T0408693

THE SCIENCE AND TECHNOLOGY OF THE ENVIRONMENT

THE SCIENCE AND TECHNOLOGY OF THE ENVIRONMENT

James G. Speight, PhD

First edition published 2023

Apple Academic Press Inc.
1265 Goldenrod Circle, NE,
Palm Bay, FL 32905 USA

760 Laurentian Drive, Unit 19,
Burlington, ON L7N 0A4, CANADA

CRC Press
6000 Broken Sound Parkway NW,
Suite 300, Boca Raton, FL 33487-2742 USA

4 Park Square, Milton Park,
Abingdon, Oxon, OX14 4RN UK

Library and Archives Canada Cataloguing in Publication

CIP data on file with Canada Library and Archives

Library of Congress Cataloging-in-Publication Data

CIP data on file with US Library of Congress

ISBN: 978-1-77463-976-4 (hbk)
ISBN: 978-1-77463-977-1 (pbk)
ISBN: 978-1-00327-751-4 (ebk)

About the Author

James G. Speight, PhD

Energy and Environmental Consultant and Author, CD&W Inc., 2476 Overland Road, Laramie, Wyoming–82070-4808, USA, Tel.: (307) 745-6069, Mobile: (307)-760-7673, E-mail: JamesSp8@aol.com, Web: https://www.drjamesspeight.com

James G. Speight, PhD, has doctorate degrees in Chemistry, Geological Sciences, and Petroleum Engineering and is the author of more than 90 books in petroleum science, petroleum engineering, biosciences, and environmental sciences, as well as books relating to ethics and ethical issues.

Dr. Speight has more than 50 years of experience in areas associated with (i) the properties, recovery, and refining of reservoir fluids, (ii) conventional petroleum, heavy oil, and tar sand bitumen, (iii) the properties and refining of natural gas, gaseous fuels, crude oil quality, (iv) the production and properties of petrochemicals from various sources, (v) the properties and refining of biomass, biofuels, biogas, and the generation of bioenergy, (vi) reactor and catalyst technology, and (vii) the environmental and toxicological effects arising from the use of energy sources. His work has also focused on safety issues, environmental effects, remediation, reactors associated with the production and use of fuels and biofuels, process economics, as well as ethics in science and engineering, and ethics in the universities.

Although he has always worked in private industry that focused on contract-based work, Dr. Speight has served as a Visiting Professor in the College of Science, University of Mosul (Iraq), Visiting Professor in Chemical Engineering at the Technical University of Denmark, and the University of Trinidad and Tobago and has had adjunct appointments at various universities. He has also served as a thesis examiner for more than 25 theses.

As a result of his work, Dr. Speight has been honored as the recipient of the following awards:

- Diploma of Honor, United States National Petroleum Engineering Society. *For Outstanding Contributions to the Petroleum Industry* (1995).
- Gold Medal of the Russian Academy of Sciences. *For Outstanding Work in the Area of Petroleum Science* (1996).
- Einstein Medal of the Russian Academy of Sciences. *In recognition of Outstanding Contributions and Service in the Field of Geologic Sciences* (2001).
- Gold Medal: Scientists without Frontiers, Russian Academy of Sciences. *In recognition of His Continuous Encouragement of Scientists to Work Together across International Borders* (2005).
- Gold Medal: Giants of Science and Engineering, Russian Academy of Sciences. *In Recognition of Continued Excellence in Science and Engineering* (2006).

- Methanex Distinguished Professor, University of Trinidad and Tobago. *In Recognition of Excellence in Research* (2007).

In 2018, he received the American Excellence Award for Excellence in Client Solutions from the United States Institute of Trade and Commerce, Washington, DC.

Contents

Abbreviations

Ag	silver
$AgNO_3$	silver nitrate
Al	aluminum
Ar	argon
As	arsenic
Au	gold
B	boron
Ba	barium
BAFs	bioaccumulation factors
BAT	best available technology
BCF	bioconcentration factor
BCT	best conventional technology
Be	beryllium
Ca	calcium
CAA	clean air act
$CaCO_3$	calcium carbonate
CaO	calcium oxide
CCl_4	carbon tetrachloride
Cd	cadmium
CEC	cation exchange capacity
CERCLA	Comprehensive Environmental Response Compensation and Liability Act
CH_3Br	methyl bromide
CH_4	methane
$CHBr_3$	bromoform
Cl	chlorine
Cl_2	chlorine
CO	carbon monoxide
Co	cobalt
CO_2	carbon dioxide
Cs	cesium
Cu	copper
CWA	clean water act
DAF	dissolved air flotation
DOC	dissolved organic carbon
DOT	Department of Transportation
EIA	Environmental impact assessment
EP	extraction procedure
EPA	Environmental Protection Agency

Fe	iron
FGD	flue gas desulfurization
Fr	francium
H_2	hydrogen
H_2CO_3	carbonic acid
H_2O_2	hydrogen peroxide
H_2S	hydrogen sulfide
H_2SO_4	sulfuric acid
HAPs	hazardous air pollutants
HCl	hydrogen chloride
HCN	hydrogen cyanide
He	helium
HF	hydrogen fluoride
Hg	mercury
HSWA	hazardous and solid waste amendments
ITCZ	intertropical convergence zone
K	potassium
KI	potassium iodide
Kr	krypton
li	lithium
LNG	liquefied natural gas
LVOCs	low-volatile organic compounds
MACT	maximum achievable control technology
MCLGs	maximum contaminant level goals
Mg	magnesium
Mn	manganese
Mo	molybdenum
MSDS	material safety data sheet
MVOCs	microbial volatile organic compounds
N_2O	nitrous oxide
Na	sodium
Na_2CO_3	sodium carbonate
NACE	National Association of Corrosion Engineers
NaCl	sodium chloride
NaOH	sodium hydroxide
Ne	neon
NEPA	National Environmental Policy Act
NH_3	ammonia
NH_4^+	ammonium
NMVOCs	non-methane volatile organic compounds
NO	nitric oxide
NO_2	nitrogen dioxide
NOM	natural organic matter
NOx	nitrogen oxides

NPDES	National Pollutant Discharge Elimination System
NPDWS	Nation Primary Drinking Water Standard
NVOCs	non-volatile organic compounds
O_2	oxygen
O_3	ozone
PAHs	polycyclic aromatic hydrocarbon
Pb	lead
$PbNO_3$	lead nitrate
PBT	bio-accumulative and toxic
PCBs	polychlorobiphenyl derivatives
PCC	pyridinium chlorochromate
PDC	pyridinium dichromate
PIP	persistent inorganic pollutant
PM	particulate matter
PMN	premanufacture notification
POPs	persistent pollutants
POTW	publicly owned treatment works
PRPs	potentially responsible parties
Pu	plutonium
Ra	radium
Rb	rubidium
RCRA	Resource Conservation and Recovery Act
SARA	Superfund Amendments and Reauthorization Act
SDWA	Safe Drinking Water Act
Se	selenium
Si	silicon
SIC	standard industrial classification
SIP	state implementation plans
SNURs	significant new use rules
SO_2	sulfur dioxide
Sox	sulfur oxides
Sr	strontium
SVOCs	semi-volatile organic compounds
TCE	trichloroethylene
TCLP	toxicity characteristic leaching procedure
TDS	total dissolved solids
TSCA	Toxic Substances Control Act
U	uranium
UIC	underground injection control
UV	ultraviolet
V	vanadium
VOCs	volatile organic compounds
WWTP	wastewater treatment plant
Zn	zinc

Preface

The history of any subject is the means by which the subject is studied so that, hopefully, the errors of the past will not be repeated. In the context of this text, environmental management and environmental awareness are not new, both having been practiced in pre-Christian times. What appears to have been available and known became lost and/or forgotten during the so-called Dark Ages and remained virtually lost until recent times.

As a result of the rebirth of environmental awareness, governments in a number of nations have passed legislation to deal with waste materials. In the United States, which is used as an example throughout this text, such legislation has included the following: (1) Toxic Substances Control Act of 1976; (2) Resource, Recovery, and Conservation Act (RCRA) of 1976 (amended and strengthened by the Hazardous and Solid Wastes Amendments (HSWA) of 1984; and (3) Comprehensive Environmental Response, Compensation, and Liability Act (CERCLA) of 1980.

It will be a surprise to many, and perhaps no surprise to a few, that environmental regulations are not new to civilization. Few people seriously discount the need for environmental regulations, but many will debate the levels at which they are proscribed-citing the cost as an unfair burden in a highly competitive economy. Regulation is, of course, a necessary step, but unfortunately, it provides no real incentive for proactive improvement. Regulation is still interpreted as a license to avoid the external costs of environmental impact. In contrast to this position, recent experience has demonstrated that systematic elimination of waste and environmental impact can provide net economic and strategic benefits such as (i) higher quality products, (ii) more efficient operations; and (iii) goodwill of an informed public who will expect a cleaner and healthier environment. As regulations become more demanding and the public more aware and concerned, this incentive for environmentally conscious actions will become increasingly apparent.

Human activities have released a large number of chemicals in the environment from industrial activities and domestic activities that are detrimental to health when coming into contact with the human body. Many of these chemicals are toxic in nature and cause serious health problems in humans. Carcinogenic chemicals cause cancer, whereas many others cause various types of allergies in the skin and respiratory system. Pesticides and a host of other chemicals enter the food chain through which it reaches the human body and affects it adversely. Metals such as lead, mercury, and arsenic are introduced into the environment by industrial and domestics. These substances are toxic and cause various ailments when they enter the human body, along with food or water.

This book is a ready-at-hand (one-stop-shopping) guide to the many issues that are related to the effects of chemicals on various ecosystems as well as to pollutant mitigation and clean-up. It is an introductory overview, with a considerable degree of detail, of the various aspects of environmental technology. The book focuses on the science and technology of environmental issues, especially chemical waste. Any chemistry in the text is used as a means of explanation of a particular point but is maintained at an elementary level.

Thus, for the purposes of the text, in general, waste is generally referred to as chemical and is only classed as hazardous when the nature of the text permits. The all too general use of the descriptor hazardous to classify wastes is often lacking in specificity and ignores the purpose of the definition. The indiscriminate use and twisting of words to describe a chemical also cannot escape some criticism. In one form or another, chemicals are harmful to the environment, especially in amounts that are in excess of the indigenous amount of chemicals that occur naturally in the environment.

The initial chapters (Chapters 1 and 2) are an introduction and a description of the various resources that can pose pollution problems. Chapters 3–5 describe the various ecosystems. The following chapters (Chapters 6–11) deal with the various aspects of waste management, and the final two chapters (Chapters 12 and 13) cover the regulations that focus on various waste streams.

Where possible, selected standard tests, as defined by the ASTM International (formerly known as the American Society for Testing and Materials), are referenced in the text. This is an aid to the reader who may wish to consult the relevant standards and necessary tests to study their application.

The literature has been reviewed up to September 2020 but to give full references for every source used while preparing this book would require a supplementary volume. The most important sources are listed, with a preference for the most easily accessible review articles and books. This provides the reader with sources that s/he can then use to build up a more comprehensive bibliography of the subject matter.

—***Editor***

PART I
Definitions and Resources

CHAPTER 1

History, Definitions, and Terminology

1.1 INTRODUCTION

The environment provides the life support system of the Earth and is the source of various resources which are essential for life. Different components of environment interact within themselves and with living organisms present over there. These interactions have great bearing on the survival of organisms (Jazib, 2018).

The air (the atmosphere), the water (the aquasphere), and the land (the geosphere) are the major parts of the environment and life on Earth depends on these interrelated components of the environment. Any disturbance to any of these components will affect the entire environmental system. Human intervention in the natural environment has already caused a great deal of changes to the system although natural causes-often ignored by some observers also contribute to climate change (Speight, 2020). Both natural and anthropogenic effects can cause changes to the living organisms (flora and fauna) in the environment.

By way of definition and description, a living organism within the environment is a product of the genetic makeup (genotype) of the organism and of the environment in which the organism lives (phenotype). Also, it is an environment which shapes the nature, distribution, and prosperity of a population. And this is where environmental science and environmental engineering play a role.

Environmental science is an interdisciplinary field that integrates the physical sciences and the biological sciences such as (alphabetically) atmospheric science, biology, chemistry, ecology, geology, mineralogy, limnology, oceanography, soil science, and zoology to the study of the environment, and the solution of environmental problems. Similarly, environmental engineering integrates the various engineering disciples to the development of processes and infrastructure for the supply of water, the disposal of waste, and the control of pollution. The activities of environmental scientists and engineers protect public health and well-being by preventing disease transmission, and they preserve the quality of the environment by averting the contamination and degradation of air, water, and land.

More specifically, environmental science and environmental engineering involve studies of the interactions between the physical, chemical, and biological components of the environment, including their effects on all types of organisms (Pfafflin and Ziegler, 2006). Earth science (also known as geoscience), is an inclusive term for all sciences related to

---

The Science and Technology of the Environment. James G. Speight (Author)

Earth (such as geology, meteorology, and oceanography). Although environmental science, environmental engineering, and earth science cover essentially the same material, environmental science places greater emphasis on the biological realm, while earth science places greater emphasis on the physical realm.

By way of further definition, limnology is the study of inland aquatic systems and includes aspects of the biological, chemicals, physical, and geological characteristics and functions of inland waters (running and standing waters, fresh, and saline, natural, and manmade). This includes the study of lakes, reservoirs, rivers, streams, wetlands, and groundwater. Limnology is closely related to aquatic ecology and hydrobiology which involve the study of aquatic organisms and the interactions of these organisms with the abiotic (non-living) environment. While limnology has substantial overlap with freshwater-focused disciplines (such as freshwater biology), it also includes the study of inland salt lakes. A more recent sub-discipline of limnology, termed landscape limnology which involves studies, management, and the conservation of these ecosystems using a landscape perspective, by explicitly examining connections between an aquatic ecosystem and the related watershed. Recently, the need to understand global inland waters as part of the system of the Earth led to the creation of a sub-discipline called global limnology, which considers processes in inland waters on a global scale, like the role of inland aquatic ecosystems in global biogeochemical cycles.

Briefly, an ecosystem works as a unit in an efficient and organized way and receives energy from that is passed on through the components of the ecosystem and, in fact, all life depends on this flow of energy. Green plants (including phytoplankton species) are able to trap the solar energy in an ecosystem and they make use of this energy for their growth and maintenance. Heterotrophic species (or consumers) obtain the energy requirements from this stored energy (in green plants) as food and use it for their development, growth, maintenance, or other life activities. All life forms in an ecosystem are linked together by the flow of energy. Besides energy, various nutrients and water, which are also required for life processes, are exchanged by the biotic components within themselves and with the abiotic components (Jazib, 2018).

Terminology is the means by which various subjects are named so that reference can be made in conversations and in writings and so that the meaning is passed on. For example, as applied in this book, the term technology refers to the sum of techniques, skills, and processes that are used in the production of products and services; technology can also be the knowledge of techniques and processes. The simplest form of technology is the development and use of basic tools. The simplest example of the term relates to: (i) the prehistoric discovery of the means by which firs could be controlled, the means to increase; (ii) the means to increase the available sources of food; and (iii) the invention of the wheel that gave humans to ability to travel greater distances and also to control the environment. Developments such as the printing press, the telephone, and (of late) the internet have removed many of the physical barriers to communication and allowed humans to interact freely (but not in all countries) on a global scale.

However, the impact of technology on environment is not uniform throughout the world, since the development and use of technology is not uniformly distributed. The development, acceptance, and use of technology by humans is uneven and varies vastly from region to region and nation to nation, depending on their economic and social conditions. This, of course, influence the environmental issues that arise in various countries as well as the sense of urgency for environmental protection.

Nevertheless, environmental issues permeate everyday life, especially the lives of workers in various occupations where hazards can result from exposure to many external influences (Lipton and Lynch, 1994). In order to combat any threat to the environment, it is necessary to understand the nature and magnitude of the problems involved (Ray and Guzzo, 1990). It is in such situations that environmental technology has a major role to play. Also, environmental issues even arise when old and outdated (non-relevant) laws are taken into modern consideration. Thus, the concept of what seemed to be a good idea at the time the action occurred no longer holds when the law influences the environment.

It is not the intent to subscribe to the notion that the state of the environment is acceptable, and that it always will be acceptable or that the many heaps of mis-understanding will correct all that is wrong. There is, however, the need to strike a balance between: (i) the definition of a pollutant; (ii) the cessation of pollutant releases; (iii) and the definition of a clean environment.

As a start to this process, a pollutant is a substance or energy introduced into the environment that has undesired effects, or adversely affects the usefulness of a resource. A pollutant may cause long or short-term damage by changing the growth rate of plant or animal species, or by interfering with human amenities, comfort, health, or property values. Some pollutants are biodegradable and therefore will not persist in the environment in the long term but can still cause damage to the environment during the lifetime of the pollutant. A pollutant can also be a naturally-occurring material that is reintroduced into the environment in an amount that exceeds the naturally occurring amount in the environment.

Following from this definition, a primary pollutant is a pollutant that is emitted directly from a source. On the other hand, a secondary pollutant is a pollutant that is produced by interaction of a primary pollutant and with another chemical or by dissociation of a primary pollutant or other effects within a particular ecosystem (Table 1.1).

Human activities have released a large number of chemicals in the environment from industrial activities and domestic activities that are detrimental for health when come into contact with human body. Many of these chemicals are toxic in nature and causes serious health problems in humans. Carcinogenic chemicals cause cancer whereas many others cause various types of allergies in the skin and respiratory system. Pesticides and a host of other chemicals enter the food chain through which it reaches the human body and affects it adversely. A variety of chemicals such as the health-threatening metals (for example, lead, mercury, and arsenic) are introduced into the environment by industrial and domestic activities (Figure 1.1) (Speight, 1996). These substances are toxic and cause various ailments when they enter the human body along with food or water.

This chapter introduces the history of environmental observations, the atmosphere, the aquasphere, the geosphere, and most important, the various ecological cycles that from key indications of the manner in which various environmental events are interrelated. Also, the purpose of this text is to examine the effects that human development has had on the environment and also to consider the methods by which the environmental impact of this development might be mitigated. Industries such as the production of minerals, the production and combustion of fossil fuels, a variety of industrial processes, and agricultural activities have all served to cause changes in the environment (Mooney, 1988). In short, the text focuses on the means by which sustainable development (Brady and Geets, 1994) might be achieved with minimum disturbance of the environment.

TABLE 1.1 Typical Environmental Pollutants*

Pollutant	Description
Mercury (Hg)	A chemical element commonly known as quicksilver, a liquid under STP**.
Ozone (O_3)	A pale blue gas with a distinctively pungent smell.
	Formed from oxygen by the action of ultraviolet light and (UV) light and electrical discharges (lightning) within the atmosphere.
	Present in low concentrations throughout the atmosphere.
	The highest concentration is in the ozone layer of the stratosphere.
Particulate matter (PM)	Microscopic particles of solid or liquid matter suspended in the air.
	Sources can be natural or anthropogenic.
Persistent organic pollutants (POPs)	Organic compounds that are resistant to environmental degradation by chemical, biological, and photolytic processes.
PAHs***	Chemical compounds containing only carbon and hydrogen in the form of multi-fused aromatic rings.
	The simplest examples are naphthalene (two fused aromatic rings) and phenanthrene (three fused aromatic rings.
VOCs****	Organic compounds that have a high vapor pressure and low boiling point at STP**.

*Listed alphabetically rather than by effect.

**Standard conditions of temperature and pressure.

***Polycyclic aromatic hydrocarbon derivatives; sometimes referred to as polynuclear aromatic hydrocarbon derivatives (PNAs).

****Volatile organic compounds.

1.2 HISTORICAL ASPECTS

There is the general belief that the environmental science and engineering are relatively new technical disciplines that arose from cause and effects in the 1960s and 1970s. In spite of this belief, the influence of environmental effects on human (and animal) populations have been realized for centuries. However, the development of human civilizations has resulted in perturbations of much of the ecosystems of the Earth leading to pollution or the ecosystems-the air, the land, and the water. In this case, pollution is defined as the introduction of a chemical (or chemicals) into the air systems, the water systems, and the land systems that were not indigenous to these systems.

In the context of this book, environmental history emerged out of the environmental movement of the 1960s and the 1970s, and much of the impetus continued until issues regarding the health of the environment because a near-global concern-there are still some countries who pay lip service for supposed attention to environment issues but little is done in practice. (Krech et al., 2003). As all history occurs in the natural world, the history of environmental issues focuses on events that influence the environment-such as the spill of a chemical liquid or the release of a noxious gas into the atmosphere- or a geographic region.

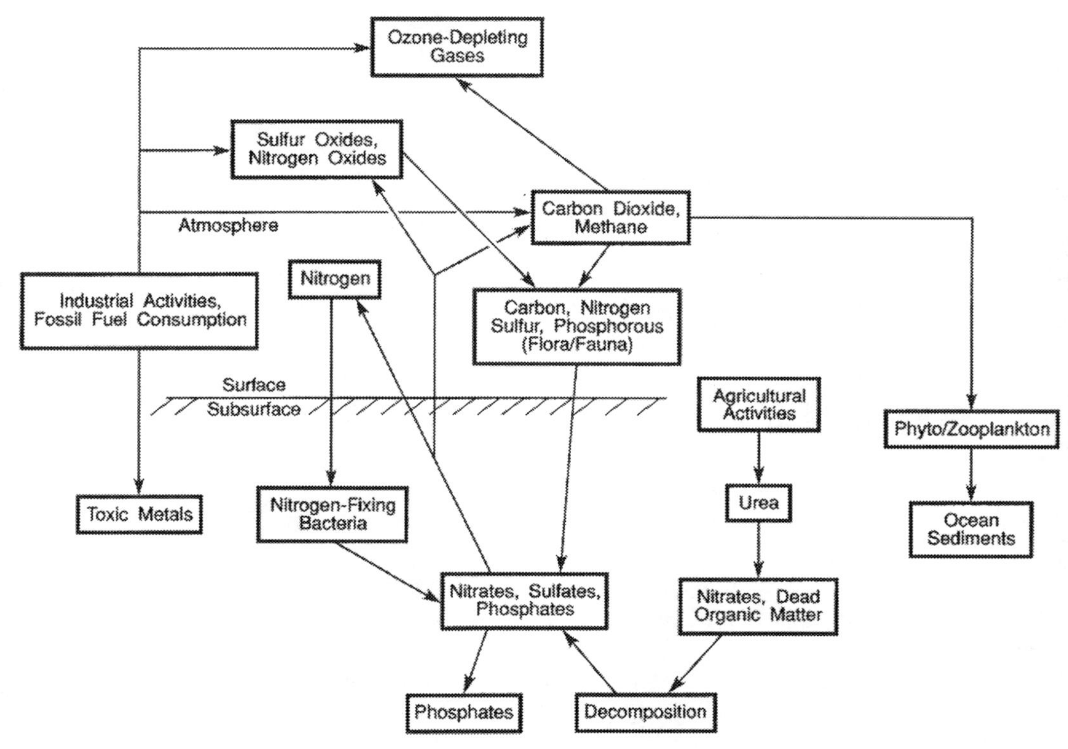

FIGURE 1.1 The industrial cycle and the domestic cycle showing pollutant entry into the environment.
Source: Reproduced with permission from: Speight (1996). © Taylor and Francis.

In a combination of historical and geological perspectives, the aggressive appearance of humans on the Earth, whether it was: (i) by evolution from some primordial soup; or (ii) through creation (Krauss, 1992; Christian Bible, 1995), has also paralleled the demand for the use of the resources of the Earth which been intimately related to the onset of environmental pollution. Whatever the origin of the human species, the environment can be endangered by any single person or by several persons participating in a variety of human activities (Mooney, 1988; Pickering and Owen, 1994). Indeed, the first use of wood for the fire as a source of warmth led to the overwhelming modern-day use of the resources of the Earth (Tester et al., 1991).

In order to put this development into the perspective of the geological time scale of the Earth, which spans approximately 4.5 billion years (4.5×10^9 years) (Table 1.2), it is necessary to remember that the Earth was virtually uninhabited for eons (eon: a division of the geologic time). For example, the Pre-Cambrian period spans approximately 3.8 billion (3.8×10^9) years, or 85% of this period the time scale.

TABLE 1.2 Age of the Earth and the Subdivision of the Years

Era	Period	Epoch	Years Ago × 106**
Precambrian	Precambrian	–	570
Paleozoic	Cambrian	–	510
–	Ordovician	–	439
–	Silurian	–	409
–	Devonian	–	363
–	Mississippian*	–	323
–	Pennsylvanian*	–	290
–	Permian	–	245
Mesozoic	Triassic	–	208
–	Jurassic	–	146
–	Cretaceous	–	65
Cenozoic	Tertiary	Paleocene	57
–	–	Eocene	35
–	–	Oligocene	23
–	–	Miocene	5
–	–	Pliocene	2
–	Quaternary	Pleistocene	0.01
–	–	Holocene	Present

*The Mississippian and Pennsylvanian periods are sometimes (collectively) referred to as the Carboniferous period.

**The numbers in years indicate the beginning and the end of the periods and epochs.

Significant effects of human activities on the environment have been observed for the 10,000 years when agriculture was started. However, the impacts were drastic in the last two centuries, during which most of the countries underwent rapid industrial development with the commencement of the industrial revolution. This development helped humans to progress along many fronts, such as health, food security, education, technology, and luxury, but at the same time, humans have exploited the resources of the Earth beyond the capacity of the planet to regenerate the depleted resources. Moreover, human activities have also resulted in contaminated air, water, soil, and other components of the environment with the by-products generated during human development. As a result, the growing pressure on ecosystems is causing habitat destruction or degradation and permanent loss of productivity, threatening both biodiversity and human well-being.

As examples, the land-borne and waterborne diseases of the pre-Christian and post Christian eras are common knowledge and are cited in many of the old texts. However, the relationship of disease to pollution is not often recognized, nor could it be expected to be recognized in those times. Furthermore, Roman armies on the march recognized the need for clean, untainted drinking water.

There are records that show that the water wagons following the Roman army carried more vinegar than they did water. Water, albeit often tainted, was plentiful in Europe but vinegar was not readily available! There are also records that show that the water wagons following the army carried more vinegar than they did water. Water, albeit often tainted, was plentiful in Europe vinegar was not readily available! The water that the Roman soldiers drank was water plus 10% v/v vinegar (chemically, acetic acid, CH_3COOH and the water-vinegar mix was often referred to in older texts as gall) which was soaked into a sponge that was carried into a leather pouch attached to the soldier's belt. As the march continued, the soldier and thirst needed his attention; the soldier would suck on the sponge without even breaking his stride, thereby allowing the army to keep moving. He would only give up his sponge if he was mortally wounded or to a mortally wounded colleague. Such was his inclination to preserve his source of aqueous nourishment.

The Romans were also aware that if the Roman army camped upstream from an enemy army, the ejection of all camp refuse into the river would cause the occurrence of sickness and disease in the enemy camp. Of course, the Roman commanders had the knowledge to draw the water requirement from even further upstream than the camp. The same line of thinking applied to time when the Roman army besieged a city through which a river ran. The Roman army would either block the river or throw waste (human and animal) into the river to spread disease at a cost to the enemy within the city and thereby preserve the health and lives of the Roman soldiers.

Whether or not it was for esthetic reasons or the disadvantages of having sewage areas close by is another matter. Following the recognition by the Romans that dumping refuse (including raw sewage) in a river upstream of an enemy encampment, the disadvantages of dumping raw sewage into rivers at places upstream of villages, towns, and cities were also recognized in the pre-Christian and post-Christian era.

In the non-military sense, there has been an awareness that the environment was becoming tainted by the presence of large bodies of the population. For example, the location of sewage areas away from buildings, rather than next to (or even in) the building, is a simple example of this awareness. There is also early evidence of flush toilets or, at least, flowing water for sewage transport away from buildings (into the nearby river) in several of the ancient cities (such as in the ancient city of Nineveh, across the river the modern city of Mosul in Iraq). Also, in ancient Rome, there was the recognition cites needed adequate water supply and sewage systems had taken place by the middle ages (James and Thorpe, 1994).

Just as tainted water was recognized as a carrier of disease, other carriers were also recognized. An example is the frequent occurrence of the bubonic plague which invaded European cities for the 14th Century. The bacterium (Yersinia Pestis) that was responsible for the plague is believed to have originated in China, in 1331, and migrated west along the silk road. The fatal disease reached Western Europe in full force in the late 1340s, striking England in 1348. Before the Black Death abated, over the next few years 'the bacteria brought death to a third of Europe's population and took half of the people in densely crowded cities, such as London. Across the English countryside, whole villages were virtually wiped out, their few remaining survivors taking refuge in neighboring communities. Over time, the physical remains of the deserted village dissolved into the Earth and the place name was erased from maps and forgotten (Huntley, 2020). The spread of the plague was eventually related to the infestation by fleas from rats is an example of a land-borne disease (Cartwright and Biddiss, 1991). The infestations were, in turn, due to unhygienic living conditions where piles of refuse gave the rats sustenance and healthy rats gave the fleas their sustenance which then turned to the human part of their life cycle for their leisure activities!

As industries developed, especially during the industrialization of the middle ages, they added their discharges to those of the community (Gimpel, 1976). When the concentration of added substances became dangerous to humans or so degraded the water that it was unfit for further use, water pollution control began. With the increasing development of land areas, pollution of surface water supplies became more critical because wastewater of an upstream community became the water supply of the downstream community. In fact, serious epidemics of waterborne diseases such as cholera, dysentery, and typhoid fever were caused by underground seepage from privy vaults into town wells. Such direct bacterial infections through water systems were the cause behind the disease, of course, the role of the organism that thrived on the garbage and sewage must also be recognized.

Another form of bacteriological contamination involved the use of severed heads as ammunition for a trebuchet as a documented event in medieval warfare (Turnbull, 1995). The trebuchet was a type of catapult that uses a long arm to throw a projectile and was a common powerful siege engine until the advent of gun powder. These unusual forms of missiles were hurled over the walls of a besieged city (Berwick, 1333) by the English in the war against the Scots (who occupied the border city) and may have been intended for something more than the usual intimidation of the besieged persons. Recognition of the head of a family member, a neighbor, a friend, or a colleague would certainly have been intimidating to the inhabitants of the besieged city. The heads, presumably some days old and subject to the heat of the summer, would also have been a source of bacteriological contamination (thus, disease) of the inner city. There are also records of the carcass of a dead horse and plague-ridden bodies being hurled over the walls and into besieged cities, presumably not for recognition by friends or relatives but for the spread of disease. Although it was not necessarily a waterborne disease in the direct sense (such as typhus and dysentery) but the infection of Europe by syphilis from the 1490s onwards seems to have been the result of the return of Columbus (by water) from the New World (Cartwright and Biddiss, 1991). Whatever the means by which the disease was brought to Europe, the French referred to syphilis as the Spanish disease and the English referred to syphilis as the French disease.

In terms of air emissions, it was well known in the Middle Ages that the air was infected with invisible spirits, some of which were benign but most were evil and dangerous. Whether this is related to a fear of airborne disease or to a superstitious population is another thought-probably the latter. However, it may have led to the early recognition that diseases could be transmitted by air. In fact, there is a well documented example of the recognition of air pollution in medieval times. In fact, a singularly important environmental event occurred in 1257 (Galloway, 1882). The event which, although unknown at the time, was perhaps a forerunner of the modern awakening of an environmental consciousness in terms of the recognition of the adverse aspects of modern (i.e., 13th century) technology.

Thus, it occurred that Eleanor, wife, and Queen of Henry III of England, was residing in the town of Nottingham whilst the King was on a military expedition into Wales. To shorten the story, the Queen was obliged to leave Nottingham due to the troublesome and noxious effects of the smoke from the coal fires that were being used for heating and cooking. Perhaps as a result of this awakening, a variety of proclamations were issued over the next several decades by Henry and by his son, Edward I. These proclamations threatened the population with the loss of various liberties, as well as the loss of a limb or two, and even life, if the consumption of coal was not seriously decreased and, in some cases, halted.

It is possible that the hasty departure of Queen Eleanor from Nottingham led to the beginning of an environmental awakening and that the consequences of burning coal were recognized at that time. However, the proclamations seem to have had little effect (there being considerable income from the sale of coal to continental Europe) and, by the last decades of the 13th Century, London had the dubious privilege of becoming the first city documented to suffer manmade pollution. In the period 1285 to 1288 (inclusive) complaints were recorded concerning the corruption of the air in the city by fumes from the wood-fired and coal-fired furnaces of the lime kilns.

The lime kilns were used for the calcination of limestone (calcium carbonate, $CaCO_3$) to produce the form of lime (quicklime, CaO) which was used in building mortar and as a stabilizer in mud floors. Thus:

$$CaCO_3 + \text{heat} \rightarrow CaO + CO_2$$

This reaction takes place at 900°C (1,650°F) (at which temperature the partial pressure of carbon dioxide is 1 atmosphere, 14.7 psi), but a temperature around 1,000°C (1,800°F; at which temperature the partial pressure of carbon dioxide is CO_2 is approximately 3.8 atmospheres, 55 psi) is usually used to make the reaction proceed quickly.

Nevertheless, the Royal positions on the pollution problem, if that is really the issue, may have solved part of the immediate problem, it did not have any lasting effect pollution from a variety of source made London (and the River Thames) one of the most polluted sites in England. In fact, in the Middle Ages, there was also pollution which seemed to continue unabated in the fledgling industrial operations. For example, mining and smelting operations were carried out without any form of protection for the workers, as evidenced from woodcut illustrations of the period (Agricola, 1556).

Also, in the Middle Ages, towns and cities suffered from industrial water pollution, principally due to the activities of butchers and tanners. Municipalities were always trying to move the butchers and tanners downstream, outside the precincts of the town, because the industry corrupted the waters of the riverside dwellers. Indeed, butchers and tanners notwithstanding, sources of drinking water have continued to be threatened by the dumping of chemical waste which is then leached (washed) from the surface into the underground water systems (aquifers). Gaseous emissions from the combustion of fossil fuels which were recognized early in their use (Chapter 2), as well as the emission of chlorofluorocarbon derivatives and other chemicals, can cause severe, often irreversible, damage to the atmosphere (Lipton and Lynch, 1994; Speight, 2019).

And so, the examples from history are numerous, especially when the source and causative agent of the disease are recognized. In many cases, the causative agent could be related to living in unsanitary conditions which, in turn, were caused by pollution of the environment. Furthermore, because of the emergence of industrial processes and the parallel emergence of fossil fuels as major sources of energy, and the emissions that are produced, the importance of protecting the environment has become a top priority within government and industry over the last two decades (Benarde, 1989; Lipton and Lynch, 1994).

The passage of the Clean Air Act (CAA) of the United States that was voted into law in 1970 and the subsequent passage of a series of amendments, the latest in 1990 (Chapter 10) (United States Congress, 1990; Stensvaag, 1991) speaks to the recognition that the emission of noxious constituents into the atmosphere must cease. Other countries have also passed similar legislation, and there are also laws to prevent pollution of the geosphere (land systems) and the aquasphere (water systems).

However, the potential for damage to the environment was not recognized until recent times, and any voices related to concerns regarding environmental issues were not heard with any great effect until after the 1960s. By that time, a variety of gaseous, liquid, and solid waste emissions had been liberally released to the environment without any understanding of any of the causative events happening.

As a start, it must be recognized that the terminology found in the various areas of industrial technology can be extremely confusing to the uninitiated; excellent examples of the confusion that abounds are to be found in the area of fossil fuel technology (Speight, 2013, 2014). Thus, it is pertinent at this point to present several definitions of the terms used in environmental technology as a means of alleviating any of the confusion that can arise.

Thus, in order to alleviate any confusion that might arise, it is necessary to define the relevant terms that will be used throughout the book.

1.3 DEFINITIONS AND TERMINOLOGY

Definitions are the means by which scientists and engineers communicate the nature of a material to each other and to the world, either through the spoken word or through the written word. Thus, the definition of a material can be extremely important and have a profound influence on the technical and public (as well as political) perception of that material. For example, water is essential to life, except in the unfortunate instance of drowning. However, a simple and constructive (perhaps destructive, although technically correct) change in the definition of water has the potential to cause a marked change in the perception of this substance. A change in the name of water to dihydrogen monoxide (H_2O) initiates thoughts of dihydrogen peroxide (H_2O_2) and even to carbon monoxide (CO), which are both dangerous chemicals, and the serious consequences that result from the release of these chemicals to the environment. Even use of the name oxygen dihydride (OH_2)—which is the same as H_2O, only written backwards—does not conjure up the familiar liquid that is essential for the support of life on the Earth.

In fact, precise definitions are an important aspect of any technology, especially environmental technology. It is in the arena of environmental technology arena where many decisions are made (and continue to be made) on the basis of emotion rather than on the basis of scientific and engineering principles (Fumento, 1993). Many activities associated with the use of chemicals have a high degree of perceived risk (O'Riordan, 1995). Therefore, the use of precise definitions should (must) be the norm.

The environment, in the context of this book, is the sum of all external factors (biotic and abiotic) that influence the life of an organism. Biotic factors include all living beings (e.g., humans, animals, plants, and microorganisms), whereas abiotic factors include all physico-chemical entities (such as air, water, soil, rocks, minerals, and mountains). The environment can be natural, human-engineered, or even abstract (non-material). Owing to such vagueness, the term environment has been used in various ways or in various perspectives. For example, terms such as natural environment, extra-terrestrial environment, human-engineered environment, socio-political-cultural environment, business environment, family environment, and workplace environment are used in general conversation.

Thus, of necessity environmental science and environmental engineering are multidisciplinary fields which involve the study of all the components or factors that make or influence life-supporting biophysical environment, such as earth processes, ecological systems, biodiversity, natural resource, alternative energy systems, climate change, and the various types of pollution. These entities or processes are guided by complex interaction of physical, chemical, and biological processes, as well as significant human intervention. Disciplines such as biology, chemistry, physics, geology, geography, sociology, economics, and management have been integrated to develop different subdivisions of environmental science. The major subdivisions of environmental science include ecology, geosciences, environmental chemistry, atmospheric science, environmental microbiology, environmental toxicology, and EIA.

In addition, there are subdivisions (such as environmental studies, environmental engineering, environmental economics, environmental ethics, environmental management, environmental sociology, and environmental biotechnology) that are generally treated as independent academic disciplines parallel to environmental science. Environmental conservation is the main emphasis for most of these disciplines, but the approaches vary. For example, environmental studies incorporate more of the social sciences for understanding human relationships, perceptions, and policies towards the environment. Environmental engineering, on the contrary, focuses on design and technology for improving environmental quality.

Finally, in the current context, there are natural cycles that must be considered that illustrate the transportation of chemical wastes throughout the environment. The main feature of a simplified cycle is the recognition that a chemical pollutant can interact with the land, the water, the air, and with the animals/plants found therein. In addition, understanding these cycles can help in the understanding of the means by which a chemical pollutant can enter the various ecosystems as a result of being produced during an industrial or domestic cycle thereby influencing the behavior of the ecosystem and thereby influencing floral and faunal activity within the ecosystem.

On the other hand, environmental engineering is a discipline that takes from broad scientific topics like chemistry, biology, ecology, geology, hydraulics, hydrology, microbiology, and mathematics to create solutions that will protect and also improve the health of living organisms and improve the quality of the environment. Environmental engineering is the application of scientific and engineering principles to improve and maintain the environment to: (i) protect beneficial ecosystems; and (ii) to improve environmental-related enhancement of the quality of human life.

Environmental engineers devise solutions for wastewater management, water, and air pollution control, recycling, waste disposal, and public health as well as the design of musical water sully and industrial water treatment systems. They also study the effect of technological advances on the environment, addressing local and worldwide environmental issues such as acid rain, global warming, ozone depletion, and water pollution as well as air pollution.

The key to the successful outcome of an environmental science and/or an environmental engineering project is teamwork in which both disciplines and other necessary professionals (as dictated by the nature of the problem) work together to solve and environmental problems.

The subject matter of any problem that is related to the environment of the Earth can be divided into two main components which are: (i) nature itself and any changes that occur over time, which includes the physical impact of humans on the atmosphere, the water systems which are referred to as the aquasphere in this book, and the land systems, which are referred to as the geosphere in this book but often falling under the term biosphere in some texts; and (ii) the means by which humans use nature, which includes the environmental consequences of an ever-increasing population and the advances in technology that is accompanies by the changing patterns of production and consumption that accompanied the transition of humans from nomadic to hunter-gatherers to settled agricultural communities along with the effects of colonialism and the environmental and human consequences of the industrial revolution as well as the succeeding technological revolutions. In the context of this text, environmental issues are not new management is not new, having been practiced in pre-Christian times (James and Thorpe, 1994). What appears to have been available became lost and/or forgotten during the so-called dark ages and remained virtually lost until recent times.

In fact, environmental science and environmental engineering are collective disciplines that are involved in the study of the environment. These studies can vary from the effects of changes in the environmental conditions on flora and fauna of a region to the more esoteric studies of animals in laboratories and can include aspects of chemistry, chemical engineering, microbiology, and hydrology as they can be applied to solve environmental problems. As a historical aside, environmental engineering (formerly known as sanitary engineering) originally developed as a sub-discipline of civil engineering.

Another extremely relevant definition, environmental technology is the application of scientific and engineering principles to the study of the environment, with the goal of the improvement of the environment. Furthermore, the issues related to the pollution of the environment are relative. In fact, the purity of the environment is in the eyes of the beholder!

Any organism is exposed to an environment, even if the environment is predominantly many members of the same organism. An example is a bacterium which, in a culture, is exposed to many members of the same species. Thus, the environment is all external influences, abiotic (physical factors) and biotic (actions of other organisms), to which an organism is exposed. The environment affects basic life functions, growth, and reproductive success of organisms, and determines their local and geographic distribution patterns. A fundamental idea in ecology is that the environment changes in time and space and living organisms respond to these changes.

Since ecology is that branch of science related to the study of the relationship of organisms to their environment, an ecosystem is an ecological community (or living unit) considered together with the nonliving factors of its environment as a unit. The environment is all of the external influences, such as abiotic (physical) factors and biotic (actions of other organisms) factors, to which an organism is exposed.

By way of brief definition, abiotic factors include such influences as light radiation (from the Sun), ionizing radiation (cosmic rays from outer space), temperature (local and regional variations), water (seasonal and regional distributions), atmospheric gases, wind, soil (texture and composition), and catastrophic disturbances. These latter phenomena are usually unpredictable and infrequent disturbances, such as fire, hurricanes, volcanic activity, landslides, and major floods, and drastically alter the environment of an area and thus change the species composition and activity patterns of the inhabitants.

On the other hand, biotic factors include natural interactions (e.g., predation, and parasitism) and anthropogenic stress (e.g., the effect of human activity on other organisms). These factors can be described as any living component that affects another organism or shapes the ecosystem. This includes both animals that consume other organisms within their ecosystem, and the organism that is being consumed. Biotic factors also include human influence, pathogens, and disease outbreaks. Each biotic factor needs the proper amount of energy and nutrition to function day to day. Because of the abiotic and biotic factors, the environment to which an organism is subjected can affect the life functions, growth, as well as the reproductive success of the organism and can determine the local and geographic distribution patterns of an organism.

Living organisms respond to changes in the environment by either adapting or becoming extinct. The basic principles of the concept that living organisms respond to changes in the environment were put forth by Darwin and Lamarck. The former (Darwin) noted the slower adaptation (evolutionary trends) of living organisms whilst the latter (Lamarck) noted, the more immediate adaptation of living organisms to the environment. Both Darwin and Lamarck developed (individually) the concept of the survival of the fittest alluding to the ability of an organism to live in harmony with the surrounding environment. This was assumed to indicate that the organism which competed successfully with environmental forces would survive. However, there is the alternate thought that the organism that can live in a harmonious symbiotic relationship with the surrounding environment has an equally favorable chance of survival.

The influence of the environment on organisms can be viewed on a large scale (i.e., the relationship between regional climate and geographic distribution of organisms) or on a smaller scale (i.e., some highly localized conditions determine the precise location and activity of individual organisms). For example, an organism may respond differently to the frequency and duration of a given environmental change. In addition, if some individual organisms in a population have adaptations that allow them to survive and to reproduce under new environmental conditions, the population will continue, but the genetic composition will have changed (Darwinism). On the other hand, some organisms have the ability to adapt to the environment (i.e., to adjust their physiology or morphology in response to the immediate environment) so that the new environmental conditions are less (certainly no more) stressful than the previous conditions. But such changes may not be genetic (Lamarckism).

In terms of anthropogenic stress (the effect of human activity on other organisms), there is the need for the identification and evaluation of the potential impacts of proposed projects, plans, programs, policies, or legislative actions upon the physical-chemical, biological, cultural, and socioeconomic components of the environment. This activity is also known as EIA and refers to the interpretation of the significance of anticipated changes related to a proposed project. The activity encourages consideration of the environment and arriving at actions that are environmentally compatible.

Identifying and evaluating the potential impact of human activities on the environment requires the identification of mitigation measures. Mitigation is the sequential consideration of the following measures: (i) avoiding the impact by not taking a certain action or parts of an action; (ii) minimizing the impact by limiting the degree or magnitude of the action and its implementation; (iii) rectifying the impact by repairing, rehabilitating, or restoring the affected environment; (iv) reducing or eliminating the impact over time by preservation and maintenance operations during the life of the action; and (v) compensating for the impact by replacing or providing substitute resources or environments.

Nowhere is the effect of anthropogenic stress felt more than m the development of natural resources of the Earth. The natural resources are varied in nature and often require definition and explanation (Chapter 2). For example, in relation to mineral resources, for which there is also descriptive nomenclature (ASTM C294), the terms related to the available quantities of the resource must be defined. In this instance, the term resource refers to the total amount of the mineral that has been estimated to be ultimately available. The term reserves refer to well-identified resources that can be profitably extracted and utilized by means of existing technology. In many countries, fossil fuel resources are often classified as a subgroup of the total mineral resources.

In some cases, environmental pollution is a clear-cut phenomenon, whereas in others, it remains a question of degree. The ejection of various materials into the environment is often cited as pollution. But there is the ejection of the so-called beneficial chemicals which can assist the air, water, and land to perform their functions. However, it must be emphasized that the ejection of chemicals into the environment, even though those chemicals are indigenous to the environment, in quantities above the naturally-occurring limits can be extremely harmful.

In fact, the timing and the place of a chemical release are influential in determining whether a chemical is beneficial, benign, or harmful. Thus, what may be regarded as a pollutant in one instance can be a beneficial chemical in another instance. The phosphates in fertilizers are examples of useful (non-destructive or beneficial) chemicals, whilst phosphates generated as by-products in the metallurgical and mining industries may, depending upon the specific industry, be considered pollutants (Chenier, 1992). In this case, the means by which such pollution can be prevented must be recognized (Breen and Dellarco, 1992).

Thus, increased use of the resources of the Earth, as well as the use of a variety of chemicals that are non-indigenous to the Earth, have put a burden on the ability of the environment to tolerate such materials. It should also be mentioned at this point that chemicals which are indigenous to the Earth can also be harmful when injected into the environment in quantities above the naturally-occurring limits.

Finally, some recognition must be made of the term carcinogen since many of the environmental effects referenced in this text can lead to cancer. Carcinogens are cancer-causing substances, and there is a growing awareness of the presence of carcinogenic materials in the environment, and there is a classification scheme provided for such materials (Table 1.3) (Zakrzewski, 1991; Kester et al., 1994; Milman and Weisburger, 1994). Because the numbers of substances with which a person comes in contact are in the tens of thousands and there is not a full understanding of the long-term effects of these substances in their possible propensity to cause genetic errors that ultimately lead to carcinogenesis.

TABLE 1.3 Classification Scheme as Determined by the US EPA

Group	Description
A	Human carcinogen
Bl	Probable human carcinogen-limited human data are available.
B2	Probable human carcinogen; carcinogen in animals.
C	Possible human carcinogen.
D	Not classifiable as a human carcinogen.
E	No carcinogenic activity in humans.

It is generally understood that pollution is the result of a non-indigenous chemical (a gaseous, liquid, or solid chemical into an ecosystem. Pollution can also involve the introduction of an indigenous chemical into an ecosystem in an amount beyond the natural abundance of the chemical in the ecosystem. Thus, a pollutant is a substance (for simplicity, most are referred to as chemicals) present in a particular location when it is not indigenous to the location or it is a greater-than-natural concentration and often being the product of human activity.

For many pollutants, the atmosphere, the aquasphere, and the geosphere have the ability to sanitize (clean) themselves within hours or days, especially when the effects of the pollutant are minimized by the natural constituents of the ecosystem. For example, the atmosphere might be considered to be self-cleaning as a result of rain. However, removal of some pollutants from the atmosphere (such as sulfate derivatives and nitrate derivatives) by rainfall results in the formation of acid rain which can and will cause serious environmental damage to ecosystems within the aquasphere and the geosphere (Johnson and Gordon, 1987; Pickering and Owen, 1994; Stensland, 2006)).

As a further aspect of this definition, the pollutant, by virtue of its name, has a detrimental effect on the environment, in part or in toto. Pollutants can also be subdivided into two classes (1) primary pollutants, and (2) secondary pollutants:

$$\text{Source} \rightarrow \text{Primary pollutant} \rightarrow \text{Secondary pollutant}$$

Thus, a pollutant is a chemical which is emitted directly from the sources. In terms of atmospheric pollutants, examples are carbon monoxide, carbon dioxide, sulfur dioxide (from which sulfur trioxide-in this case, a secondary pollutant is derived), and nitrogen oxides from combustion of carbonaceous fuels:

$$2C_{fuel} + O_2 \rightarrow 2CO$$
$$C_{fuel} + O_2 \rightarrow CO_2$$
$$2N_{fuel} + O_2 \rightarrow 2NO$$
$$N_{fuel} + O_2 \rightarrow NO_2$$
$$S_{fuel} + O_2 \rightarrow SO_2$$
$$2SO_2 + O_2 \rightarrow 2SO_3$$

The question related to the classification of nitrogen dioxide and sulfur trioxide as primary pollutants often arises, as does the origin of the nitrogen. In the former case, these higher oxides can be formed in the upper levels of the combustors. The nitrogen, from which the nitrogen oxides are formed (Chapter 2), does not originate solely from the fuel but may also often originate from the air used for the combustion.

On the other hand, secondary pollutants are produced by interaction of primary pollutants and with another chemical or by dissociation of a primary pollutant, or other effects within a particular ecosystem. Again, using the atmosphere as the example, the formation of the constituents of acid rain is an example of the formation of secondary pollutants:

$$SO_2 + H_2O \rightarrow H_2SO_3$$
sulfurous acid

$$SO_2 + H_2O \rightarrow H_2SO_4$$
sulfuric acid

$$NO + H_2O \rightarrow HNO_2$$
nitrous acid

$$3NO_2 + 2H_2O \rightarrow HNO_3$$
nitric acid

In many cases, these secondary pollutants can have significant environmental effects, such as the formation of acid rain and smog (Chapter 10).

The source of the pollutant is as important as the pollutant itself because the source is generally the logical place to eliminate pollution. On the understanding that the conversion of a pollutant to a contaminant, and vice versa, can occur and there may be the need to understand the means by which this can happen.

Briefly, a contaminant, which is not usually classified as a pollutant unless it has some detrimental effect, can cause deviation from the normal composition of an environment. On the other hand, a receptor is an object (animal, vegetable, or mineral) or a locale that is affected by the pollutant.

A chemical waste (Chapter 6) is any solid, liquid, or gaseous waste material (often a mixture of two or more chemicals) which, if improperly managed or disposed of, may pose substantial hazards to human health and the environment. At any stage of the management process, a chemical waste may be designated, by law, a hazardous waste. Improper disposal of these waste streams in the past has created a need for expensive cleanup operations (Chapter 10) (Tedder and Pohland, 1993). Correct handling of these chemicals (National Research Council, 1981) as well as dispensing with many of the myths related to chemical processing (Kletz, 1990) can mitigate some of the environmental problems that will occur when incorrect handling is the norm.

1.4 THE EARTH SYSTEM

Finally, for convenience as a means of differentiation, in this book, the earth system consists of three main systems (often referred to as spheres). The first system, the atmosphere is an envelope of gas that keeps the planet warm and provides oxygen for breathing and carbon dioxide for photosynthesis. The second system comprises the areas of Earth that are covered with amounts of water, called the aquasphere and the sub-system often referred to as the hydrosphere, which excludes that part of the aquasphere that contains large quantities of ice at the poles and elsewhere is referred to as the cryosphere. Then there is the geosphere which consists of the interior and the surface of Earth, both of which are made up of rocks. The limited part of the planet that can support living flora and fauna comprises part of the geosphere that may also be referred to as the biosphere. Furthermore, the three systems (the aquasphere, the aquasphere, and the geosphere) and the various sub-systems (the hydrosphere, the cryosphere, and the biosphere) are complex systems that interact to maintain the earth system as it currently exists.

1.4.1 THE ATMOSPHERE

The atmosphere (Chapter 3) is the envelope of gases surrounding the Earth and it is subdivided into regions depending on the altitude (Parker and Corbitt, 1993). The constituents of the atmosphere are primarily nitrogen (N_2, 78.08% v/v), oxygen (O_2, 20.95% w/w) and water vapor (0 to 0.25% w/w), although the concentration of water vapor (H_2O) is highly variable, especially near the surface, where volume fractions can be as high as 4% in the tropics. There are many minor constituents or trace gases such as (alphabetically rather than by abundance, which is variable) argon (Ar), carbon dioxide (CO_2), helium (He), hydrogen (H_2), krypton (Kr), methane (CH_4). (neon, Ne), nitrous oxide (N_2O), ozone (O_3), and water vapor (H_2O) (Table 1.4) (Prinn, 1987).

In addition to the gaseous constituents, the atmosphere also contains suspended solid and liquid particles. Aerosols are particulate matter (PM) usually less than 1 micron in diameter (also called 1 micrometer which is equivalent to 1 meter × 10^{-6} in diameter, i.e., one-millionth of a meter or one-thousandth of a millimeter, 0.001 mm, or approximately 0.000039 of an inch) that are created by gas-to-particle reactions and are lifted from the surface by the winds. A portion of these aerosols can become centers of condensation or deposition in the growth of water and ice clouds. Cloud droplets and ice crystals are made primarily of water with some trace amounts of particles and dissolved gases. Their diameters range up to 100 micrometers. Water or ice particles larger than approximately 100 microns begin to fall because of gravity and may result in precipitation at the surface.

Ozone is found in trace quantities throughout the atmosphere, the largest concentrations being located in a layer in the lower stratosphere between the altitudes of 9 and 18 mi (15 and 30 km). This ozone results from the dissociation by solar ultraviolet (UV) radiation of molecular oxygen in the upper atmosphere and nitrogen dioxide in the lower atmosphere. Ozone also plays an important role in the formation of photochemical smog and in the purging of trace species from the lower atmosphere (Chapters 3, Chapter 8).

TABLE 1.4 Approximate Composition of the Atmosphere

Component, % v/v*	Amount, % v/v
	Major Components:
Nitrogen (N_2)	78.08
Oxygen (O_2)	20.95
	Minor/Trace Components:
Argon (Ar)	0.93
Carbon dioxide (CO_2)	0.035
Helium (He)	0.00052
Hydrogen (H_2)	0.00005
Krypton (Kr)	0.00010
Methane (CH_4)	0.00014
Neon (Ne)	0.018
Nitrous oxide (N_2O)	0.00005
Ozone (O_3)	0.0000007
Water vapor (H_2O)	0.025**

*Listed alphabetically

**Variable; can be as high as 4% v/v in humid areas.

The chemistry of ozone formation can be explained in relatively simple terms, although the reactions are believed to be much more complex (Chapter 3). Thus, above approximately 19 miles (30 km), oxygen is dissociated during the daytime by energy (hv) from UV light:

$$O_2 + hv \rightarrow O + O$$

The oxygen atoms produced then form ozone:

$$O + O_2 + M \rightarrow O_3 + M$$

In this equation, M is an arbitrary molecule required to conserve energy and momentum in the reaction that produces ozone.

Although present in only trace quantities (Table 1.4), atmospheric ozone plays a critical role for the biosphere by absorbing the UV radiation with a wavelength from 240 to 320 nanometers (nm, 1 nm = 1 meter × 10^9, which would otherwise be transmitted to the surface of the Earth.

The atmospheric ozone should not be confused with the ozone layer, which acts as a region of the stratosphere of the Earth (Chapter 3) that absorbs most of the UV radiation from the Sun. The ozone layer is mainly found in the lower portion of the stratosphere, from approximately 9 to 22 miles above the Earth, although the thickness of the layer varies seasonally and geographically. This layer is so-named because it contains a high concentration of ozone (O_3) in relation to other parts of the atmosphere, although still small in relation to other gases in the stratosphere. The ozone layer contains up to 10 parts per million of ozone, while the average ozone concentration in the atmosphere of the Earth as a whole is on the order of 0.3 parts per million.

The UV radiation is lethal to simple unicellular organisms (algae, bacteria, protozoa) and to the surface cells of higher plants and animals. It also damages the genetic material of cells (deoxyribonucleic acid, DNA) and is responsible for sunburn in human skin. In addition, the incidence of skin cancer has been statistically correlated with the observed surface intensities of the UV wavelengths from 290 to 320 nm, which are not totally absorbed by the ozone layer.

An important effect noted as a result of the changes in the constituents of the atmosphere is the tendency for the temperature close to the surface of the Earth to rise, a phenomenon referred to as the greenhouse effect. This term is used to describe the rise in the temperature of the Earth, analogous to the rise in temperature in a greenhouse when the energy from the Sun is trapped and cannot escape from the enclosed space (Figure 1.2) (Speight, 1996).

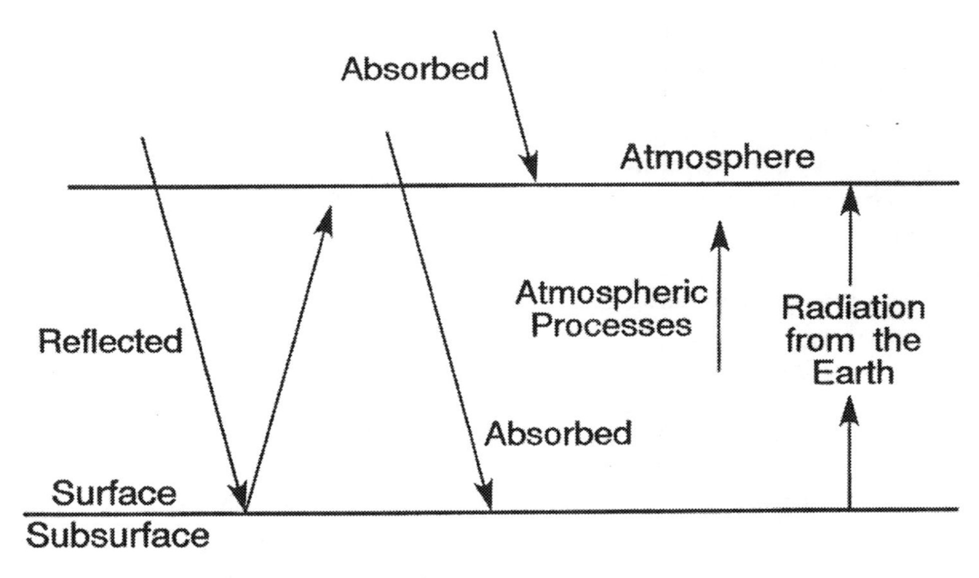

FIGURE 1.2 Illustration of the greenhouse effect.
Source: Reproduced with permission from: Speight (1996). © Taylor and Francis.

Although the term greenhouse effect has generally been used for the role of the whole atmosphere (mainly water vapor and clouds) in keeping the surface of the Earth warm, it has been increasingly associated (perhaps erroneously) with the contribution of carbon dioxide. However, there are various types of gases that are the result of industrial and domestic activities that can contribute to this effect and, thus, the continual rise in the surface temperature of the Earth (Mannion, 1991; Pickering and Owen, 1994).

The arguments related to the magnitude of the greenhouse effect have range back and forth for some time with carbon dioxide the main (if not the sole) cause without recognition of other causes, several of which are natural causes or events (Speight, 2020). There are those who believe that the Earth is doomed to a rise in temperature and serious harm to the human race is imminent. On the other hand, there are those observers who believe that there is no cause for concern related to the environment and humans can go on merrily, as has been the case for centuries and pollute the atmosphere without any concerns related to the consequences of such actions.

Whatever the correct side on which to base an argument, there is no doubt that the emissions, which can give rise to such an effect must be limited (Bradley et al., 1991). The continuous pollution of the atmosphere with the so-called greenhouse gases can be of no advantage to life on Earth, even if the effects of these gases are not manifested in a temperature rise but in the form of aggravating pollutants to flora and fauna.

Both of these opposite opinions are of some concern because they may mask the reality of the situation. It is analogous to other situations that have arisen in the last several decades. For example, in the late 1960s and early 1970s, there were warnings related to an approaching ice age. In fact, they were those observers who would have us believe that, upon looking out of the window, an observer(s) would see glaciers approaching from the end of the street! In fact, it is entirely likely that several important analysts who then warned of global cooling and imminent glaciation of northern societies are now warning of global warming (Easterbrook, 1995; Speight, 2020). The glaciers did not arrive and now, in a little more 50 years later, the frantic warnings are no related to a rise in global temperature! There are also advisories that the rise in temperature that is the basis of global warming is being accompanied (perceived or real) by the emergence (or reemergence) of a variety of infectious diseases. If, as has been suggested, the Earth has warmed 0.3 to 0.6°C (approximately l°F, or less) during this century, the perception is that a higher rise in the temperature of the Earth may lead to a series of global catastrophes.

These types of contradictory reports and arguments add much confusion to an already difficult area of technology. It seems that every time a government appropriates money to study an issue, the heretofore unheard-of-experts spring into action. What is really needed is a careful study of the data, the generation of new data, and less enthusiasm for catching the headlines.

Obviously, a major challenge to the atmospheric scientist, or for that matter to any environmental scientist and engineer, is to ensure that the assessment of the influence of any chemical on the environment, through the accurate measurement of the effects is well understood (Newman, 1993).

1.4.2 THE AQUASPHERE

The Earth is a unique planet insofar as there is an abundance of water that is necessary to sustaining life on the Earth, and helps tie together the atmosphere, the land (the geosphere), the oceans and rivers (the aquasphere) into an integrated system. Precipitation, evaporation, freezing, and melting and condensation are all part of the hydrological cycle, which is-a never-ending global process of water circulation from clouds to land, to the ocean, and back to the clouds. This cycling of water is intimately linked with energy exchanges among the atmosphere, ocean, and land that determine the climate of the Earth and cause much of natural climate variability. The impacts of climate change and variability on the quality of human life occur primarily through changes in the water cycle.

The water systems of the Earth (Chapter 4), often referred to as the aquasphere or the hydrosphere, refers to water in various forms: oceans, lakes, streams, snowpack, glaciers, the polar ice caps, and water under the ground (groundwater) (Parker and Corbitt, 1993). An important aspect of the water system is an aquifer which is a water-bearing (water-rich) subsurface formation (a subsurface zone) that yields water to wells. An aquifer may be porous rock, unconsolidated gravel, fractured rock, or cavernous limestone. Aquifers are important reservoirs storing large amounts of water which, in theory, should be relatively free from evaporation loss or pollution. However, in practice, this is not always the case.

The oceans play a key role in the water cycle insofar as the oceans hold 97% v/v of the total water on the Earth and 78% v/v of the global precipitation that occurs over the oceans, and it is the source of 86% v/v of global evaporation. Besides affecting the amount of atmospheric water vapor and hence rainfall, evaporation from the sea surface is important in the movement of heat in the climate system. Water evaporates from the surface of the ocean, mostly in warm, cloud-free subtropical seas. This continuing event cools the surface of the ocean, and a large amount of heat absorbed by the ocean partially buffers the greenhouse effect from increasing carbon dioxide and other gases. Water vapor carried by the atmosphere condenses as clouds and falls as rain, mostly in the intertropical convergence zone (ITCZ), far from where it evaporated. Condensing water vapor releases latent heat and which drives much of the atmospheric circulation in the tropics. This latent heat release is an important part of the heat balance of the Earth, and it couples the planet's energy and water cycles.

By way of explanation, the ITCZ, known by sailors as the doldrums or the calms because of the monotonous, windless weather, is the area where the northeast and southeast trade winds converge. The zone encircles Earth near the thermal equator, although the specific position of the zone can vary on a seasonal basis. When the zone lies near the geographic equator, it is referred to as the near-equatorial trough. When the ITCZ is drawn into and merges with a monsoonal circulation, the zone is sometimes referred to as a monsoon, a usage that is more common in Australia and parts of Asia.

The major physical components of the global water cycle include the evaporation from the ocean and land surfaces, the transport of water vapor by the atmosphere, precipitation onto the ocean and land surfaces, the net atmospheric transport of water from land areas to ocean, and the return flow of freshwater from the land back into the ocean. The additional components of oceanic water transport are few, including the mixing of freshwater through the oceanic boundary layer, transport by ocean currents, and sea ice processes.

On land, the situation is more complex, and includes the deposition of rain and snow on land; water flow in runoff; infiltration of water into the soil and groundwater; storage of water in soil, lakes, and streams, and groundwater; polar and glacial ice; and use of water in vegetation and human activities. Processes labeled include precipitation, condensation, evaporation, evapotranspiration (from tree into atmosphere), radiative exchange, surface runoff, groundwater and streamflow, infiltration, percolation, and soil moisture. Furthermore, in the water systems (particularly in the lakes), the term eutrophication becomes important (Chapter 4). Eutrophication is the deterioration of the esthetic and life supporting qualities of lakes and estuaries, caused by excessive fertilization from effluents high in phosphorus, nitrogen, and organic growth substances. Algae and aquatic plants become excessive, and when they decompose, a sequence of objectional features arises.

Water for human consumption (and, in many cases as on farms for animal consumption) from such lakes must be filtered and treated. Diversions of sewage, better utilization of manure, erosion control, improved sewage treatment and harvesting of the surplus aquatic crops alleviate the symptoms.

1.4.3 THE GEOSPHERE

Land systems (Chapter 5) are those components that contribute to the earth system (Skinner and Porter, 1987; Parker and Corbitt, 1993). The land systems (also called the geosphere in this book) constitute the terrestrial component of the systems of the Earth and encompass all processes and activities related to the human use of land, including technological and organizational investments and arrangements, as well as the benefits gained from land and the unintended social and ecological outcomes of societal activities. Changes in land systems have large consequences for the local environment and human well-being and are at the same time pervasive factors of global environmental change. In more specific terms, the lithosphere refers to the minerals in the rust of the Earth, whereas the term geosphere is often more broad in coverage and refers to the complex and variable mixture of minerals, organic matter, water, and air which make up the soil.

The land provides vital resources to society, such as food and fuel and many other ecosystem services that support production functions, regulate risks of natural hazards, or provide cultural services. By using the land, changes are the direct result of human decision making which range from local landowners decisions to national scale planning and trade agreements. The aggregate impact of many local land system changes has far reaching consequences for the earth system, that feedback on ecosystem services, human well-being, and decision making. As a consequence, land system change is both a cause and consequence of socio-ecological processes.

Briefly, as used in this book, the term earth system refers to the interacting physical, chemical, and biological processes that occur on the Earth. The system consists of the land, oceans, atmosphere, and poles. It includes the planet's natural cycles-the carbon, water, nitrogen, phosphorus, sulfur, and other cycles and deep earth processes. In addition, life is also an integral part of the earth system. Life affects the carbon, nitrogen, water, oxygen, and many other cycles and processes. Two other definitions include the terms global change and global climate change.

Global change refers to planetary-scale changes in the earth system. More completely, the term global change encompasses planetary-scale changes to atmospheric circulation, ocean circulation, climate, the carbon cycle, the nitrogen cycle, the water cycle and other cycles, sea-ice changes, sea-level changes, food webs, biological diversity, pollution, health, fish stocks, and more. Global climate change refers to the long-term average of the aggregation of all components of weather of the components of the climate of the Earth: precipitation, temperature, and cloudiness, for example. The climate system includes processes involving ocean, land, and sea ice in addition to the atmosphere. Many changes in earth system functioning directly involve changes in climate. However, the earth system includes other components and processes, biophysical, and human, that are important for its functioning. Some earth system changes, natural or driven by humans, can have significant consequences involving changes in climate (Speight, 2020).

The lithosphere is the outermost shell of the Earth that is composed of the crust and the portion of the upper mantle that behaves elastically on time scales of thousands of years or greater; defined on the mineralogy. The lithosphere is subdivided into tectonic plates which are the uppermost part of the lithosphere that chemically reacts to the atmosphere, the hydrosphere and the biosphere through the soil-forming process (called the pedosphere). There are two types of lithosphere which are: (i) the oceanic lithosphere, which is associated with the oceanic crust and exists in the oceanic basins and has a density of approximately of about 2.9 grams per cubic centimeter; and (ii) the continental lithosphere, which is associated with the continental crust and has an order of 2.7 grams per cubic centimeter.

The oceanic lithosphere consists mainly of a mafic mantle (mafic: an adjective describing a silicate mineral or igneous rock that is rich in magnesium and iron) crust and ultramafic mantle (over 90% mafic) and is denser than continental lithosphere. It thickens as it ages and moves away from the mid-ocean ridge. This thickening occurs by conductive cooling, which converts the hot asthenosphere into the lithospheric mantle. It was less dense than the asthenosphere for tens of millions of years, but after this became increasingly denser. The gravitational instability of mature oceanic lithosphere has the effect that when tectonic plates come together, the oceanic lithosphere invariably sinks underneath the overriding lithosphere. New oceanic lithosphere is constantly being produced at mid-ocean ridges and is recycled back to the mantle at subduction zones, so the oceanic lithosphere is much younger than its continental counterpart. The oldest oceanic lithosphere is about 170 million years old compared to parts of the continental lithosphere which are billions of years old.

On the other hand, the continental lithosphere (also called the continental crust) is the layer of igneous, sedimentary rock that forms the continents and the continental shelves. This layer consists mostly of granitic rock. Continental crust is also less dense than oceanic crust, although it is considerably thicker. Approximately 40% of the surface of the Earth is covered by continental crust, but continental crust makes up about 70% v/v of the crust of the Earth. The continental crust is believed to have been derived from the fractional differentiation of oceanic crust over geologic time (hundreds of millions of years) and was primarily a result of volcanism and subduction.

The lithosphere is underlain by the asthenosphere, which is the weaker, hotter, and deeper part of the upper mantle. The lithosphere-asthenosphere boundary is defined by a difference in response to stress in which the lithosphere remains rigid for long periods of geologic time in which it deforms elastically and through brittle failure. On the other hand, the asthenosphere undergoes considerable deformation and accommodates strain through this deformation.

The term biosphere refers to living organisms and their environments on the surface of the Earth (Manahan, 1991). Included in the biosphere are all environments capable of sustaining life above, on, and beneath the surface of the Earth, as well as in the oceans. Consequently, the biosphere includes virtually all of the water systems (the aquasphere, the hydrosphere) as well as portions of the atmosphere and the upper lithosphere (land systems).

In addition, there are relationships between the atmosphere, the water systems, the land systems, and the biosphere. These relationships are physical and are also caused by environmental forces. It is worth noting here that such relationships, perhaps not in this exact concept, were noted more than four centuries ago. Paracelsus (Philippus Aureolus Theophrastus Bombast von Hohenheim; 1493–1541) taught that the macrocosm (the heavens) and the microcosm (the Earth and all of the living creatures) were linked together and that the macrocosm directed the growth and development of the microcosm (Huser, 1589).

In the modern relationships between the atmosphere, the water systems, and the land systems, acid rain (Chapters 3–5) is the precipitation phenomenon that incorporates anthropogenic acids (i.e., those acids which are the result of human activities) and other acidic materials. The deposition of acidic materials into the water systems and onto the land occurs in both wet and dry forms as rain, snow, fog, dry particles, and gases.

The effect of acid rain (acid deposition) on a particular ecosystem depends largely on the sensitivity of the ecosystem to acid deposition (Johnson and Gordon, 1987). There is also the ability of the ecosystem to neutralize the acid, as well as the concentration and composition of acid reaction products and the amount of acid added to the system.

The trace element is a term that refers to those elements that occur at low levels in a given system. The somewhat ambiguous term probably arose from the inadequacy of earlier analytical techniques-before modem methods such as atomic absorption, plasma emission, neutron-activation analysis, gas chromatography, and mass spectrometry extended the limits of detection to the low levels currently attainable. In many early investigations, it was only possible to detect the presence of an element, as it was said to be present at a trace level. A reasonable definition of a trace element is that the element occurs at a level of a few parts per million or less. The term trace substance is a more general definition that is applied to both elements and chemical compounds.

There are a variety of trace elements encountered in natural waters, some of which are the nutrients required for animal and plant life. Of these, many are essential at low levels but toxic at higher levels. This is typical for many substances in the aquatic environment, a point that must be kept in mind in judging whether a particular element is beneficial or detrimental. Some of these elements, such as lead or mercury, have such toxicological and environmental significance that they are discussed in detail in separate sections.

Some of the heavy metals are among the most harmful elemental pollutants. These metals include essential elements like iron as well as toxic metals like lead (Pb), cadmium (Cd), and mercury (Hg). Most of them have a tremendous affinity for sulfur and attack sulfur bonds in enzymes, thus immobilizing the enzymes. Protein carboxylic acid (–COOH) derivatives and amino ($-NH_2$) groups are also chemically bound by heavy metals. Cadmium, copper, lead, and mercury ions bind to cell membranes, hindering transport processes through the cell wall. Heavy metals may also precipitate phosphate compounds or catalyze their decomposition.

1.4.4 THE EARTH SYSTEM

Not only do the Earth systems overlap, but they are also interconnected insofar as whatever affects one of the Earth systems can affect another of the Earth systems. For example, when part of the air in the atmosphere becomes saturated with water, precipitation, such as rain or snow, can fall to the surface of the Earth. That precipitation connects the hydrosphere with the geosphere by promoting erosion and weathering of the rocks, and these two surface processes (erosion and weathering) slowly break down large rocks into smaller ones. Over time, the erosion and weathering processes change large pieces of rocks into sediments, such as sand or silt. The cryosphere can also be involved in erosion, as large glaciers scour bits of rock from the bedrock beneath them. The geosphere includes all the rocks that makeup Earth, from the partially melted rock under the crust, to ancient, towering mountains, to grains of sand on a beach.

The many interactions between Earth systems are complex and are happening constantly, but the effects are not always obvious. There are some extremely dramatic examples of Earth systems interacting, such as volcanic eruptions and tsunamis, but there are also slow, nearly undetectable changes that alter ocean chemistry, the content of the atmosphere, and the microbial biodiversity in soil. Each part of the Earth, from the inner core to the top of the atmosphere, has a role in making Earth home to billions of lifeforms.

Both the geosphere and hydrosphere provide the habitat for the biosphere, a global ecosystem that encompasses all the living things on Earth. The biosphere refers to the relatively small part of the environment of the Earth in which living things (the flora and fauna) can survive. Thus, the biosphere contains a wide range of organisms, including fungi, plants, and animals, that live together as a community. Biologists and ecologists refer to this variety of life as biodiversity. All the living things in an ecosystem are often referred to as the biotic factors while the non-living things that organisms require to survive, such as water, air, and light are often referred to as the abiotic factors.

1.5 ECOLOGICAL CYCLES

The ecological cycles are the various self-regulating processes that recycle the limited resources of the Earth (water, carbon, nitrogen, and other elements) that are essential to sustain life. Understanding how local cycles fit into global cycles is essential to make the best possible management decisions to maintain ecosystem health and productivity for now and the future.

There are biogeochemical cycles for the chemical elements (such as calcium, carbon, hydrogen, mercury, nitrogen, oxygen, phosphorus, selenium, and sulfur) as well as; molecular cycles for water and silica; macroscopic cycles such as the rock cycle as well as human-induced cycles for synthetic compounds such as polychlorinated biphenyl derivatives (PCBs). In some cycles there are reservoirs where a substance remains for a long period of time.

There are several ecological cycles that also need a clearer and more detailed some definition as each cycle plays an important role in the interrelationship of the various ecological systems to the environment as a whole (Prinn, 1987; Clark, 1989; Graedel and Crutzen, 1989; Schneider, 1989; Maurits la Riviere, 1989; Frosch and Gallopoulos, 1989). These cycles can be perturbed by anthropogenic stress as well as by natural occurrences.

For example, the oxygen cycle (Figure 1.3) (Speight, 1996) illustrates the role of oxygen in the land water and air either as oxygen per se or combined with carbon as carbon dioxide and as mineral carbonate derivatives. The cycle also includes industrial activities (such as fossil fuel burning) and natural activities (represented by volcanic activity). The nitrogen cycle (Figure 1.4) (Speight, 1996) relates not only to nitrogen in the atmosphere but also to nitrogen fixation in the soil and its use in plant growth.

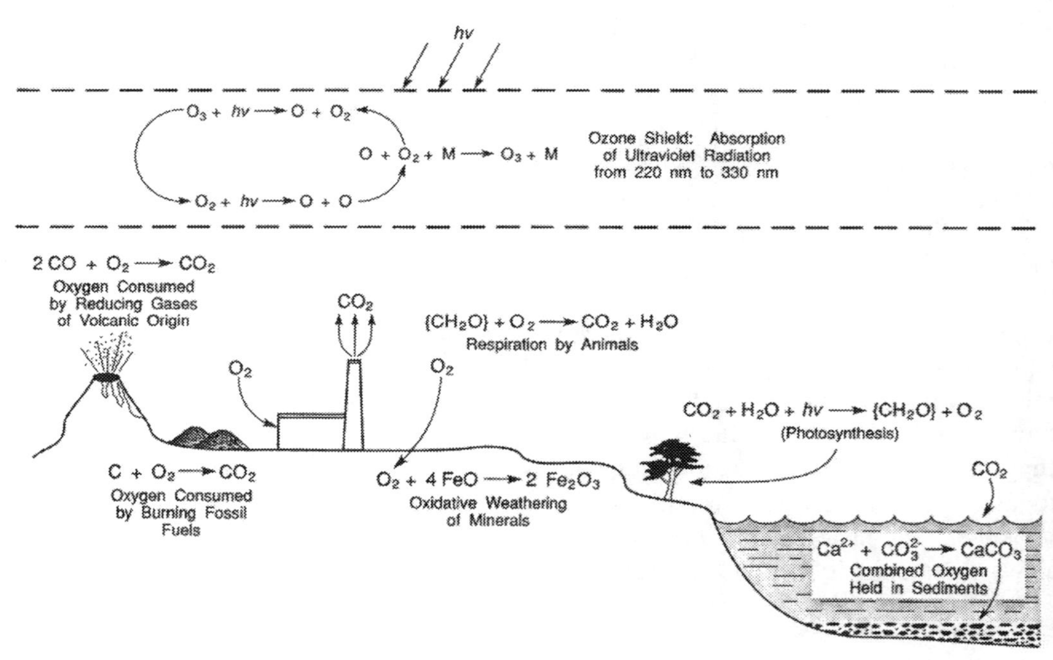

FIGURE 1.3 The oxygen cycle.
Source: Reproduced with permission from: Speight (1996). © Taylor and Francis.

The water cycle (also referred to as the hydrologic cycle) (Chapter 4) is the means by which water is transported throughout several ecosystems. In theory, there should be a harmonious balance between the water movement on the Earth. Indeed, there should be a harmonious balance between the atmosphere, the water systems, and the land systems (Gore, 1993). But the existence of contrasting desert areas and tropical rain forests speaks to the apparent imbalance in these cycles, particularly in the hydrological cycle. A subcategory of the hydrological cycle involves the dispersion of water in various groundwater systems. These systems are responsible for the availability of water to the various flora and fauna in a particular region. The water availability of water from groundwater systems is considered, perhaps, to be more important than water availability from lakes and rivers. However, the interrelationships of water in the whole hydrological cycle are the most important features of this cycle.

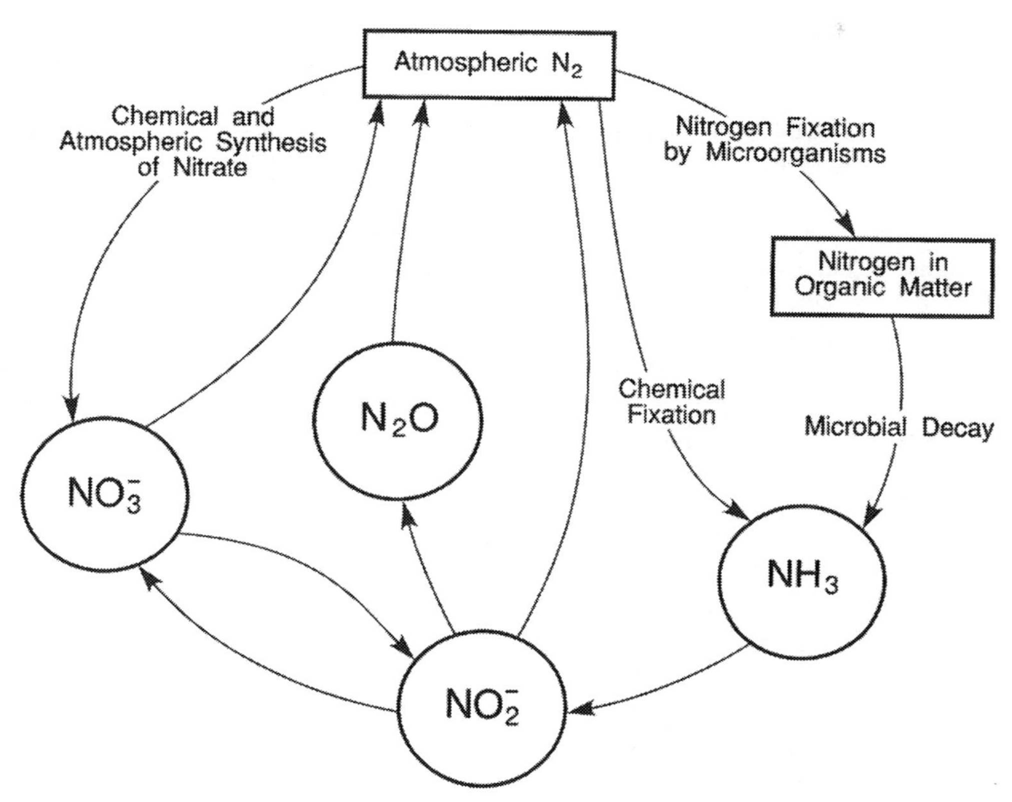

FIGURE 1.4 The nitrogen cycle.
Source: Reproduced with permission from: Speight (1996). © Taylor and Francis.

For example, the biogeochemical cycle (also called the substance turnover or cycling of substances) is a pathway by which a chemical substance moves through biotic compartments the biosphere) and through abiotic compartments of the Earth (such as the atmosphere, the hydrosphere, and the lithosphere).

The Earth system contains several 'great cycles' in which key materials are transported through the environment. In general, cycles occur in closed systems; at the global scale, many systems may be assumed to be closed because the Earth receives negligible quantities of minerals from space (as a result of meteorite impacts) and because only limited quantities of materials can escape the atmosphere of the Earth.

The biogeochemical cycles operate at the global scale and involve all of the main components of the Earth System in which materials are transferred continually between the atmosphere, the aquasphere, and the geosphere. However, since the biogeochemical cycles involve elements that are essential for life, organisms play a vital part in those cycles. Typically then, the biogeochemical cycles involve an inorganic component (the abiotic part of the cycle, including sedimentary and atmospheric phases) and an organic component (comprising plants and animals, both living and dead). Like other environmental systems, biogeochemical cycles involve the flow of substances between stores (also known as reservoirs) in the geosphere, atmosphere, hydrosphere, and biosphere. Water plays a vital role in mediating many of the flows between stores. Three of the key biogeochemical cycles are the nitrogen, carbon, and sulfur cycles.

The nitrogen cycle is a relatively fast and complex cycle. Most of the atmosphere consists of gaseous nitrogen which is fixed (in other words, made available for use by plants) biologically in soils. Soil bacteria convert nitrogen to ammonia; this, together with inorganic nitrate, is absorbed by plant roots and converted to organic compounds (such as proteins) in plant tissues. These compounds are eaten and consumed by herbivorous animals. In turn, nitrogenous compounds are passed to carnivores, and they are ultimately returned to the soil in the form of nitrogenous waste products (such as urine and fecal matter) and as a result of the death and decomposition of organisms. Bacteria then convert the organic nitrogen compounds into ammonia and ammonium compounds, which are then converted by bacteria into nitrites and then nitrates, which are then available for re-uptake by plants.

Some of the nitrogenous compounds that are not absorbed by plants are leached from the soil into groundwater, surface water and ultimately into seas and oceans. Of that nitrogenous material, some is used by aquatic plants, some accumulates as organic sediment, and some evaporates into the atmosphere. The cycle is completed by denitrifying bacteria which eventually convert nitrates and nitrites to ammonia, nitrogen, and nitrogen oxides.

The carbon cycle is a cycle in which carbon is stored in the atmosphere in the form of carbon dioxide, which is absorbed by plants and converted to carbohydrates by the process of photosynthesis. The cycle then follows food chains, with carbohydrates being consumed by herbivores and then carnivores, being metabolized during the process of respiration. Carbon dioxide is returned to the atmosphere as animals exhale and when organic waste and dead organisms' decay. Vegetation and animals are thus important stores of carbon, although that carbon may be rapidly returned to the atmosphere if vegetation is burned. Soils are also important reservoirs for carbon. Atmospheric carbon dioxide is soluble in water, in which, it forms carbonic acid, which forms bicarbonate ions and carbonate ions, which in turn form salts (such as the insoluble calcium carbonate, which accumulates in marine sediments, marine organisms and carbonate rocks, such as limestone). Carbon is typically stored in these forms until it is released to the atmosphere by chemical weathering.

The sulfur cycle is a cycle in which sulfur is released into the atmosphere during volcanic eruptions (in the forms of sulfurous gas, dust, and particles) and as a result of the weathering of rocks. The oceans also play an important role in the sulfur cycle, as marine phytoplankton produce dimethyl sulfide, some of which enters the atmosphere and is converted to sulfur dioxide and sulfate aerosols. These compounds are ultimately converted to sulfuric acid and are deposited on the surface of the Earth as part of the precipitation. In terrestrial ecosystems, bacteria break down sulfurous compounds and release the sulfur to the atmosphere again, mainly in the form of hydrogen sulfide.

Finally, it must be noted that the biogeochemical cycles have been modified substantially by human activities, which is necessary for consideration to understand the various environmental issues related to human activities.

1.6 ENVIRONMENTAL ETHICS

Last but certainly not the least, there is the evolving sub-discipline of environmental ethics, which involves the study of the moral relationship of human beings with the environment and its non-human contents. This is especially important to scientists and engineers who are working on issues related to maintaining an environment that is as close as possible to a pristine environment.

The concept of environmental ethics developed into a specific philosophical discipline in the 1970s due to the increasing awareness in the 1960s of the effects that technology, industry, economic expansion and population growth were having on the environment (Carson, 1962; Attfield, 1983; Taylor, 1986; Warren, 2000). However, pollution and the depletion of natural resources have not been the only environmental concerns since that time and issue such as: (i) dwindling plant and animal biodiversity; (ii) the loss of wilderness areas; (iii) the degradation of ecosystems; and (iv) climate change are all part of a raft of issues that have implanted themselves into both public consciousness and public policy over subsequent years.

Thus, environmental ethics deals with the issues related to the rights and duties of individuals that are fundamental to life and well-being of present human society, future generations (of human), as well as of other living beings present on the Earth. Environmental ethics evolved in the 1970s as a sub-discipline of environmental science and environmental engineering. The subject differs from traditional ethics, which is concerned with relationships among humans people only. The need of environmental ethics has arisen as a result of the following three major factors: (i) modern technological civilization has been affecting nature greatly; therefore, there is a need to analyze the consequences of human actions; (ii) as a result of the advancement of science, human understanding about nature and environmental problems is increasing; and (iii) there is a concern- or belief-that other living species have equal rights to live on the Earth is also raising the need for environmental ethics.

Therefore, it is necessary to understand and accept that the life-supporting environment of the Earth is the outcome of complex interaction of innumerable physical, chemical, and biological factors. With the advancement in the methods of scientific investigation, the human understanding of these factors is also increasing.

The term conservation ethics is an extension of use-value into the non-human biological world, which focuses only on the worth of the environment in terms of its utility or usefulness to humans. The concept generally advocates preservation of the environment on the basis that it has extrinsic value insofar as it is instrumental to the welfare of human beings. Conservation is, therefore, a means to an end and purely concerned with mankind and inter-generational considerations.

Conservation ethics is related to anthropocentrism, which advocates that humans are the most important or critical element in any given situation; that the human race must always be its own primary concern. Detractors of the anthropocentrism concept argue that the Western tradition biases homo sapiens when considering the environmental ethics of a situation and that humans evaluate their environment or other organisms in terms of utility for them. Many argue that all environmental studies should include an assessment of the intrinsic value of non-human beings. In fact, based on this very assumption, a philosophical article has recently explored the possibility of humans' willing extinction as a gesture toward other beings.

Thus, environmental ethics is the part of environmental philosophy which considers extending the traditional boundaries of ethics from solely including humans to including the nonhuman world. It exerts influence on a large range of disciplines including law, sociology, theology, economics, ecology, and geography. Environmental ethics says that humans should base their behavior on a set of ethical values that guide the human approach toward the other living beings in nature. Even if the human race is considered the primary concern of society, plants, and animals (flora and fauna) are also important have the right of existence.

KEYWORDS

- **carbon monoxide**
- **environmental impact assessment**
- **hydrogen peroxide**
- **intertropical convergence zone**
- **polychlorinated biphenyl derivatives**

REFERENCES

Agricola, G., & Bauer, G., (1556). De Re Metallica. Froben, Basel, Switzerland.

ASTM C294, (1995). *Descriptive Nomenclature of Constituents of Natural Mineral Aggregates.* Annual Book of ASTM Standards, ASTM International, West Conshocken, Pennsylvania.

Attfield, R., (1983). *The Ethics of Environmental Concern.* Blackwell books, Oxford, United Kingdom.

Benarde, M. A., (1989). *Our Precarious Habitat: Fifteen Years Later.* John Wiley & Sons Inc., New York.

Bradley, R. A., Watts, E. C., & Williams, E. R., (1991). Limiting net greenhouse gas emissions in the United States. *Volume I: Energy Technologies and Volume II: Energy Responses.* United States Department of Energy, Washington, D.C.

Brady, G. L., & Geets, P. C. F., (1994). *International Journal of Sustainable Development and World Ecology, 1*(3), 189.

Breen, J. J., & Dellarco, M. J., (1992). *Pollution Prevention in Industrial Processes.* Symposium Series No. 508. American Chemical Society, Washington, D.C.

Carson, R., (1962). *Silent Spring.* Houghton Mifflin, Boston, Massachusetts.

Cartwright, F. F., & Biddis, M. D., (1991). *Disease and History.* Dorset Press, New York.

Chenier, P. J., (1992). *Survey of Industrial Chemistry* (2nd edn.). VCH Publishers Inc., New York.

Easterbrook, G., (1995). *A Moment on the Earth: The Coming Age of Environmental Optimism.* Viking Press, New York.

Fumento, M., (1993). *Science Under Siege: Balancing Technology and the Environment* (p. 372). William Morrow and Company Inc., New York.

Galloway, R. L., (1882). *A History of Coal Mining in Great Britain.* Macmillan & Co., London, United Kingdom.

Gimpel, J., (1976). *The Medieval Machine: The Industrial Revolution of the Middle Ages.* Holt, Reinhart, and Winston, New York. Chapter 4.

Gore, A., (1993). *Earth in the Balance: Ecology and the Human Spirit*. Penguin Books USA Inc., New York.
Huntley, D., (2020). *The Plague Village* (Vol. 41, No. 5, pp. 21–23). British Heritage Magazine.
Huser, (1589). *Paracelsus: Opera Omnia*. Huser, Basel, Switzerland.
James, P., & Thorpe, N., (1994). *Ancient Inventions*. Ballantine Books, New York.
Jazib, J., (2018). *Basics of Environmental Sciences*. Iqra Publishers, New Delhi, India.
Johnson, R. W., & Gordon, G. E., (1987). *The Chemistry of Acid Rain: Sources and Atmospheric Processes*. Symposium Series No. 349. American Chemical Society, Washington, DC.
Kester, J. E., Hattemer-Frey, H. A., & Krieger, G. R., (1994). In: Ayers, K. W., Deb, K., & Hattemer-Freyer, H. A., (eds.), *Environmental Science and Technology Handbook*.
Kletz, T. A., (1990). *Improving Chemical Engineering Practices*. Hemisphere Publishing Corp., London, United Kingdom.
Krauss, L. M., (1992). In: Shore, W. H., (ed.), *Mysteries of Life and the Universe* (p. 47). Harcourt Brace and Co., New York.
Krech, S., McNeill, J. R., & Merchant, C., (2003). *Encyclopedia of World Environmental History* (Vol. 1–3). Routledge, Taylor & Francis Group, London, United Kingdom.
Lipton, S., & Lynch, J., (1994). *Handbook of Health Hazard Control in the Chemical Process Industry*. John Wiley & Sons Inc., New York.
Lopez, B. H., (1978). *Of Wolves and Men*. Charles Scribner and Sons Inc., New York.
Manahan, S. E., (1991). *Environmental Chemistry*. Lewis Publishers Inc., Chelsea, Michigan.
Mannion, A. M., (1991). *Global Environmental Change: A Natural and Cultural Environmental History*. Longman Scientific and Technical Publishers, Harlow, Essex, United Kingdom.
Milman, H. A., & Weisburger, E. K., (1994). *Handbook of Carcinogen Testing* (2nd edn.). Noyes Data Corp., Park Ridge, New Jersey.
Mooney, H., (1988). *Towards an Understanding of Global Change*. National Academy Press, Washington, D.C.
Mowat, F., (1963). *Never Cry Wolf Little*. Brown and Co., New York.
National Research Council, (1981). *Prudent Practices for Handling Hazardous Chemicals in Laboratories*. National Academy Press, Washington, D.C.
Newman, L., (1993). *Measurement Challenges in Atmospheric Chemistry*. Advances in Chemistry Series No. 232. American Chemical Society, Washington, D.C.
O'Riordan, T., (1995). *Perceiving Environmental Risks*. Academic Press Inc., San Diego, California.
Parker, S. P., & Corbitt, R. A., (1994). *Encyclopedia of Environmental Science and Engineering*. McGraw-Hill, New York.
Pfafflin, J. R., & Ziegler, E. N., (2006). *Environmental Science and Engineering* (5th edn.). CRC Press, Taylor & Francis Group, Boca Raton, Florida.
Pickering, K. T., & Owen, L. A., (1994). *Global Environmental Issues*. Routledge Publishers, New York.
Prinn, R. G., (1987). In: Parker, S. P., (ed.), McGraw-Hill Encyclopedia of Science and Technology (Vol. 2. pp 171 & 185). McGraw-Hill, New York.
Ray, D. L., & Guzzo, L., (1990). *Trashing the Planet: How Science Can Help Us Deal with Acid Rain, Depletion of the Ozone, and Nuclear Waste (Among Other Things)*. Regnery Gateway, Washington, D.C.
Rittenberg, S. C., (1987). In: Parker, S. P., (ed.). *McGraw-Hill Encyclopedia of Science and Technology* (Vol. 19. p. 317). McGraw-Hill, New York.
Skinner, B. J., & Porter, S. C., (1987). *Physical Geology*. page. John Wiley & Sons, Hoboken, New Jersey.
Speight, J. G., (1996). *Environmental Technology Handbook*. Taylor & Francis, Washington, DC.
Speight, J. G., (2013). *The Chemistry and Technology of Coal* (3rd edn.). CRC Press, Taylor & Francis Group, Boca Raton, Florida.
Speight, J. G., (2014). *The Chemistry and Technology of Petroleum* (5th edn.). CRC Press, Taylor & Francis Group, Boca Raton, Florida.
Speight, J. G., (2019). *Natural Gas: A Basic Handbook* (2nd edn.). Gulf Publishing Company, Elsevier, Cambridge, Massachusetts
Speight, J. G., (2020). *Global Climate Change Demystified*. Scrivener Publishing, Beverly, Massachusetts.
Stensland, G. J., (2006). Acid rain: In: Pfafflin, J. R., & Ziegler, E. N., (eds.), *Environmental Science and Engineering* (5th edn., Vol. 1. Pp. 1–14). CRC Press, Taylor & Francis Group, Boca Raton, Florida.

Stensvaag, J. M., (1991). *Clean Air Act Amendments: Law and Practice*. John Wiley and Sons Inc., New York.
Taylor, P. W., (1986). *Respect for Nature: A Theory of Environmental Ethics.* Princeton University Press, Princeton University, Princeton, New Jersey.
Tedder, D. W., & Pohland, F. G., (1993). *Emerging Technologies in Hazardous Waste Management III.* Symposium Series No. 518. American Chemical Society, Washington, D.C.
Tester, J. W., Wood, D. O., & Ferrari, N. A., (1991). *Energy and the Environment in the 21st Century.* The MIT Press, Cambridge, Massachusetts.
Turnbull, S., (1995). *The Book of the Medieval Knight.* Arms and Armor Press, London, United Kingdom.
United States Congress, (1990). Public law 101-549. *An Act to Amend the Clean Air Act to Provide for Attainment and Maintenance of Health Protective National Ambient Air Quality Standards, and for Other Purposes.*
Warren, M. A., (2000). *Moral Status: Obligations to Persons and Other Living Things.* Oxford University Press, Oxford university, Oxford, United Kingdom.
Zakrzewski, S. F., (1991). *Principles of Environmental Toxicology.* American Chemical Society, Washington, DC.

CHAPTER 2

Resources and Resource Utilization

2.1 INTRODUCTION

Resources are those usable materials that occur naturally within the Earth and are often referred to as natural resources, which are usually either: (i) non-renewable; or (ii) renewable (Table 2.1). The former category refers to those resources, as the name implies, are those that are no longer available once the available resource at a site is exhausted. In general terms, the development of resources is usually concerned with the production or use of energy. These resources have been useful to humans in a variety of ways, either directly in the original form (such as coal) or indirectly after certain modifications (such as crude oil products). At one time, the Earth was believed to have an inexhaustible supply of natural resources. However, the fossil fuel resources are not unlimited and inexhaustible, but each resource has a limited lifetime that is dependent upon the rate of use and, in fact, have been depleted to an irreparable extent (Jazib, 2018).

By way of definition and clarification, the term mineral resources, which are non-renewable (i.e., the fossil fuels-natural gas, crude oil, and coal) are carbonaceous resources which have been formed over geological time from the remains of buried prehistoric organisms and currently (in the first two decades of the 21st Century) account for the majority of the total energy consumption of the world and this usage (based on the rate of current consumption and the amount of each resource that remains) is projected to continue for the next several decades (Speight, and Islam, 2006). On the other hand, natural renewable resources, such as wind and water, have been used throughout recorded history, and until recently were a predominant energy source. In fact, in many developing countries, biomass is the primary energy source for the majority of the population (Speight, 2011b).

The resources that are the subject of this chapter include the fossil fuels (natural gas, crude oil, and coal), mineral resources, wood, and other renewable energy resources (Figure 2.1) (Cassedy and Grossman, 1990; Johansson et al., 1993; Pickering and Owen, 1994; Speight, 1996, 2020a). In terms of energy production, it is the fossil fuel resources that are have been by far, and will continue to be for the next three-to-five decades which are the most important resources insofar as they constitute the majority of the energy produced in the major industrialized nations. Moreover, the relationship of energy resource development and the effects of this development and use on the environment is well known (Pickering and Owen, 1994). In addition, mineral ores which yield radioactive materials

The Science and Technology of the Environment. James G. Speight (Author)

on processing are also included under the general term minerals. In addition, there are also questions relating to the policies which relate to resource use (Cooper, 1994). The issues from such questions, although not the subject of this text, must be considered when resource development is planned.

TABLE 2.1 Categorization of Natural Resources

Category	Type	Description
Non-renewable	Fossil fuel	Natural gas
–	–	Crude oil
–	–	Heavy crude oil
–	–	Extra heavy crude oil
–	–	Tar sand bitumen
–	–	Coal
–	–	Oil shale
–	Nuclear	Radioactive metals such as plutonium, thorium, and uranium
Renewable	Biomass	Wood
–	–	Grasses
–	–	Other vegetation
–	Geothermal energy	Hot water
–	–	Natural steam
–	Hydro energy	River
–	Ocean energy	Ocean heat
–	–	Ocean current
–	–	Wave energy
–	Solar energy	Solar thermal conversion
–	–	Photoelectric energy
–	–	Photochemical conversion
–	–	Stored solar hear
–	Tidal energy	Tidal energy
–	Wind energy	Windmill energy

The use of natural resources has continued in an as-needed (often haphazard and uncontrolled) manner for several centuries (and even millennia) as well as into prehistoric times, if the use of wood for fires and the use of specific ores to produce tools are included (Singer et al., 1958). Industrial operations, and there are many (Mooney, 1988; Austin, 1984), which produce products from a natural resource must make (and are making) every attempt to ensure that natural resources such as air, land, and water remain unpolluted. The environmental aspects of such an operation need to be carefully, and continually, addressed (Chenier, 1992).

Whilst the focus of this text is on the environmental aspects of resource utilization, it is necessary to understand the means by which pollution can occur. An important aspect, therefore, is an understanding of the various cycles that exist in the atmosphere, the aquasphere, and the geosphere (Chapter 1). For the present purposes, it is more pertinent to consider the use of natural resources since the onset of the Industrial Revolution in the late 1700s when the use of the natural resources of the Earth, which has focused on the production of energy. Therefore, it is necessary to understand the relationships that exist between a natural resource, the commodity produced from that resource, and the environment, which can be represented by a simple equation of progression from one to the other:

$$\text{Natural Resource} \rightarrow \text{Commodity} \rightarrow \text{Environment}$$

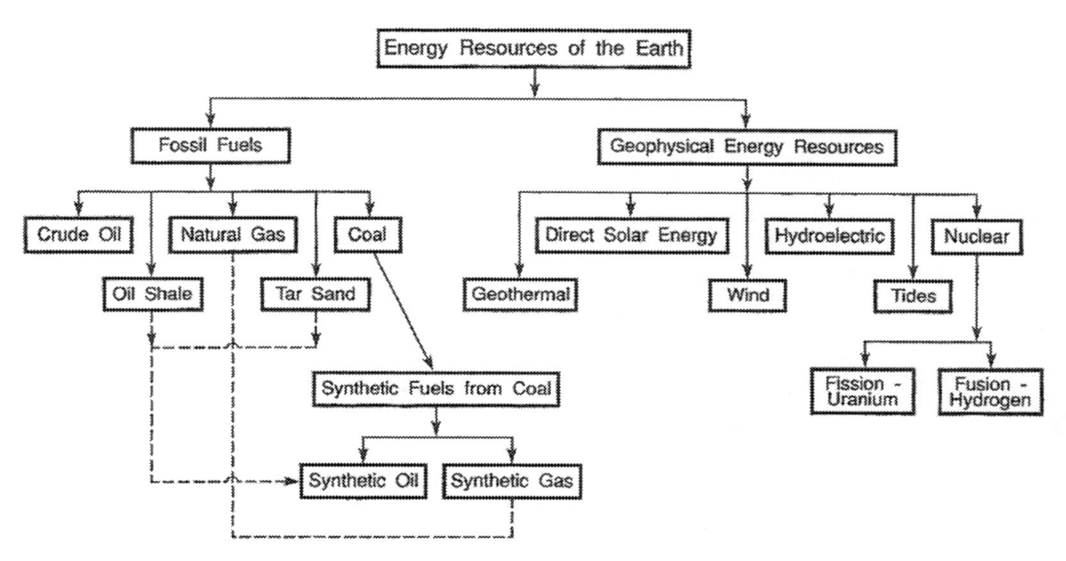

FIGURE 2.1 Energy Resources of the Earth.
Source: Reproduced with permission from: Speight (1996). © Taylor and Francis.

Perturbations in one aspect of this equation typically usually cause perturbations in one or both of the other two aspects. For example, the availability of many metals depends upon the quantity of energy used to recover the metals for the respective ores and, more important, the amount of environmental damage that results from the extraction of the metals as well as the environmental damage that can be tolerated when valuable metals are extracted low-grade (low-metal-containing) ores. Mining and processing of resources offer humans the world valuable commodities suitable for a variety of industrial and domestic uses but also results in the generation of a variety of contaminants. In addition, production, and use may also involve major environmental concerns, including disturbance of land by mining operations as well as the high potential for pollution of the air from the dust generated by mining and land transportation as well as the disruption of aquifers leading to water pollution. This problem is aggravated by the depletion of the higher-grade ores and a necessary trend to the utilization of the lower-grade ores which is analogous to the acceptance by natural gas and crude oil refineries of lower quality natural gas and lower quality feedstocks such as heavy crude oil, extra-heavy crude oil, and tar sand bitumen (Table 2.2) (Speight, 2014, 2017).

TABLE 2.2 Simplified Descriptions of Conventional Crude Oil, Tight Oil, Heavy Crude Oil, Extra Heavy Crude Oil, and Tar Sand Bitumen*

Conventional Resources
Conventional Crude Oil
Mobile in the reservoir; API gravity: >25°
High-permeability reservoir
Tight Oil
Similar properties to the properties of conventional crude oil
Immobile in the reservoir
Low-to no permeability reservoir
Heavy Crude Oil
More viscous than conventional crude oil; API gravity: 10–20°
Mobile in the reservoir
High-permeability reservoir
Secondary recovery
Tertiary recovery (enhanced oil recovery-EOR, e.g., steam stimulation)
Unconventional Resources
Extra Heavy Crude Oil
Similar properties to the properties of tar sand bitumen; API gravity: <10°
Mobile in the reservoir
High-permeability reservoir
Elevated temperature in reservoir-maintains oil in liquid state
Tar Sand Bitumen
Immobile in the deposit; API gravity: <10°
High-permeability reservoir
Near solid to solid in low-temperature reservoir

*This list is intended for general descriptive purposes only and not for use as a means of classification.

However, before launching into a description of the various natural resources, it is necessary to understand how the reserves of these natural resources are defined. A resource (such as any fossil fuel resources) is the whole commodity that exists in the subterranean sediments and strata, but the reserves are only a part of the resource and represent the fraction of a commodity that can be recovered-typically based on the economics of the recovery operation. Further, the definition states that the reserves have a defined size (or amount). Factors that affect whether or not reserves can be extracted economically include: (i) the demand; (ii) market price; (iii) mining costs, transportation costs; (iv) new technologies that can extract the material at a lower price; (v) taxes; (vi) environmental laws; and (vii) government price controls. Understandably, the economic value of a resource can change with time as these factors change.

Thus, at some time in the future, certain resources may become reserves or, because of unforeseen issues, reserves may fall into the category of a non-recoverable resource. In the first case, a reclassification can arise as a result of improvements in recovery technology which may either: (i) render the resource more accessible; or (ii) result in lower costs of the recovery operations. In addition, other uses may also be found for the finished products resulting in an increase in demand which leads to the need to recover more of the resource that was frequently an uneconomical prospect. On the other hand, a resource may become exhausted and unable to produce at the necessary recovery rate thereby forcing the recovery operations to focus on a lower-grade, high-recovery-cost resource. Thus, caution is advised when the term reserves is used as a descriptor of the resource which leads to more specific definitions such as: proved or proven reserves, probable reserves, possible reserves, and potential reserves (Figure 2.2) (Speight, 1996, 2014; Speight and Islam, 2016).

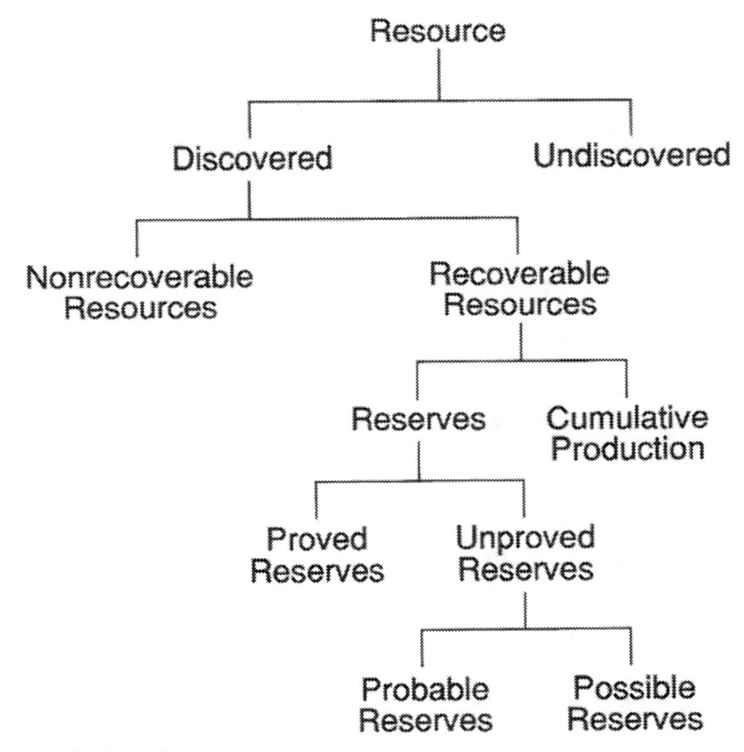

FIGURE 2.2 Categorization of resources and reserves.
Source: Reproduced with permission from: Speight (1996). © Taylor and Francis.

Proved reserves are those reserves of the mineral that have been positively identified as recoverable with current technology. Probable reserves are those reserves of the mineral that are nearly certain but a slight doubt does exist. Possible reserves are those reserves where there is a higher degree of uncertainty about the available amounts of the resource. Potential reserves are based upon geological features of the sediments or strata in which the resource occurs, but the derived data are based on limited information and may, at best, only be an educated guess. Also, there are the so-called undiscovered reserves which are lacking specific data and are often subject to speculation-some observers would say that these types of reserves are little more than figments of the imagination! In reality, the proved reserves may be a small part of the total resource and the data may be hypothetical or speculative.

2.2 FOSSIL FUEL RESOURCES

As defined above, a natural-occurring resource is the natural concentration of mineral (fossil fuels are classified as naturally occurring mineral resources-that is, or may become, of potential economic interest due to the inherent properties. The resource will also be present in sufficient quantity to make it of intrinsic economic interest. In addition, a resource that might previously have remained unworked, because of their poor quality or was beneath excessive overburden thickness are increasingly being considered as potential sources of supply. Current examples are the natural gas reserves and the crude oil reserves that are contained the so-called tight formations. These formations are low-permeability siltstone formations, sandstone formations, and carbonate formations that occur in close association with a shale source rock (US EIA, 2011, 2013; Mayes, 2015).

Fossil fuels are those fuels, namely coal, crude oil (including heavy crude oil and bitumen), natural gas, and oil shale are produced by the decay of plant remains over geological time. These fuels are carbon-based and represent a vast source of energy. In fact, at the present time, the majority of the energy consumed by humans is produced from the fossil fuels, coal, crude oil, and natural gas with smaller amounts of energy coming from nuclear and hydroelectric sources (Figure 2.2) (Speight, 1996, 2020a).

As an aside, the nuclear power industry is truly an industry where the future is uncertain or, at least, at the crossroads (Lowinger and Hinman, 1994). As a result, fossil fuels are projected to be the major sources of energy for the next 50 years (Speight and Islam, 2016; Speight, 2020b).

Planning for the future use of these materials requires (as it does for any resource) an estimate of the amounts of fossil fuels that remain. However, the estimates of the fossil fuels available for future use differ considerably. The issue lies with the use of so-called undiscovered reserves (or reserves-yet-to-be-discovered) of the fossil fuels but suppositions that there are yet-to-be-discovered reserves of fossil fuels is open to much debate and many questions!

In terms of resource availability, it is a matter of understanding: (i) the amount of a resource that is available; (ii) recovery of the resources with the maximal efficiency; and (iii) the application of recovery methods that will cause minimal damage to (disturbance of) the environment. Estimates of the quantities of recoverable fossil fuels in the world vary and are subject to much speculation because of the loose definition of the term resource (see also Chapter 1) (Speight, 2013a, 2014).

Apart from providing the major source of energy for industrial and domestic consumption, fossil fuels also provide, in many cases the means (energy) by which other resources can be recovered and converted to products for consumer use. Therefore, it is worthwhile at this point giving consideration to the impact of fossil fuels not only on the evolution of science and engineering (in fact, on the progress of technology in general) but also on the environment with some indication of the steps that will be necessary to ensure reductions in emissions from the use of such fuels (Walker and Wirl, 1994).

First, and to reemphasize, fossil fuels are not the only source of pollutants, but they are a major source of energy. As such, the use of fossil fuels has been the cause for considerable criticism but it is not the fuel that is at fault. It is the manner in which the fuels are used and has been due to lack of environmental-related oversight during the production of energy from the fuel. Thus, it is necessary to reconsider the strategies for fossil fuel use whereupon it may be discovered (again!) that fossil fuels can be used in an environmentally acceptable manner. Hence, it is necessary to understand, by way of background, the various types of fossil fuels that generate energy and the evolution of their use in this present time. This will also aid in understanding the means by which fossil fuels can produce noxious emissions. Such an understanding will then lead to means by which the emissions can be reduced, if not eliminated completely.

In summary, understanding the nature of fossil fuels and their recovery from the formation in which they exist can be equated to the recovery of any mineral and its use as a commodity. There is also the need to understand the influence of the various processes on the ecological cycles (Chapter 1) as well as, in many cases, the potential for the production of acid rain and the potential for global warming, or any similar effect (Easterbrook, 1995; Speight, 2020c).

Thus, understanding the resource is a benefit to an understanding the recovery processes and the conversion processes through an evaluation of the products and the effects on the environment of the recovery process and manufacture of the products.

2.2.1 NATURAL GAS

Natural gas, like crude oil, does not have a universally acceptable classification scheme (Speight, 2014, 2019a). The term natural gas is intended to mean the gaseous components that often occur in reservoirs with crude oil. There are sources of natural gas which occur without the associated presence of crude oil. Even though the gas may, to all intents and purposes, be characterized as methane, there are those constituents of natural gas which present the potential for pollution and must be removed (Speight, 2014, 2019a).

Just as crude oil and its derivatives were used in antiquity, natural gas was also known in antiquity and was used for lighting in medieval China (James and Thorpe, 1994). Gas wells were an important aspect of religious life in ancient Persia and exhibitions of burning pillars of fire must have been awe-inspiring, to say the least since, as the old name *varishnak* implies they needed no food (Forbes, 1958). As another example, (Christian Bible, Book of Daniel, Chapter 3) note is made of the eternal fires as they existed in the time of King Nebuchadnezzar. These fires, which have been quoted as varying between nine cubits (approximately 12 feet) and forty cubits (approximately 60 feet) high, have since been reported as being a self-igniting natural gas seepage located close to where the Tigris and the Euphrates rivers meet.

There is also a very distinct possibility that the voices of the gods, as referenced in old texts, was natural gas forcing its way through fissures in the surface of the Earth (Scheil and Gauthier, 1909; Schroder, 1920). There is a passage in the records of the Assyrian king Tukulti Ninurta (ca. 885 B.C.) in which the voices of the gods are heard arising from rocks near to Hit, the place of the bitumen deposits which became so well known as a building mastic. Similar sounds attributed to be the voices of the gods are noted to have also occurred in the region around Kirkuk. In classical times these wells were often flared (Lockhart, 1939; Forbes, 1958, 1964), environmental aspects notwithstanding. They are depicted as burning near the shrine of Phoebus Apollo at Delphi (Greece) and on coins of Apollonia (Crete) as well as coins of Selenizza (Albania) (Forbes, 1958). Plutarch mentions that Alexander the Great saw burning gas wells near to Ecbatana which he (Plutarch) described as a gulf of fire streaming from an inexhaustible source!

Historical records indicate that the use of natural gas, other than for religious purposes, dates back to approximately 250 AD when it was used as a fuel in China. The gas was obtained from shallow wells and was distributed through a piping system constructed from hollow bamboo stems. Gas wells were also known in Europe in the middle ages and were reputed to eject oil from the wells such as the phenomena observed at the site near to the town of Mineo in Sicily (Forbes, 1958). Natural gas was used on a small scale for heating and lighting in northern Italy during the early 17th Century.

In a more modern context, there is the record of a burning spring in 1775 near Charleston, West Virginia, as well as on land owned by George Washington (Lincoln, 1785). The possibility also exists that the fire may have been caused by the ignition of naphtha seepages but there is the very strong, if not the only, possibility that the fire was caused by a gas seepage that was ignited by the flaming torch held by one of the investigators as he brought it near to the place of the egress of the gas from the reservoir to the surface. There is also the first record of a natural gas well being drilled (to a depth of 27 feet) in the United States occurred near a burning spring at Fredonia, New York, in 1821.

In the years following this discovery, natural gas use was restricted to the local environs since the available technology for storage and transportation (bamboo pipes notwithstanding!) was not well developed and, at that time, natural gas had little or no commercial value. In fact, in the 1930s when crude oil refining was commencing an expansion in technology that still continues, gas was not considered to be a major fuel source and was only produced as an unwanted by-product of crude oil production.

The principal gaseous fuel source at that time (i.e., the 1930s) was the gas produced by the surface gasification of coal. In fact, each town of any size had a plant for the gasification of coal (hence, the use of the term town gas). Most of the natural gas produced at the crude oil fields was vented to the air or burned in a flare stack. Only a small amount of the natural gas from the crude oil fields was pipelined to industrial areas for commercial use. It was only in the years after World War II that natural gas became a popular fuel commodity leading to the recognition that it has at the present time.

Currently, natural gas production is predicted to increase, and gas fired generators are also predicted to overtake nuclear power as the secondary electricity source due to relatively low capital costs, high efficiency, and low emissions). One of the reasons for the increase in natural gas usage as a power source is the predicted increase in electricity demand as well as growth in the industrial sector for cogeneration and other uses.

2.2.2 CRUDE OIL

Liquid fossil fuels are usually given one or more of the names: oil, crude oil, crude oil, and heavy crude oil. Unlike coal, there is no universally acceptable system by which crude oil can be (or has been) classified. Several attempts at classification of crude oil, using a physical property, have been attempted (Speight, 1991) but much of the classification nomenclature tends to be focused on general descriptions. For example, heavy crude oil is so-called because of its tendency to contain lesser amounts of the distillable fractions and have a density (0.90 to 1.00) close to the density of water (1.00). Thus, at the temperature of density measurement, heavy crude oil is lighter than water. On the other hand, the near solid bitumen, is often referred to as natural asphalt, is heavier than water. In addition, the term natural asphalt is incorrect since asphalt is a product of refinery processing and maybe an altered, rather than natural, material (Figure 2.3) (Speight, 1996, 2014, 2017). In summary, more complete descriptions of crude oil would be necessary in any universally acceptable classification scheme.

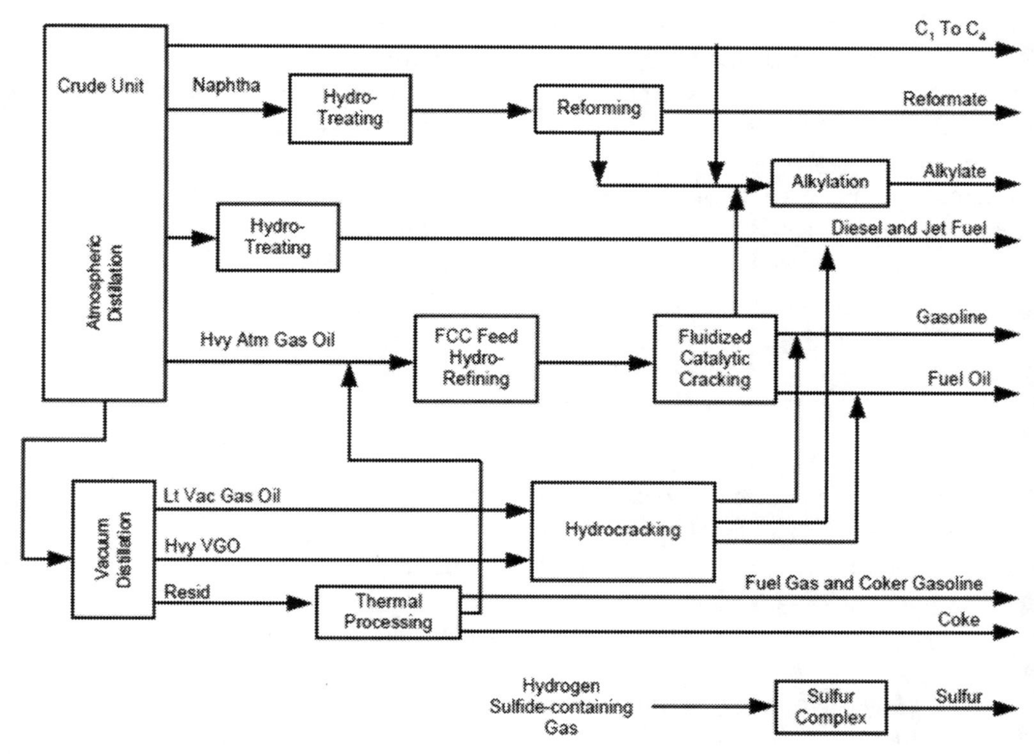

FIGURE 2.3 Schematic of a refinery.

Source: Reproduced with permission from: Speight (1996). © Taylor and Francis.

The recorded use of crude oil and the nonvolatile derivative of crude oil, asphalt, have been known for approximately 6,000 years (Agricola, 1556; Abraham, 1945; Speight, 1991; James and Thorpe, 1994). There are references to oil seepages, the product being frequently referred to as pitch (and referred to as slime in the Bible) in the ancient world of the Greeks and Persians (Herodotus, 447 B.C.) and Alexander the Great is reputed to have discovered crude oil on the banks of the River Oxus (Abraham, 1945).

There are also records of the use of mixtures of pitch and sulfur as a weapon of war during the Battle of Palatea, Greece, in the year 429 B.C. (Forbes, 1959). There are references to the use of a liquid material, naft (presumably the volatile fraction of crude oil (naphtha) and which is used as a solvent or as a precursor to gasoline), as an incendiary material during various battles of the pre Christian era (James and Thorpe, 1994). This is the so-called Greek fire, a precursor and distant cousin to napalm, of which much is known and the effects recognized by those on the delivery end and on the receiving end, with drastic consequences, as well as by historians.

With the onset of the industrial revolution in the seventeenth century, the use of fossil fuels as sources of energy became well established. First coal, which carried the revolution through the 19th Century, and then crude oil. It is crude oil that has fueled the military vehicles of many nations becoming a prime commodity in the 20th Century and, in the thoughts of some, the prime cause of all wars since the Franco-Prussian disagreements of 1870–1871 (Yergin, 1992).

After World War II, there was a phenomenal rise in the use of crude oil and its various products, mostly through the development of the internal combustion engine (the automobile) and the accompanying expansion of industrial operations. At the same time, the growth of the use of electricity during the last four decades is related to the rise in the use of household consumer products such as refrigerators and stoves.

Currently, crude oil refining is a complex sequence of events that result in the production of a variety of products (Figure 2.3) (Speight, 1996). In fact, crude oil refining might be considered as a collection of individual, but yet related, processes that are each capable of producing effluents. The refining industry, as it is currently known, will continue at least for the next three-to-five decades (Speight, 2020b).

2.2.3 COAL

Coal is a very abundant fossil fuel and is classified by rank (degree of metamorphism/ progressive alteration) as a natural series from lignite to anthracite (ASTM D388). Coal forms a major part of the fossil fuel resources of the Earth (Speight, 2013a, 2020a), the amount available being subject to the method of estimation and to the definition of the resources. In fact, coal has been referred to as the keystone to a variety of energy scenarios (Speight, 2013a, 2020a).

Coal is, perhaps, the most familiar of the fossil fuels not necessarily because of its usage throughout the preceding centuries (Galloway, 1882; James and Thorpe, 1994) but more because of its extremely common usage during the nineteenth century. Coal was, in large part, the fuel which allowed the industrial revolution to proceed. And because of this, as well as the continuing popularity of coal, it is extremely important in the present context because of the associated, and very necessary, cleanup of the emissions when it is burned.

Coal has probably been known and used for a considerable time, but the records are somewhat less than complete. There are frequent references to coal in the Christian Bible (Cruden, 1930) but all in all, the recorded use of coal in antiquity is well documented (Galloway, 1882; James and Thorpe, 1994).

Over the centuries, coal use increased substantially. In fact, in the late 1500s, an increasing shortage of wood in Europe resulted in coal becoming an exploitable resource in the United Kingdom, France, Germany, and Belgium. In the mid-to-late 1700s, the use of coal increased dramatically in Britain with the successful development of coke smelters and the ensuing use of the coke to produce steam power. By the 1800s, coal was supplying most of the energy requirements of Britain and the use of gas from coal (usually referred to as town gas because of the generation site and use) for lighting was also established.

In contrast, in the United States, wood was more plentiful, the population density was much lower than in Europe; any coal required for energy was imported from Britain and Nova Scotia. But after the Revolutionary War (1775–1781), coal entered the picture as an increasing source of energy; as an example, the state of Virginia supplied coal to New York City. However, attempts to open the market to accept coal as a fuel were generally ineffective in the United States and the progress was extremely slow. It was not until the period from 1850 to 1885 that coal use in the United States increased, spurred by the emerging railroad industry both for the manufacture of steel rails and as a fuel for locomotives.

Coal occurs in various forms, defined by rank or type (Speight, 2013a); it is not only a solid hydrocarbon material with the ability to produce energy on demand but it also has the potential to produce considerable quantities of carbon dioxide, nitrogen oxides, and sulfur oxides as a result of combustion (Speight, 1993).

Coal production is predicted to increase 1% per year to meet the domestic and foreign demand for coal. Currently, coal consumption for electricity generation accounts for more than 56% of the total electricity generation and this is expected to rise by the year 2050. In addition, by that year, electricity generation will account for nearly 90% of the coal use. As a result of the increased use of coal for power generation, electricity generators are seen as being responsible for one-third of the carbon emissions. However, there are other aspects of coal use that requires monitoring.

Coal combustion, for example, in coal-fired power plants, is also responsible for the production of bottom ash and fly ash (Williamson and Wigley, 1994). Both types of ash have the ability to promote erosion and corrosion of equipment as well as pollution of the environment, and the leachability of trace elements from the ash is cause for environmental concern.

2.2.4 BIOMASS

Non-fossil fuels are alternative sources of energy (often referred to as renewable source of energy) that do not rely on continued consumption of the limited supplies of crude oil, coal, and natural gas. Examples of the non-fossil fuel energy sources include: (i) biomass; (ii) geothermal; (iii) hydrogen; (iv) (v) nuclear; (vi) solar; (vii) tidal; and (viii) wind sources (Nersesian, 2007; Speight, 2020a, b)-in the current context, the focus is on bio-sources. Such resources are considered to be extremely important to the future of energy generation because they are renewable energy sources that could be exploited continuously and not suffer depletion.

Renewable energy sources include (alphabetically rather than by common use or preference) include: (i) biomass; (ii) geothermal energy, (iii) hydrogen energy, (iv) nuclear energy, (v) solar energy, (vi) tidal energy, and (vii) wind energy. However, while it is necessary to indicate the potential sources of alternate fuel (or alternate energy), the focus of this book remains on biomass.

Biomass is a term used to describe any material of recent biological origin, including plant materials such as trees, grasses, agricultural crops, and even animal manure. Other biomass components, which are generally present in minor amounts, include: (i) triglyceride derivatives, which are ester derivatives derived from glycerol and three fatty acids; (ii) sterol derivatives, which are also known as steroid alcohol derivatives); (iii) alkaloid derivatives, which form a class of naturally occurring organic compounds that mostly contain basic nitrogen atoms; (iv) terpene derivatives, which are the primary constituents of the essential oils of many types of plants and flowers; (v) terpenoid derivatives, which are sometimes referred to as isoprenoids and form a large and diverse class of naturally occurring organic chemicals derived from terpenes-most are multicyclic structures with oxygen-containing functional groups); and (vi) wax derivatives, which are a diverse class of organic compounds that are lipophilic, malleable solids near ambient temperatures and include higher molecular weight alkane derivatives and lipids, typically with melting points above about 40°C (104°F) and melt to give low-viscosity liquids. Waxes are insoluble in water but soluble in organic, nonpolar solvents. Natural waxes of different types are produced by plants and animals and occur in crude oil).

This list (above) includes everything from primary sources of crops and residues harvested/collected directly from the land, to secondary sources such as sawmill residuals, to tertiary sources of post-consumer residuals that often end up in landfills. A fourth source, although not usually categorized as such, includes the gases that result from anaerobic digestion of animal manures or organic materials in landfills (Wright et al., 2006).

Thus, knowledge of the composition of a biomass feedstocks is critical to the selection of the varieties with optimized properties for downstream conversion (De Jong, 2014). This can be partially achieved by selecting varieties with biomass composition that are better suited to the conversion process. Lignocellulosic biomass displays considerable recalcitrance to biochemical conversion because of the inaccessibility of its polymer components to enzymatic digestion and the release or production of fermentation inhibitors during pretreatment. If the ratio of hemicellulose, cellulose, and lignin in a woody biomass feedstock was optimized for the specific biochemical conversion method, then the pretreatment methods could be reduced or avoided.

By definition, biomass is material that comes from plants. Plants use the light energy from the sun to convert water and carbon dioxide to sugars that can be stored, through a process called photosynthesis. Some plants, such as sugar cane and sugar beets, store the energy as simple sugars, which are mostly used for food. Other plants store the energy as more complex sugars (starches) and include grains such as corn and are also used for food. Another type of plant matter-cellulosic biomass-is made up of complex sugar polymers (complex polysaccharides), and is not generally used as a food source. This type of biomass is increasing in use as a feedstock for ethanol (sometimes, in this case, referred to as bioethanol) production. Other feedstocks of interest include agricultural and forestry residues, organic urban wastes, food processing and other industrial wastes, and energy crops.

Biomass is produced by a photosynthetic process (photosynthesis) which involves chemical reactions occurring on the Earth between sunlight and green plants within the plants in the form of chemical energy. In the process, solar energy is absorbed by green plants and some microorganisms to synthesize organic compounds from low energy carbon dioxide (CO_2) and water (H_2O). For example:

$$6CO_2 + 6H_2 \rightarrow 6CH_2O + 3O_2$$

(Biomass)

For convenience here, the general formula of the organic material produced during photosynthesis process is $(CH_2O)_n$ which is mainly carbohydrate material. Some of the simple carbohydrates involved in this process are the simple carbohydrates glucose ($C_6H_{12}O_6$) and sucrose ($C_{12}H_{22}O_{11}$) and constitute biomass, which is a renewable energy source due to its natural and repeated occurrence in the environment in the presence of sunlight. The amount of biomass that can be grown certainly depends on the availability of sunlight to drive the conversion of carbon dioxide and water into carbohydrates. In addition to limitations of sunlight, there is a limit placed by the availability of appropriate land, temperature, climate, and nutrients, namely nitrogen, phosphorus, and trace minerals in the soil.

Biomass is clean for it has negligible content of sulfur, nitrogen, and ash, which give lower emissions of sulfur dioxide, nitrogen oxides, and soot than conventional fossil fuels. Biomass resources and many and varied and, include: (i) forest and mill residues; (ii) agricultural crops and wastes; (iii) wood and wood wastes; (iv) animal wastes; (v) livestock operation residues; (vi) aquatic plants; (vii) fast-growing trees and plants; and (viii) municipal and industrial wastes. The role of wood and forestry residues in terms of energy production is as old as fire itself, and in many societies, wood is still the major source of energy. In general, biomass can include anything that is not a fossil fuel that is based on bio-organic materials other than natural gas, crude oil, heavy crude oil, extra-heavy crude oil and tar sand bitumen (Lucia et al., 2006; Lee and Shah, 2013; Speight, 2020a, b).

There are many types of biomass resources that can be used and replaced without irreversibly depleting reserves, and the use of biomass will continue to grow in importance as replacements for fossil fuel sources and as feedstocks for a range of products. Some biomass materials also have particular unique and beneficial properties which can be exploited in a range of products including pharmaceuticals and certain lubricants. In this context, the increased use of biofuels should be viewed as one of a range of possible measures for achieving self-sufficiency in energy, rather than a panacea to completely replace the fossil fuels.

Last, but by no means least, wood is one of the major natural resources of the Earth and it has the potential to be a major source of energy in addition to being a source of chemicals (Tillman, 1978). Wood is a renewable resource, but there are caveats that must be applied to the unlimited use of wood. One of these is the deforestation of large tracts of land and the ensuing imbalance in the carbon dioxide cycle. Wood ranks first in the world as a raw material for the manufacture of other products which, in addition to the conventional wood products, include cellophane, rayon, paper, methanol, plastics, and turpentine.

Wood is composed of polysaccharides such as cellulose, part of which can be extracted for chemical use. A wide variety of organic compounds and tannins, pigments, sugars, starch, cyclitols, gums, mucilages, pectins, galactans, terpenes, hydrocarbon derivatives, acid derivatives, ester derivatives, fatty acid derivatives, aldehyde derivatives, sterol derivatives, and waxes. Substantial amounts of methanol (often referred to as wood alcohol) are extracted from wood. Methanol, once a major source of liquid fuel, is again being considered for that use (Skrzypek et al., 1994).

Energy generation utilizing biomass and municipal solid wastes (MSW) are also promising in regions where landfill space is limited. Technological advances in the fields have made this option efficient and environmentally safe, possibility even supplementing refinery feedstocks as sources of energy through the installation of gasification units.

The various forms of biomass are often referred to as carbon neutral fuels insofar as any carbon dioxide emitted (for example, during combustion or other forms of processing) is negated by the carbon dioxide taken by the biomass during the lifetime of the plant(s). However, this is a paper calculation and must be treated with caution. In reality, biomass. While biomass is carbonaceous feedstock that is composed of a variety of organic constituents that contain carbon, hydrogen, oxygen, often nitrogen and also varying amounts of other atoms, including alkali metals, alkaline earth metals, and heavy metals (i.e., metals with relatively high densities, atomic weights, or atomic numbers-all of which are detrimental to process catalysts (Speight, 2014, 2017) and potentially toxic to the various forms of flora and fauna (Pourret and Hursthouse, 2019).

Briefly, the alkali metals consist of the chemical elements lithium (li), sodium (Na), potassium (K), rubidium (Rb), cesium (Cs), and francium (Fr). Together with hydrogen they make up Group I of the Periodic Table (Table 2.3). On the other hand, the alkaline earth metals are the six chemical elements in Group 2 of the Periodic Table and are beryllium (Be), magnesium (Mg), calcium (Ca), strontium (Sr)barium (Ba), and radium (Ra). These elements have similar properties-they are shiny, silvery-white, and are somewhat reactive at standard temperature and pressure. The common transition metals such as copper (Cu), lead (Pb), and zinc (Zn) are often classed as heavy metals but the criteria used for the definition and whether metalloids (types of chemical elements which have properties in between, or that are a mixture of, those of metals and nonmetals) are included, vary depending on the context. These metals are often found in functional molecules such as the porphyrin molecules which include chlorophyll and which contains magnesium.

TABLE 2.3 The Periodic Table of the Elements

1	2	3	4	5	6	7	8	9	10	11	12	13	14	15	16	17	18
H																	He
Li	Be											B	C	N	O	F	Ne
Na	Mg											Al	Si	P	S	Cl	Ar
K	Ca	Sc	Ti	V	Cr	Mn	Fe	Co	Ni	Cu	Zn	Ga	Ge	As	Se	Br	Kr
Rb	Sr	Y	Zr	Nb	Mo	Tc	Ru	Rh	Pd	Ag	Cd	In	Sn	Sb	Te	I	Xe
Cs	Ba	57-71	Hf	Ta	W	Re	Os	Ir	Pt	Au	Hg	Tl	Pb	Bi	Po	At	Rn
Fr	Ra	89-103	Rf	Db	Sg	Bh	Hs	Mt	Ds	Rg	Cn	Nh	Fl	Mc	Lv	Ts	Og

La	Ce	Pr	Nd	Pm	Sm	Eu	Gd	Tb	Dy	Ho	Er	Tm	Yb	Lu
Ac	Th	Pa	U	Np	Pu	Am	Cm	Bk	Cf	Es	Fm	Md	No	Lr

2.2.5 OTHER ENERGY RESOURCES

It is predictable that fossil fuels will be the primary source of energy for the next several decades, well into the next century and therefore the message is clear: until other energy sources supplant coal, natural gas, and crude oil the challenge is for the development of technological concepts that will provide the maximum recovery of energy from these fossil fuel resources. Thus, it is absolutely essential that energy from such resources be obtained not only cheaply but also efficiently and with minimal detriment to the environment.

To complete the energy resource scenario and to put the fossil energy resources into perspective, there are non-fossil fuel energy resources (Figure 2.1) (Speight, 1996) which are derived from the sun and are generically referred to as non-fossil resources, or renewable resources, or/and geophysical energy resources (Golob and Brus, 1993; Mills and Diesendorf, 1994; Pickering and Owen, 1994). And even though it is not widely recognized, since renewable energy resources are assumed to be "natural" and therefore in harmony with nature, there are environmental issues that need to be addressed when renewable energy resources are to be employed.

Beyond this, there is much doubt related to the time when humans can look forward to the generation of energy from the unconventional sources. The time scale for the generation of electrical energy from such sources is generally unknown or, at best, open to much speculation and criticism. The resources include direct solar energy, hydroelectric energy, wind energy, geothermal energy, energy from biomass, and elements of minerals from which nuclear energy may be derived (Jones and Radding, 1980; Robinson, 1980; Dickson and Fanelli, 1990; Himmel et al., 1994; Speight, 2020a).

Nuclear energy, which can be as hazardous as any fossil fuel in terms of destruction of the environment, deserves some comment insofar as an appreciation of its use may assist in the general background and aid in putting fossil fuel resources into a more complete context. Energy from nuclear sources has shown potential to be a significant energy source in the future. The technology is known (Knief, 1992) but has suffered some setbacks. Accidents and dubious claims of the ease with which energy may be derived from nuclear sources have reduced the credibility of the nuclear industry. Nevertheless, the potential still exists for the extraction of energy from nuclear sources.

As a brief summary, energy from nuclear sources may be the result of nuclear fission, which uses uranium ore or refined uranium-235 as its basic energy source. The ore must be concentrated into fissionable isotopes through nuclear processing. Eventually, the spent material must be replaced, and the resources of uranium ore may be limited as determined by a specific test (ASTM E901). Hence the need to explore the potential for generating more of the fissionable isotopes. However, the use of such materials has become the core of a major controversy which threatens to be, at least for the time being, the death-knell of the nuclear industry.

Nuclear fusion, unlike nuclear fission, which involves the use of fissionable material, involves the fusion of hydrogen nuclei. The concept has been suggested as having the potential to provide a clean and virtually unlimited supply of energy. However, it is generally felt (Halpern, 1980) that the technology is at the stage where several decades appear to be needed for the performance of believable laboratory-scale experiments which are critical to the development of controlled energy from fusion.

Hence, the continued use of the so-called conventional fuel resources derived from the remains of ancient plants and animals. Those same resources which have been instrumental in the phenomenal expansion of the industrialized world. And also those same resources which can have serious adverse effects on the flora and fauna (including man) of the world.

2.3 RESOURCE UTILIZATION

Mineral resources have been known and developed throughout historical time, especially ores containing copper (James and Thorpe, 1994). Mineral deposits, including ore bodies and potential ore, can be developed to provide economically valuable commodities (Bain, 1987). Therefore, extraction and maximum use must be planned for the benefit of the largest number of people. Such a policy is known as conservation, and it is influenced by many economic and political factors. As costs increase, people tend to use less; they retain materials for more essential uses, thus practicing conservation.

For convenience, raw materials are classified as fuels (see above), metals (including ferrous metals, nonferrous metals, light metals, as well as precious metals), and industrial minerals or nonmetallic elements and have a variety of uses. It is because of these various uses that metals find their way into the environment. The increased demand for a particular metal, coupled with the necessity to utilize lower grade ores, has a significant effect upon the amount of ore that must be mined and processed to produce a unit of a product. Typically, in many industries for every unit of the product, a leaner ore will produce more units of by products, some of which can be used and some of which cannot because of the tendency for the by-products to be unusable thereby requiring discharge to the land and/or water systems and the accompanying environmental consequences.

A number of minerals other than those used to produce metals are also important sources of industrial and domestic products. For example, clays, which occur as suspended and sedimentary matter in water and as secondary minerals in soil, are also used for clarifying oils, as catalysts in crude oil processing, as fillers and coatings for paper, and in the manufacture of firebrick, pottery, sewer pipe, and floor tile.

Fluorine compounds are widely used in industry; for example, fluorspar (CaF_2) 1s used in large quantities as a flux in steel manufacture. Freon-12 (difluoro dichloromethane, CF_2Cl_2) is widely used as a refrigeration and air-conditioning fluid and is considered to be a major stratospheric pollutant. Sodium fluoride is used for water fluoridation.

Mica is a complex aluminum silicate mineral which is transparent, tough, flexible, and elastic. Better grades of mica are cut into sheets and used in electronic apparatus, capacitors, generators, transformers, and motors. Finely divided mica is widely used in roofing, paint, welding rods, and many other applications. Phosphorus, along with nitrogen and potassium, is one of the major fertilizer elements. Many soils are deficient in phosphate. Its non-fertilizer applications include supplementation of animal feeds, synthesis of detergent builders, and preparation of chemicals such as pesticides and medicines.

Sulfur can exist in many forms which undergo a complicated cycle (Figure 2.4) (Speight, 1996). Sulfur may be expelled from volcanoes as sulfur dioxide or as hydrogen sulfide, which is oxidized to sulfur dioxide and sulfates in the atmosphere. Highly insoluble metal sulfides are oxidized upon exposure to atmospheric oxygen to relatively soluble metal sulfates.

In the ground and in water, sulfate (SO_4^{2-}) is converted to organic sulfur by plants and bacteria. Bacteria mediate transitions among sulfates, elemental sulfur, organic sulfur, and hydrogen sulfide. Sulfur may either escape to the atmosphere (usually as the oxides) and reappear in the form of sulfurous and sulfuric acids in rainfall (acid rain) or be retained s metal sulfides. In fact, the four most important sources of sulfur are (in decreasing order): deposits of elemental sulfur, hydrogen sulfide recovered from sour natural gas, organic sulfur recovered from crude oil, and pyrite (FeS_2).

The development of large machines for handling huge tonnages of rock and the use of refined explosives have improved mining methods, as illustrated in many coal mining operations (Speight, 2013a) and in the mining of oil sand (Speight, 2014, 2017). Open-pit and strip mines are larger and more efficient than ever before, and lower grades of ore are being recovered. It is common for a mine to produce 50,000 tons (45,000 metric tons) of ore a day, and some yield more than 100,000 tons (90,000 metric tons).

Improved methods of transportation, concentration, and smelting permit less costly handling of large tonnages and improve recoveries. Newer smelting methods are cleaner, have as good or better recovery records, and may prove to be less expensive to operate. And the environmental aspects of mining many of these minerals need attention. Even the simplest mining procedure such as the recovery of gravel for asphalt highways can have important environmental effects.

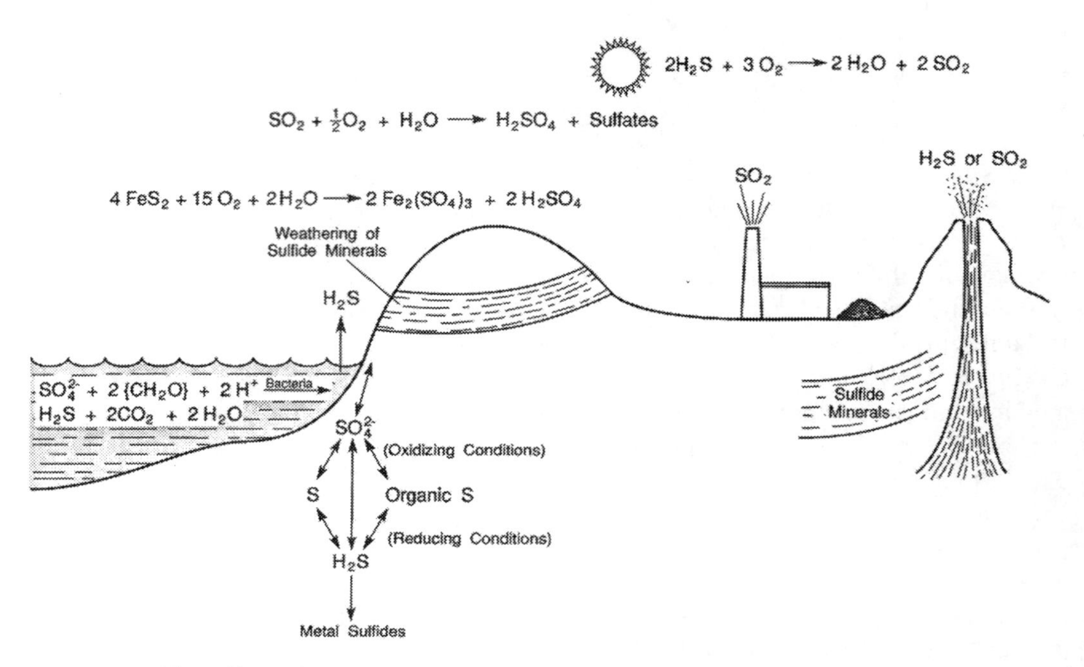

FIGURE 2.4 The sulfur cycle.
Source: Reproduced with permission from: Speight (1996). © Taylor and Francis.

The increased use and popularity of mineral resources are due, no doubt, to the relative ease of accessibility which has, in many cases, remained relatively unchanged over the centuries. There are exceptions to the ready availability (crude oil is an example) because of various physical and political reasons, although the prognosis for a continuing crude oil industry remains optimistic though cooperation and planning (Al-Sowayegh, 1984; Kubursi and Naylor, 1985; Niblock and Lawless, 1985; Speight, 2011a, 2020b).

The relatively simple means by which many mineral resources can be utilized has also been a major factor in determining their popularity. And many of the commodities produced from mineral resources:

$$\text{Ore} \rightarrow \text{mineral} \rightarrow \text{commodity}$$

It may be interchangeable on a purely physical basis insofar as one commodity can be converted to another, thereby finding supplemental uses that were not available in the early days of resource development.

Thus, the ease with which mineral resources (such as the fossil fuels) were available to the world did, indeed, play a major role in their increased use (Cassedy and Grossman, 1990; Blunden and Reddish, 1991). The conversion of these natural products to usable products and, in some cases, to valuable chemicals served to increase their popularity, if coal, crude oil, and natural gas can be used as examples. More specific examples are the coal chemicals industry of the 19th Century and the petrochemical industry of the 20th Century (Speight, 2013a, 2014, 2019b).

There are projections that the era of fossil fuels (gas, crude oil, and coal) will be almost over when cumulative production of the fossil resources reaches 85% of their initial total reserves (Hubbert, 1973). Should the same apply to mineral resources in general, then there may be some cause for concern, perhaps alarming and the issue is worthy of some attention. And a major issue is, naturally, the scarcity of the fossil fuel resources that are used as sources of energy as a means of producing valuable products from a variety of non-fossil fuel resources. For example, the scarcity of crude oil (relative to a few decades ago) is real, but it seems more than likely that the remaining reserves of crude oil, coal, and natural gas make it likely that there will be an adequate supply of energy for several decades.

Therefore, technologies to ameliorate the effects of the technologies that are employed in developing the mineral resources must be pursued vigorously. There is a need for online analysis of the various product streams (Breen and DeMarco, 1992) to determine the nature of the effluents before they are released to the environment. There is a challenge that must not be ignored, and the effects of, for example, acid rain on soil and water leave no doubt related to the need to control its causes. Indeed, recognition of the need to address these issues is the driving force behind recent energy strategies as well as a variety of research and development programs (Katz, 1984; United States Department of Energy, 1990).

The solution to the issue of acid rain deposition from the atmosphere (Chapter 3) lies in the control of sulfur oxide emissions (usually sulfur dioxide, SO_2) as well as the nitrogen oxide (NOx) emissions (Chapter 6). These gases react with the water in the atmosphere and the result is an acid:

$$SO_2 + H_2O \rightarrow H_2SO_3$$

Sulfurous acid

$$2SO_2 + O_2 \rightarrow 2SO_3$$

$$SO_3 + H_2O \rightarrow H_2SO_4$$

Sulfuric acid

$$2NO + H_2O \rightarrow 2HNO_2$$

Nitrous acid

$$2NO + O_2 \rightarrow 2NO_2$$

$$NO_2 + H_2O \rightarrow HNO_3$$

Nitric acid

In fact, the combustion of any sulfur-containing fuel or nitrogen-containing fuel during the conversion of feedstock energy to electrical energy can account for the substantial amounts of the sulfur oxides (SO_x) and nitrogen oxides (NO_x) released to the atmosphere (Speight, 2013a, b).

Whichever technologies succeed in reducing the amounts of these gases in the atmosphere should also succeed in reducing the amounts of urban smog, those notorious brown and gray clouds that are easily recognizable at some considerable distances from urban areas, not only by their appearance but also by their odor.

This kind of visible air pollution is composed of nitrogen oxides, sulfur oxides, ozone, smoke, and PM is derived from the emissions from the combustion of carbonaceous fuels, vehicular emissions, industrial emissions, forest fires, and agricultural fires. The oxides of carbon (carbon monoxide, CO, and carbon dioxide, CO_2) are also of importance insofar as all carbon-based materials produce either or both of these gases during use. And both gases have the potential for harm to the environment.

A reduction in the emissions of these gases, particularly carbon dioxide, which is the final combustion product of carbon-based mineral resources, has little chance of being reduced without a traumatic switch to non-carbon resources. However, such emissions can be moderated by trapping and recovering the carbon dioxide at the time of resource processing. It is also necessary to note here, on the positive side for natural gas, that methane produces less carbon dioxide per unit of energy than coal.

Natural gas is far less abundant than coal and the inadvertent release of natural gas into the atmosphere (methane, CH_4, amongst other gases, is a greenhouse gas) (Wuebbles and Edmonds, 1988; Graedel and Crutzen, 1989; Hileman, 1992) may tend to offset any of the advantages of its use. Indeed, the release of other gases resulting from fossil fuel usage into the atmosphere has been projected to cause environmental perturbations such as the deposition of acid rain (Graedel and Crutzen, 1989).

Resource development is a necessary part of the modern world, hence the need for stringent controls over the amounts and types of emissions from such development. The necessity for the cleanup of process effluents is real. Indeed, the emissions resulting from the use of the various mineral resources have had deleterious effects on the environment and promise further detriment unless adequate curbs are taken to control not only the nature but also the amount of effluents being released into the environment.

Effluent products and by-products are produced in various industries (Austin, 1984; Probstein and Hicks, 1990; Speight, 2013a, 2014). These products have the potential to contain quantities of noxious materials that are a severe detriment to the environment. Not to admit to this would be a serious omission of fact. All such noxious products need to be removed before use or discharge to the environment.

2.4 ENVIRONMENTAL ASPECTS

The capacity of the environment to absorb the effluents and other impacts of energy technologies is not unlimited, as some would have us believe. The environment should be considered to be an extremely limited resource, and discharge of chemicals into it should be subject to severe constraints. Indeed, the declining quality of many raw materials dictates that more material must be processed to provide the needed products. And the growing magnitude of the effluents from industrial processes has moved above the line where the environment has the capability to absorb such effluents without disruption. Thus, the general prognosis for environmental protection through an increased awareness of environmental issues is not pessimistic and can be looked upon as being quite optimistic.

In terms of gaseous emissions, the control of nitrogen oxides is achieved, to date, mainly through modification of the combustion process, and other methods, including various flue gas treating processes are being developed. Indeed, it is anticipated that flue gas control technologies may achieve significantly higher control of gaseous emissions than currently being recorded. In fact, the current flue gas desulfurization (FGD) techniques are capable of removing approximately 90% w/w of the sulfur (Speight, 2014, 2017). However, it is extremely important that there is an understanding of the various types of emissions that escape to the atmosphere, as well as to the other constituents of the surrounding environment.

A number of inorganic emissions enter the atmosphere as the result of human activities. The most common of these emissions are sulfur dioxide (SO_2), nitric oxide (NO), and nitrogen dioxide (NO_2). Other inorganic gaseous emissions include ammonia (NH_3), hydrogen sulfide (H_2S), chlorine (Cl_2), hydrogen chloride (HCl), and hydrogen fluoride (HF), as well as other oxides of nitrogen. Because of carbon monoxide emissions from internal combustion engines, the highest levels of this gas tend to occur in congested urban areas at times specific times, such as during heavy road traffic periods (rush hours). Carbon monoxide emissions from automobiles may be lowered by employing a leaner air-fuel mixture, i.e., an air-fuel in which the weight ratio of air to fuel is relatively high. Current automobiles use catalytic exhaust reactors to cut down on carbon monoxide emissions and nitrogen oxide emissions. But, in more general terms, there is the need to increase the efficiency of automobile transportation (Nivola and Crandall, 1995) which, hopefully, would reduce the noxious emissions into the atmosphere.

If the carbon monoxide escapes to the atmosphere, it is generally agreed that it is removed from the atmosphere by reaction with hydroxyl radical:

$$CO + HO^{-} \rightarrow CO_2 + H$$

This reaction accompanied by other chemical reactions which lead to the regeneration of more hydroxyl radicals:

$$O_2 + H \rightarrow HOO$$

$$HOO^{-} + NO \rightarrow HO^{-} + NO_2$$

$$HOO^{-} + HOO^{-} \rightarrow H_2O_2$$

$$H_2O_2 + hv \rightarrow 2HO^{-}$$

Soil microorganisms act to remove carbon monoxide from the atmosphere.

The sulfur cycle (Figure 2.4) (Speight, 1996) involves primarily hydrogen sulfide, sulfur dioxide, sulfur trioxide, and sulfates. Sulfur compounds enter the atmosphere to a large extent through human activities, primarily as sulfur dioxide from the combustion, or processing, of sulfur-containing fuels. Non-anthropogenic sulfur enters the atmosphere largely as hydrogen sulfide from volcanoes and from the biological decay of organic matter and reduction of sulfate.

Organic emissions may have a strong effect upon atmospheric quality. The effects of such emissions in the atmosphere may be divided into two major categories which are: (i) direct effects, such as effects caused by exposure to the emissions; and (ii) the formation of secondary pollutants, especially smog (a mixture of smoke and fog). Because of their widespread use in fuels, hydrocarbon derivatives predominate among organic atmospheric pollutants.

Crude oil products, primarily gasoline, are the source of most of the anthropogenic hydrocarbon emissions found in the atmosphere. The Clean Air Act (CAA) of 1990 has already impacted the gasoline markets of the United States (Miller, 1994). Additives and reformulation have already been implemented in some cases to meet regulations and emission standards required for compliance, and gasoline will continue to be reformulated.

The regulations for reformulated gasoline have been divided into Phase I and II, starting in 1995 and 2000, respectively, with each phase having more stringent requirements. Phase I is divided into two compliance models which are designed to facilitate the determination of fuel formulation compliance without the expense of extensive motor performance testing. The simple model can be use from 1995 to 1997 and the complex model from 1997 on and will be used for Phase II.

Alkanes are among the more stable hydrocarbon derivatives in the atmosphere. Because of their high vapor pressures, alkanes with 6 or fewer carbon atoms are normally present as gases, alkanes with 20 or more carbon atoms are present as aerosols or are adsorbed onto atmospheric particles. Alkanes with 6 to 21 carbon atoms per molecule may be present either as vapor or particles, depending upon conditions.

Alkenes enter the atmosphere from a variety of processes, including emissions from internal combustion engines and turbines, foundry operations, and crude oil refining. In addition to the direct release of alkenes, these hydrocarbon derivatives are commonly produced by the partial combustion and cracking (thermal decomposition) at high temperatures of alkanes, particularly in the internal combustion engine. Unlike alkanes, alkenes are highly reactive in the atmosphere, especially in the presence of nitrogen oxides, hydroxyl radicals, and sunlight. In addition, the reaction of molecular oxygen to the resulting radical results in the formation of the extremely reactive hydroperoxyl radical. These radicals then participate in chain reactions such as those involved in the formation of smog (Chapter 6, Chapter 8).

Single-ring aromatic compounds are important constituents of lead-free gasoline, which has largely replaced leaded gasoline. In fact, derivatives of naphthalene (which has two fused, conjugated, aromatic rings) and several compounds containing two or more unconjugated rings have been detected as atmospheric pollutants (Table 2.4).

Polynuclear aromatic hydrocarbon derivatives (Table 2.4) are present as aerosols in the atmosphere because of their extremely low vapor pressure. The partial combustion of coal, which has an atomic hydrogen-to-carbon ratio close to or less than 1, is a source of polynuclear aromatic compounds which have considerable toxicity (Lee et al., 1981).

Carbonyl compounds, consisting of aldehyde derivatives (RCHO) and ketone derivatives (R^1COR^2 where R^1 and R^2 are the same or different hydrocarbon groups), enter the atmosphere from a large number of sources and processes. These include direct emissions from internal combustion engine exhausts, incinerator emissions, spray painting, polymer manufacture, printing, petrochemicals manufacture, and lacquer manufacture. Formaldehyde (HCHO) and acetaldehyde (CH_3CHO) are produced by microorganisms and acetaldehyde is emitted by various types of vegetation.

TABLE 2.4 Examples of Polynuclear Aromatic Hydrocarbon Derivatives

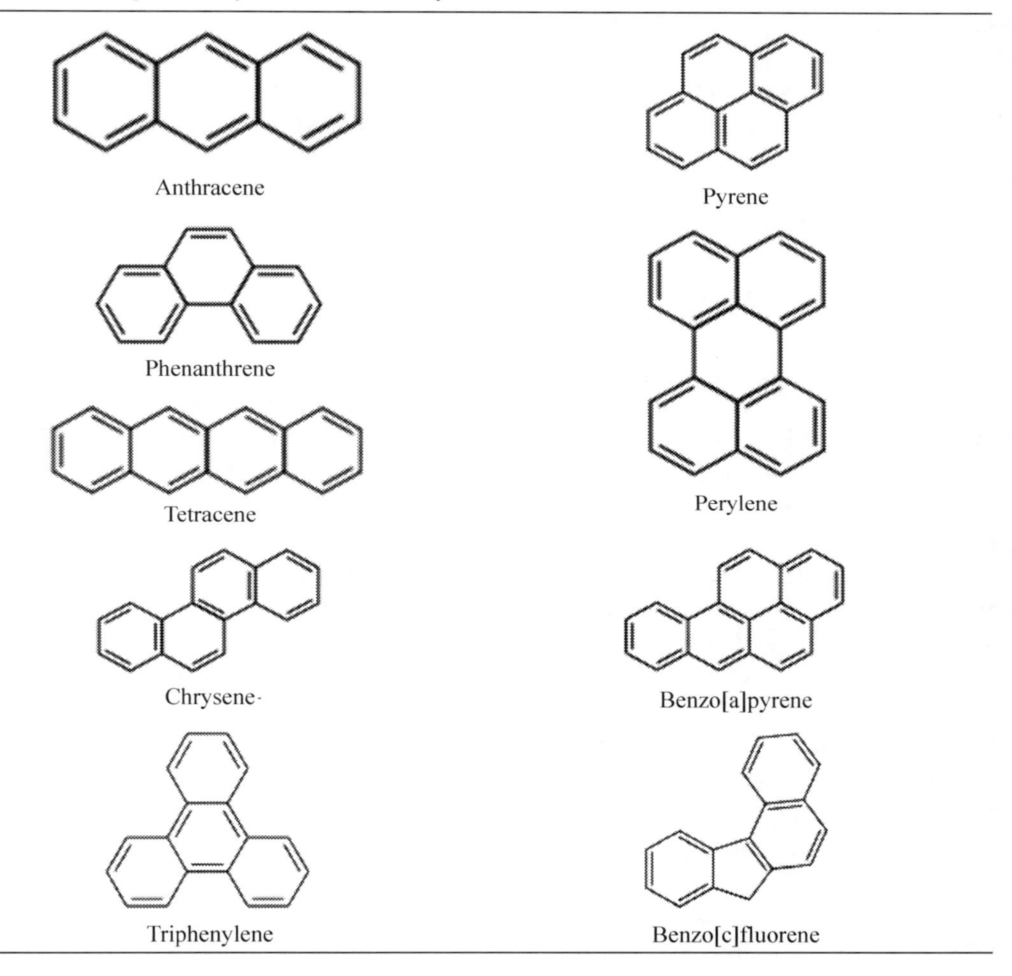

Naphthalene

Biphenyl

As expected from their low vapor pressures, the lower-boiling organic halogen compounds are the most likely to be found in the atmosphere. The three most abundant organic chlorine compounds in the atmosphere are methyl chloride (CH_3Cl) and carbon tetrachloride (CCl_4). Also found are methylene chloride (CH_2Cl_2), methyl bromide (CH_3Br), bromoform ($CHBr_3$), assorted chlorofluorocarbons and halogen-substituted ethylene compounds, such as trichloroethylene ($CHCl{=}CCl_2$), vinyl chloride ($CH_2{=}CHCl$), perchloroethylene ($CCl_2{=}CCl_2$), and ethylene dibromide (CH_2BrCH_2Br). Many of these types of chlorocarbon derivatives and chlorofluorocarbon derivatives have been implicated in the halogen-atom-catalyzed destruction of atmospheric ozone, which filters out cancer-causing ultraviolet (UV) radiation from the sun.

$$\text{Chlorofluorocarbon} \rightarrow Cl\cdot$$

$$Cl\cdot + O_3 \rightarrow ClO + O_2$$

Although inert in the lower atmosphere, chlorofluorocarbon derivatives undergo photo-decomposition by the action of high-energy UV radiation in the stratosphere to release chlorine atoms which then react with ozone to produce chlorine monoxide.

While organic sulfur derivatives are not highly significant as atmospheric contaminants on a large scale, organic sulfur compounds can cause local pollution problems-not the least of which is the highly objectionable odor. Major sources of organic sulfur compounds in the atmosphere include microbial degradation, wood pulping, volatile matter evolved from plants, animal wastes, packing house and rendering plant wastes, starch manufacture, sewage treatment, and crude oil refining.

Aromatic amines are widely used as chemical intermediates, antioxidants, and curing agents in the manufacture of polymers (rubber and plastics), drugs, pesticides, dyes, pigments, and inks. A large number of heterocyclic nitrogen compounds have been reported in tobacco smoke, and it is inferred that many of these compounds can enter the atmosphere from burning vegetation. Coke ovens are another major source of these compounds. In addition to the derivatives of pyridine, some of the heterocyclic nitrogen compounds are derivatives of pyrrole (Speight, 2013a).

PM is a term that has come to stand for (liquid or solid) particles in the atmosphere which may be organic or inorganic and originate from a wide variety of sources and processes. As one example, PM is produced from mineral matter in a carbonaceous fuel that is converted during combustion to finely divided inorganic material referred to as fly ash. This low density ash can be carried out of the stack with the hot exhaust gases and also has the ability to adsorb polynuclear aromatic hydrocarbon derivatives. Furthermore, the practice of burning any finely divided carbonaceous fuel can contribute to fly ash emissions which are composed of several chemical species (such as arsenic, cadmium, mercury, thallium, and zinc compounds) that are of environmental significance in soil, in water, and in the atmosphere (Kothny, 1973). Indeed, the United States CAA of 1990 (Chapter 10) requires the evaluation of several trace elements with a view to the banning such emissions.

The particulate matter (PM, also referred to as particulates, atmospheric aerosol particles, atmospheric PM, or suspended particulate matter, SPM) is a collection of microscopic particles of solid particles (or even liquid matter) suspended in the air. The term aerosol commonly refers to the particulate/air mixture, as opposed to the PM alone. The source of the PM can be natural or anthropogenic. The PM can have an impact on climate and precipitation that adversely affect human health in ways additional to direct inhalation.

The PM is often categorized by size of the particles such as: (i) coarse particles, which are designated PM_{10} with a diameter of 10 micrometers (10 microns, 10 μm; (1 micron = 1 meter × 10^{-6}) or less; (ii) fine particles, designated $PM_{2.5}$ with a diameter of 2.5 μm or less, (iii) ultrafine particles, $PM_{0.1}$; and (iv) soot. All of the aforementioned particles are inhalable to congregate in the lungs and, therefore, are a hazard to human health.

Soot as an airborne contaminant in the environment has many different sources, all of which are results of thermal decomposition, particularly pyrolysis. The sources include soot from coal combustion, internal-combustion engines, power-plant boilers as well as a variety of boilers, waste incineration local field (agricultural) burning, house fires, forest fires, fireplaces, and furnaces. These exterior sources also contribute to the indoor environment sources such as smoking of plant matter, cooking, oil lamps, candles, quartz/halogen bulbs with settled dust, and defective furnaces.

The formation of soot depends strongly on the fuel composition. The rank ordering of the soot forming tendency (also referred to as the propensity for soot formation) of fuel components is:

Naphthalene derivatives > benzene derivatives > aliphatic derivatives

However, the order of the soot-forming propensity within any of the above chemical hydrocarbon groups varies dramatically and is dependent on the structural type of the hydrocarbon derivative and the heteroatom (N, O, S) content.

Thus, PM, which ranges in size from approximately one-half millimeter down to molecular dimensions, may consist of either solids or liquid droplets. Atmospheric aerosols are solid or liquid particles smaller than 100 microns) in diameter. Pollutant particles in the 0.001 to 10 micron range are commonly suspended in the air near sources of pollution, such as the urban atmosphere, industrial plants, highways, and power plants. The small, solid particles include carbon black, silver iodide (AgI), combustion nuclei, and sea-salt nuclei and the larger particles include cement dust, windblown soil dust, foundry dust, and any pulverized solid fuel-thus, the need for dust control is strong (Mody and Jakhete, 1988). On the other hand, the liquid PM (mist) includes raindrops, fog, and sulfuric acid mist and some particles are of biological origin, such as viruses, bacteria, bacterial spores, fungal spores, and pollen. Atmospheric particles undergo a number of processes in the atmosphere. Sedimentation and scavenging by raindrops and other forms of precipitation are the major mechanisms for particle removal from the atmosphere; PM also reacts with atmospheric gases.

Dispersion aerosols, such as dusts, formed from the disintegration of larger particles are usually above 1 micron in size. Typical processes for forming dispersion aerosols include evolution of dust from coal grinding, formation of spray in cooling towers, and blowing of dirt from dry soil. Many dispersion aerosols originate from natural sources, such as sea spray, windblown dust, and volcanic dust. However, a vast variety of human activities break up material and disperse it to the atmosphere.

Metal oxides constitute a major class of inorganic particles in the atmosphere. These are formed during the combustion of fuels that contain metals. For example, particulate iron oxide is formed during the combustion of pyrite-containing coal:

$$3FeS_2 + 8O_2 \rightarrow Fe_3O_4\ (FeO + Fe_2O_3) + 6SO_2$$

Organic vanadium in residual fuel oil is converted to particulate vanadium oxide. Part of the calcium carbonate in the ash fraction of coal is converted by heat to calcium oxide and is emitted to the atmosphere through the stack:

$$CaCO_3 \rightarrow CaO + CO_2$$

A common process for the formation of aerosol mists involves the oxidation of atmospheric sulfur dioxide to sulfuric acid, a hygroscopic substance that accumulates atmospheric water to form small liquid droplets:

$$2SO_2 + O_2 + 2H_2O \rightarrow 2H_2SO_4$$

In the presence of basic air pollutants, such as ammonia (NH_3) or calcium oxide (CaO), the sulfuric acid (H_2SO_4) reacts to form the ammonium or calcium salts. Under conditions of low humidity, water is lost from these droplets and a solid aerosol is formed.

A significant portion of organic PM is produced by internal combustion engines. These products may include nitrogen-containing compounds and oxidized hydrocarbon polymers. Lubricating oil and its additives may also contribute to organic PM.

Particulate carbon as soot, carbon black, coke, and graphite originates from auto and truck exhausts, heating furnaces, incinerators, power plants, and steel and foundry operations and composes one of the more visible and troublesome particulate air pollutants. Because of its good adsorbent properties, carbon can be a carrier of gaseous and other particulate pollutants. Particulate carbon surfaces may catalyze some heterogeneous atmospheric reactions, including the important conversion of sulfur dioxide to sulfate.

Another mineral worthy of note as a pollutant is asbestos which is of considerable concern as an air pollutant because when inhaled the fibrils can lodge in human tissue and cause asbestosis (a pneumonia condition), mesothelioma (tumor of the mesothelial tissue lining the chest cavity adjacent to the lungs), and bronchogenic carcinomas (cancer originating with the air passages in the lungs).

By way of definition, asbestos is the name given to a group of six fibrous silicate minerals, typically those of the serpentine group. Chemically, asbestos is a group of impure magnesium silicate minerals and colors may be white, gray, green, or brown, with a specific gravity on the order of and is noncombustible. Serpentine asbestos is the mineral chrysotile, a magnesium silicate with strong and flexible fibers. Spinning is possible with the longer fibers. Amphibole asbestos includes various silicates of magnesium, iron, calcium, and sodium. The fibers are generally brittle and cannot be spun, but are more resistant to chemicals and to heat than serpentine asbestos. Because asbestos has long been indicated in asbestosis (similar to silicosis) and, in more recent years, considered a carcinogen, several countries have issued regulations that restrict its use. For many years, it has been used in fireproof fabrics, brake lining, gaskets, roofing, insulation, paint fillers, reinforcing agents in rubber and plastics, and in electrolytic diaphragm cells.

As a side note, talc is a clay mineral that is composed of hydrated magnesium silicate [$Mg_3Si_4O_{10}(OH)_2$] and has been known to occur with asbestos.

Some of the metals found predominantly as PM in polluted atmospheres are known to be hazardous to human health. All of these except beryllium (Be) are so-called heavy metals. Lead is the toxic metal of greatest concern in the urban atmosphere because it comes closest to being present at a toxic level; mercury ranks second. Others include beryllium, cadmium, chromium, vanadium, nickel, and arsenic (a metalloid). Arsenic can have a significant effect on the environment, whether the metalloid is released to the land, the water, or the air (Nriagu, 1994a, b).

Atmospheric mercury (Hg^o) is of concern because of its toxicity, volatility, and mobility. Some atmospheric mercury is associated with PM. One means by which mercury can enter the atmosphere is as volatile elemental mercury from coal combustion (ASTM D3684). Volatile organic mercury compounds such as dimethyl mercury [$(CH_3)_2Hg$] and methyl mercury salts, such as methyl mercury bromide (CH_3HgBr), are also encountered in the atmosphere. With the reduction in the use of leaded fuels, atmospheric lead is of less concern than it used to be.

A significant natural source of radionuclides in the atmosphere is radon, a noble gas product of radium decay. Radon may enter the atmosphere as either of two isotopes, ^{222}Rn (half-life: 3.8 days) and ^{220}Rn (half-life: 54.5 seconds). Both are alpha emitters in decay chains that produce a series of daughter products and terminate with stable isotopes of lead. The initial decay products, isotopes of polonium (^{218}Po and ^{216}Po) are nongaseous and adhere readily to atmospheric PM. Therefore, some of the radioactivity detected in these particles is of natural origin.

One of the more serious problems in connection with radon is that of radioactivity originating from uranium mine tailings that have been used in some areas as backfill, soil conditioner, and a base for building foundations. Radon produced by the decay of radium exudes from foundations and walls constructed on tailings. Some medical authorities have suggested that the rate of birth defects and infant cancer in areas where uranium mill tailings have been used in residential construction are significantly higher than normal.

The combustion of fossil fuels introduces radioactivity into the atmosphere in the form of radionuclides contained in fly ash. Large coal-fired power plants lacking ash-control equipment may introduce up to several hundred millicuries of radionuclides into the atmosphere each year, far more than either an equivalent nuclear or an equivalent oil-fired power plant.

The radioactive noble gas krypton (^{85}Kr) has a half-life on the order of 10.3 years and is emitted into the atmosphere by the operation of nuclear reactors and the processing of spent reactor fuels. In general, other radionuclides produced by reactor operation are either chemically reactive and can be removed from the reactor effluent or have such short half-lives that a short time delay prior to emission prevents their leaving the reactor.

The above-ground detonation of nuclear weapons can add large amounts of radioactive PM to the atmosphere. Although there are general agreements between several nations to cease and desist from testing nuclear devices in the atmosphere, not all nations agree to this proposition.

It is considered likely that most of the environmental impact of mineral resource development (including the hazards of coal mining, gaseous emissions, acid precipitation) could be substantially abated. A considerable investment in retrofitting or replacing existing facilities and equipment might be needed. However, replacing coal with natural gas, which releases less carbon dioxide per unit of energy, is at best a short-term solution.

Minimizing the carbon dioxide emissions from many mineral resources (which must also include emissions of hydrogen sulfide) would require revamping of the current technology. But it is possible, and a conscious goal must be to improve the efficiency with which the resources are transformed and consumed and by shifting to alternative products.

In terms of the production of transportation fuels from biomass sources (Sheehan, 1994), some comment is required because of the recognition that biomass fuels can also produce hydrocarbon emissions during combustion.

Ethanol (ethyl alcohol; grain alcohol) is a major biomass fuel (Speight, 2020a) but in the current market conditions it remains significantly more expensive than gasoline. The widespread use of biomass (alcohol) fuel is constrained by the size of the resource base from which they are produced. There is also the argument that methyl alcohol (methanol) produced from natural gas releases as much carbon dioxide as does gasoline. The counter argument is that methanol-fueled engines have a greater potential for improvements than gasoline engines do (for example, they can function at higher levels of compression than have heretofore been attempted).

There is the need for the systematic identification and evaluation of the potential impacts of proposed projects, plans, programs, policies, or legislative actions upon the physical-chemical, biological, cultural, and socioeconomic components of the environment. Also known as environmental impact assessment (EIA), it also includes the consideration of measures to mitigate undesirable impacts. The primary purpose of EIA is to encourage consideration of the environment in planning and decision making and ultimately to arrive at actions that are environmentally compatible.

Impact assessment refers to the interpretation of the significance of anticipated changes related to the proposed project. Impact interpretation can be based upon the systematic application of the definition of significance, as described earlier. For some types of anticipated impacts, there are specific numerical standards or criteria that can be used as a basis for impact interpretation. Examples include air quality standards, environmental noise criteria, surface-water, and groundwater quality standards, and wastewater discharge standards for particular facilities. This latter is especially true since the quality of wastewater can vary with the nature of the facility.

In this context, it is necessary to consider the impacts related to the biological environment and the potential significance of the loss of particular habitats, including wetland areas (Dennison and Berry, 1993). Another basis for assessment is public input, which could be received through the scoping process or through public participation programs.

Identifying and evaluating potential impact mitigation measures should also be an activity in the process of EIA. Mitigation can be defined as the sequential consideration of measures such as: (i) avoiding the impact by not taking a certain action or parts of an action; (ii) minimizing impacts by limiting the degree or magnitude of the action and its implementation; (iii) rectifying the impact by repairing, rehabilitating, or restoring the affected environment; (iv) reducing or eliminating the impact over time by preservation and maintenance operations during the life of the action; and (v) compensating for the impact by replacing or providing substitute resources or environments. Examples of mitigation measures include pipeline routing to avoid archeological resources, inclusion of pollution-control equipment on airborne and liquid discharges, reductions in project size, revegetation programs, wildlife protection plans, erosion control measures, remediation activities, and creation of artificial wetlands.

A key activity in EIA is associated with selecting the proposed action from alternatives that have been evaluated (Lnhaber, 1982). In public projects, there is considerable emphasis on the evaluation of alternatives; in fact, various regulations (Majumdar, 1993) indicate that the analysis of alternatives represents the heart of the impact assessment process. Conversely, for many private developments, the range of alternatives may be limited. Even so, there are still potential alternative measures that could be evaluated, including those relating to project size and design features even if location alternatives are not available.

Environmental impact studies need to address a minimum of two alternatives, and can include more than fifty alternatives. Typical studies address three to five alternatives. The minimum number usually represents a choice between construction and operation of a project versus project rejection. The alternatives may encompass a wide range of alternate considerations such as: (i) site location; (ii) design of a site; (iii) construction, operation, and decommissioning options; (iv) project size; (v) phasing in of operational units; (vi) timing of the construction, operation, and decommissioning activities; and (vii) the option that related to a non-project.

All of the aforementioned arguments for and against the use of resources do not belie the obvious. Mineral resources, including fossil fuel resources, will continue to be sources of valuable products for the next several decades (Speight, 2011a; Speight and Islam, 2016). They must be used judiciously insofar as they are sources of pollutants, many of which are gases and which must not be released to the atmosphere.

Obviously, much work is needed to accommodate the change to a different fuel source. In the meantime, there must be serious efforts to improve efficient usage and working to ensure that there is no damage to the environment. Such is the nature of resource development and use and the expectancy of environmental protection.

KEYWORDS

- **biomass**
- **fossil fuels**
- **mineral resources**
- **natural gas**
- **predominant energy**

REFERENCES

Abraham, H., (1945). *Asphalts and Allied Substances*. Van Nostrand and Co., New York.
Al-Sowayegh, A., (1984). *Arab Petro-Politics*. Croom Helm Ltd., London, United Kingdom.

ASTM D3684, (2020). *Test Method for Total Mercury in Coal by the Oxygen Bomb Combustion/Atomic Absorption Method. Annual Book of Standards*. ASTM International, West Conshohocken, Pennsylvania.
ASTM D388, (2020). *Classification of Coals by Rank*. Annual Book of Standards, ASTM International, West Conshohocken, Pennsylvania.
ASTM E901, (2020). *Classification System for Uranium Resources*. Annual Book of Standards, ASTM International, West Conshohocken, Pennsylvania.
Austin, G. T., (1984). *Shreve's Chemical Process Industries*. McGraw-Hill, New York.
Blunden, J., & Reddish, A., (1991). *Energy, Resources, and Environment*. Hodder and Stoughton, London.
Breen, J. J., & DeMarco, M. J., (1992). *Pollution Prevention in Industrial Processes: The Role of Process Analytical Chemistry.* Symposium Series No. 508. American Chemical Society, Washington, D.C.
Cassedy, E. S., & Grossman, P. Z., (1990). *Introduction to Energy Resources, Technology, and Society.* Cambridge University Press, Cambridge, United Kingdom.
Chenier, P. J., (1992). *Survey of Industrial Chemistry* (2nd edn.). VCH Publishers Inc., New York. Chapter 25.
Cooper, R. N., (1994). *Environment and Resource Policies for the World Economy*. The Brookings Institution, Washington, D.C.
Cruden, A., (1930). *Complete Concordance to the Bible*. Butterworth Press, London, United Kingdom.
De Jong, W., (2015). Biomass composition, properties, and characterization. In: De Jong, W., & Van, O. J. R., (eds.), *Biomass as a Sustainable Energy Source for the Future: Fundamentals of Conversion Processes*. American Institute of Chemical Engineers Inc., Washington, DC.
Dennison, M. S., & Berry, J. F., (1993). *Wetlands: Guide to Science, Law, and Technology*. Noyes Data Corp., Park Ridge, New Jersey.
Dickson, M. H., & Fanelli, M., (1990). *Small Geothermal Resources: A Guide to Development and Utilization.* UNITAR/UNDP Centre on Small Energy Resources, Rome, Italy.
Easterbrook, G., (1995). *A Moment on the Earth: The Coming Age of Environmental Optimism*. Viking Press, New York.
Forbes, R. J., (1958). *Studies in Early Petroleum Chemistry.* E.J. Brill, Leyden, The Netherlands.
Forbes, R. J., (1959). *More Studies in Early Petroleum History*. E.J. Brill, Leyden, The Netherlands.
Forbes, R. J., (1964). *Studies in Ancient Technology*. E.J. Brill, Leyden, The Netherlands. Volume I.
Galloway, R. L., (1882). *A History of Coal Mining in Great Britain*. Macmillan & Co., London.
Gricola, A., & Bauer, G., (1556). De Re Metallica. Froben, Basel, Switzerland.
Halpern, G. M., (1980). Fusion energy. In: *Encyclopedia of Chemical Technology* (Vol. 11, p. 590). John Wiley & Sons Inc., New York.
Herodotus, (1956). 447 BC. Historia. In: R. George and M. Kamroff (eds.), *The History of Herodotus*. Tudor Publishing Co., New York.
Himmel, M. E., Baker, J. O., & Overend, R. P., (1994). *Enzymatic Conversion of Biomass for Fuels Production.* Symposium Series No. 566. American Chemical Society, Washington, DC.
James, P., & Thorpe, N., (1994). *Ancient Inventions*. Ballantine Books, New York.
Jazib, J., (2018). *Basics of Environmental Sciences*. IQRA Publishers, New Delhi, India.
Johansson, T. D., Kelly, H., Reddy, A. K. N., & Williams, R. H., (1993). *Renewable Energy for Fuels and Electricity.* Earthscan Publications, United Nations, New York.
Jones, J. L., & Radding, S. B., (1980). *Thermal Conversion of Solid Wastes and Biomass.* Symposium Series No. 130. American Chemical Society, Washington, D.C.
Katz, J. E., (1984). *Congress and National Energy Policy.* Transaction Books. New Brunswick, New Jersey.
Knief, R. A., (1992). *Nuclear Engineering: Theory and Technology of Commercial Nuclear Power.* Taylor & Francis, Washington, D.C.
Kothny, E. L., (1973). *Trace Elements in the Environment*. Advances in Chemistry Series No. 123. American Chemical Society, Washington, D.C.
Kubursi, A. A., & Naylor, T., (1985). *Cooperation and Development in the Energy Sector*. Croom Helm, London.
Lee, M. L., Novotny, M. V., & Bartle, K. D., (1981). *Analytical Chemistry of Polycyclic Aromatic Compounds*. Chapter 3. Academic Press Inc., New York.
Lee, S., & Shah, Y. T., (2013). *Biofuels and Bioenergy: Processes and Technologies*. CRC Press, Taylor & Francis Group, Boca Raton Florida.

Lee, W. M. G., Yuan, Y. S., & Chen, J. C., (1993). *J. Environ. Sci. Health, Part A., A28*, 1017.
Lincoln, B., (1785). *Memoirs of the American Academy of Arts and Sciences* (Vol. I. p. 372).
Lowinger, T. C., & Hinman, G. W., (1994). *Nuclear Power at the Crossroads: Challenges and Prospects for the Twenty-First Century*. International Research Center for Energy and Economic Development, Boulder Colorado.
Luciani, G., (2013). *Security of Oil Supplies: Issues and Remarks*. Claeys & Casteels, Deventer, Netherlands.
Majumdar, S. B., (1993). *Regulatory Requirements for Hazardous Materials*. McGrawHill, New York.
Mayes, J. M., (2015). What are the possible impacts on US refineries processing shale oils? *Hydrocarbon Processing, 94*(2), 67–70.
Mody, V., & Jakhete, R., (1988). *Dust Control Handbook*. Noyes Data Corp., Park Ridge, New Jersey.
Mooney, H., (1988). *Towards an Understanding of Global Change*. National Academy Press, Washington, DC.
Nersesian, R. L., (2007). *Energy for the 21st Century: A Comprehensive Guide to Conventional and Alternative Fuel Sources*. M.E. Sharpe Inc., Armonk, New York.
Niblock, T., & Lawless, R., (1985). *Prospects for the World Oil Industry*. Croom Helm. London.
Nivola, P. S., & Crandall, R. W., (1995). *The Extra Mile: Rethinking Energy Policy for Automotive Transportation*. The Brookings Institute, Washington, D.C.
Nriagu, J. O., (1994a). *Arsenic in the Environment, Part I: Cycling and Characterization*. John Wiley & Sons Inc., New York.
Nriagu, J. O., (1994b). *Arsenic in the Environment, Part II: Human Health and Ecosystem Effects*. John Wiley & Sons Inc., New York.
Pickering, K. T., & Owen, L. A., (1994). *Global Environmental Issues*. Routledge Publishers Inc., New York.
Pourret, O., & Hursthouse, A., (2019). It's time to replace the term "heavy metals" with "potentially toxic elements" when reporting environmental research. *Int. J. Environ. Res. Public Health, 16*(22), 4446–4451.
Probstein, R. F., & Hicks, R. E., (1990). *Synthetic Fuels*. pH Press, Cambridge, Massachusetts.
Robinson, J. S., (1980). *Fuels from Biomass: Technology and Feasibility*. Noyes Data Corp., Park Ridge, New Jersey.
Sheehan, J. J., (1994). In: Himmel, M. E., Baker, J. O., & Overend, R. P., (eds.), *Enzymatic Conversion of Biomass for Fuels Production*. Symposium Series No. 566. American Chemical Society, Washington, D.C. Chapter 1.
Singer, C., Holmyard, E. J., Hall, A. R., & Williams, T. I., (1958). *A History of Technology* (Vol. I–V). Clarendon Press, Oxford, United Kingdom.
Skrzypek, J., Sloczynski, J., & Ledakowicz, S., (1994). *Methanol Synthesis*. Polish Academy of Sciences, Gliwice, Poland.
Speight, J. G., & Islam, M. R., (2016). *Peak Energy: Myth or Reality*. Scrivener Publishing, Beverly, Massachusetts.
Speight, J. G., (1996). *Environmental Technology Handbook*. Taylor & Francis, Washington, DC.
Speight, J. G., (2011a). *An Introduction to Petroleum Technology, Economics, and Politics*. Scrivener Publishing, Beverly, Massachusetts.
Speight, J. G., (2011b). *The Biofuels Handbook*. Royal Society of Chemistry, London, United Kingdom.
Speight, J. G., (2013a). *The Chemistry and Technology of Coal* (2nd edn.). Marcel Dekker Inc., New York.
Speight, J. G., (2013b). *Coal-Fired Power Generation Handbook*. Scrivener Publishing, Beverly, Massachusetts.
Speight, J. G., (2014). *The Chemistry and Technology of Petroleum* (5th edn.). CRC Press, Taylor & Francis Group, Boca Raton, Florida.
Speight, J. G., (2017). *Handbook of Petroleum Refining*. CRC Press, Taylor & Francis Group, Boca Raton, Florida.
Speight, J. G., (2019a). *Natural Gas: A Basic Handbook* (2nd edn.). Gulf Publishing Company, Elsevier, Cambridge, Massachusetts.
Speight, J. G., (2019b). *Handbook of Petrochemical Processes*. CRC Press, Taylor & Francis Group, Boca Raton, Florida.
Speight, J. G., (2020a). *Synthetic Fuels Handbook: Properties, Processes, and Performance* (2nd edn.). McGraw-Hill, New York.
Speight, J. G., (2020b). *Refinery of the Future* (2nd edn.). Gulf Professional Publishing, Elsevier, Cambridge, Massachusetts.
Speight, J. G., (2020c). *Global Climate Change Demystified*. Scrivener Publishing, Beverly, Massachusetts.

Tillman, D. A., (1978). *Wood as an Energy Resource*. Academic Press Inc., New York.
United States Department of Energy, (1990). *Gas Research Program: Implementation Plan.* DOE/FE-0187P. United States Department of Energy, Washington, D.C. April.
US EIA, (2011). *Review of Emerging Resources.* US Shale Gas and Shale Oil Plays. Energy Information Administration, United States Department of Energy, Washington, DC.
US EIA, (2013). *Technically Recoverable Shale Oil and Shale Gas Resources: An Assessment of 137 Shale Formations in 41 Countries Outside the United States.* Energy Information Administration, United States Department of Energy, Washington, DC.
Williamson, J., & Wigley, F., (1994). *The Impact of Ash Deposition on Coal Fired Plants.* Taylor and Francis, Washington, D.C.
Wright, L., Boundy, R., Perlack, R., Davis, S., & Saulsbury, B., (2006). *Biomass Energy Data Book* (1st edn.). Office of Planning, Budget, and Analysis, Energy Efficiency and Renewable Energy, United States Department of Energy. Contract No. DE-AC05-00OR22725. Oak Ridge National Laboratory, Oak Ridge, Tennessee.
Wuebbles, D. J., & Edmonds, J., (1988). *A Primer on Greenhouse Gases.* Report No. DOE/NBB-0083. Office of Energy Research, United States Department of Energy, Washington, D.C.
Yergin, D., (1992). *The Prize: The Epic Quest for Oil, Money, and Power.* Simon & Schuster, New York.

PART II
Ecosystems

CHAPTER 3

The Atmosphere

3.1 INTRODUCTION

As defined in Chapter 1, the term Earth system refers to the three interacting physical, chemical, and biological systems in which the Earth allows the Earth to act as a livable planet in which life-supporting processes occur and interact. The systems (by definition for this book) consist of: (i) the atmosphere; (ii) the aqua sphere, which consists of the oceans, the rivers, the lakes, and all other water systems; and (iii) the geosphere, which consists of the landmasses (Table 3.1). These systems act together and include the natural cycles, such as the carbon cycle, water cycle, nitrogen cycle, sulfur cycle, and many other cycles that involve the deep Earth processes. Life (including human life) is an integral part of the Earth

TABLE 3.1 The Main Components of the Earth System

Component	Description
Atmosphere	The gaseous layer surrounding the Earth.
	Held to the surface by gravity.
	Receives energy from solar radiation, which warms the surface of the Earth.
	Absorbs water from the surface of the Earth via evaporation.
	Redistributes heat and moisture across the surface of the Earth.
	Contains substances that are essential for life, including carbon, nitrogen, oxygen, and hydrogen.
Aquasphere	The parts of the earth system composed of water in its liquid, gaseous (vapor) and solid (ice) phases.
	Includes oceans, seas, ice sheets, sea ice, glaciers, lakes, rivers, streams, atmospheric moisture, ice crystals, and areas of permafrost.
	Also includes the moisture in the soil (soil water) and within rocks (groundwater).
	Can be sub-divided into the fluid water systems (the hydrosphere) and solid water (ice) systems (the cryosphere).
Geosphere	Composed of rocks and minerals.
	Includes the solid crust, the molten mantle, and the liquid and the core of the Earth.
	Includes the soil in which nutrients become available to living organisms.
	Provides an important ecological habitat and the basis of life forms.
	Subject to the processes of: (i) erosion; (ii) weathering; and (iii) transport, as well as to tectonic forces and volcanic activity.

The Science and Technology of the Environment. James G. Speight (Author)

system, and in many cases, the human interactions with the Earth system are also drivers of change (Jazib, 2018).

Global change refers to planetary-scale changes in the Earth system and involves changes to the atmosphere (atmospheric circulation), the aqua sphere (ocean circulation), and the geosphere (which includes biological diversity and pollution. Climate is a result of the aggregation and interaction of all components of weather such as, for example, precipitation, temperature, and the formation and presence of clouds). Although changes in the Earth system involve changes in climate, the Earth system includes other components and changes to the Earth system, whether natural or driven by humans, can have significant consequences without involving changes in climate. However, global change should not be confused with climate change which is part of a much larger challenge and is not always solely due to human interaction with the Earth system (Speight, 2020).

As the first the systems to be described in this book, the atmosphere is concentrated at the surface of the Earth and rapidly thins with altitude and blends with space at approximately 100 miles above sea level. The atmosphere is essential to life and serves a vital protective function to living species and is intimately related to the other systems of the Earth: the aqua sphere and the geosphere. On occasion, a fourth component-the biosphere- and a fifth component-the cryosphere-are is added as a part of the geosphere. These components are also systems in their own right and they are intimately tightly interconnected (Table 3.1) (Chapter 4, Chapter 5). These components are also systems in their own right and they are tightly interconnected. For example, the biosphere (which involves the participation of all three of the main systems (the atmosphere, the aqua sphere, and the geosphere) is the system that contains all living organisms (including humans), and it is intimately related to the other three systems. Most living organisms require gases from the atmosphere, water from the aquasphere, as well as nutrients and minerals from the geosphere and, as a result, inhabit one or more of the three main spheres-viz the atmosphere, the aqua sphere, and the geosphere. Much of the biosphere exists within the surface layer of the Earth, which includes the lower part of the atmosphere, the surface of the geosphere and approximately 300 feet of the top layer of the ocean.

The cryosphere is the collective term for the parts of the Earth system in which water exists in the frozen state (Table 3.1). The cryosphere comprises components such as snow, river ice, lake ice, sea ice, ice sheets, glaciers, and ice caps which exist, both on land and beneath the oceans. The lifespan of each component is different-for example, river ice and lake ice are features that generally do not survive from winter to summer while sea ice advances and retreats with the seasons and in the Arctic region, the sea ice can survive to become multi-year ice lasting several years.

While the Earth system is subject to several divisions and sub-divisions, for the purposes of this book the three predominant divisions of the Earth system are: (i) the atmosphere; (ii) the aqua sphere; and (iii) the geosphere.

The atmosphere is the region above the surface of the Earth that is, in fact, the envelope of gases which is subdivided into regions depending on the altitude (Parker and Corbitt, 1993). The atmosphere is life-supporting by virtue of the relatively high proportion of molecular oxygen which exists in spite of the presence of gases such as nitrogen, methane, hydrogen, and other gases that are capable of existing in various oxide forms. The chemical disequilibrium is maintained by the continuous production of gases from biological processes. Thus, floral, and faunal chemistry plays a major role in maintaining the composition of the atmosphere (Anderson, 2006).

Below approximately 60 miles in altitude, the proportion of the major constituents in the atmosphere is very uniform and is known as the homosphere (the lower part of the atmosphere where the bulk gases are mixed due to turbulent mixing or diffusion), to distinguish it from the heterosphere above the approximate 60 miles of altitude, where the relative amounts of the major gaseous constituents change with height. In the homosphere, there are sufficient atmospheric motions and a short enough molecular free path to maintain uniformity in composition. Above the boundary between the homosphere and the heterosphere (known as the homopause or the turbopause), the collision frequency (mean free path) of an individual molecule is sufficiently long to allow a partial separation (by gravity) of the (relatively) lighter (less dense) molecules from the heavier (more dense) molecules. Not surprisingly, the average molecular weight of the heterosphere decreases with height because the lighter (less dense) molecules dominate the composition at higher elevation.

In addition to the gaseous constituents, the atmosphere contains suspended solid particles and liquid particles in which a gas, frequently air is the continuous medium and particles of solids or liquids are dispersed in it (Hidy et al., 1979). The particle size distribution of atmospheric aerosols is multi-modal in character, usually with a bimodal mass, volume, or surface area distribution.

The nuclei mode (<0.03 micron) (1 micron, 1 μm, 1 meter × 10^{-6}, also called 1 micrometer which is equivalent to 1 meter × 10^{-6} in diameter, i.e., 1 millionth of a meter or one thousandth of a millimeter, 0.001 mm, or approximately 0.000039 of an inch) is formed by condensation of vapor or by gaseous reaction products. The intermediate mode (0.1–1.0 micron) is formed by coagulation of nuclei. The larger particles (>1.0 micron) fall out, whereas the fine particles (smaller than 0.1 micron) then agglomerate to form larger particles which remain suspended.

Aerosols are created by gas-to-particle reactions and are lifted from the surface by winds. A portion of these aerosols can become centers of condensation or deposition in the growth of water and ice clouds. Cloud droplets and ice crystals are made primarily of water with some trace amounts of particles and dissolved gases. The diameters of the particles range from several microns to 100 microns. Water or ice particles larger than approximately 100 microns begin to fall because of gravity and may result in precipitation at the surface.

Water is the only naturally occurring compound that exists in three physical forms or physical states (gas, liquid, and solid) on the surface of the Earth and is an important component of the three Earth systems. Heat energy (which is absorbed in processes such as melting, sublimation, and evaporation) is transferred through the atmosphere as water changes from one physical state to another. On the other hand, heat is lost to the atmosphere during freezing, condensation, and precipitation.

The interaction between temperature, pressure, and moisture in the atmosphere makes it possible for meteorologists to predict short-term changes in local weather conditions (weather forecasts). However, the changes that such systems undergo on a regular basis (often a daily basis) emphasize the complexity of the atmospheric system.

3.2 COMPOSITION

The atmosphere of the Earth protects life on Earth by creating pressure that: (i) allows water to be retained as a liquid; (ii) absorbing dangerous or unhealthy ultraviolet (UV) solar radiation; (iii) warming the surface through heat retention; and (iv) reducing the extremes of temperature between day and night temperatures, which is known as the diurnal temperature variation. The study of the atmosphere (atmospheric science or aerology includes subfields, such as climatology and atmospheric physics and (most important in the current context of environmental issues, global warming) paleoclimatology which is a study of the ancient atmosphere and the change made by humans.

The major constituents of the atmosphere are: (i) nitrogen, N_2, 78.08% v/v; (ii) oxygen, O_2, 20.95% w/w, and water vapor, 0 to 0.25% w/w, although the concentration of water vapor (H_2O) is highly variable, especially near the surface, where volume fractions can be as high as 4% in the tropics. There are many minor constituents or trace gases such as (alphabetically rather than by abundance, which is variable) argon (Ar), carbon dioxide (CO_2), helium (He), hydrogen (H_2), krypton (Kr), methane (CH_4). (neon, Ne), nitrous oxide (N_2O), ozone (O_3), and water vapor (H_2O) (Table 3.2) (Prinn, 1987). In addition, the atmosphere also contains organic species other than methane (Westberg and Zimmerman, 1993) that can play an important role in nucleation and in the formation of smog and as well as participate in the greenhouse effect (Chapter 1).

TABLE 3.2 Approximate Composition of the Atmosphere

Component, % v/v*	Amount, % v/v
Major Components:	
Nitrogen (N_2)	78.08
Oxygen (O_2)	20.95
Minor/Trace Components:	
Argon (Ar)	0.93
Carbon dioxide (CO_2)	0.035
Helium (He)	0.00052
Hydrogen (H_2)	0.00005
Krypton (Kr)	0.00010
Methane (CH_4)	0.00014
Neon (Ne)	0.018
Nitrous oxide (N_2O)	0.00005
Ozone (O_3)	0.0000007
Water vapor (H_2O)	0.025**

*Listed alphabetically.

**Variable; can be as high as 4% v/v in humid areas.

The atmosphere has changed considerably since, in the early days of the Earth, it was primarily an atmosphere composed of hydrogen. For example, 2.4 billion (2.4×10^9) years ago (at a time known as the great oxidation event), there was an increase of oxygen in the atmosphere from almost zero oxygen to amounts closer to present-day amounts. Humans have also contributed to significant changes in the composition of the atmosphere through air pollution, especially since the onset of the Industrial Revolution in the 18th Century that have led to changes such as ozone depletion and the onset of global climate change. It is, in fact, easy to place the blame for these changes squarely on the shoulders of humans but there are also several natural events (which are often ignored in the discussions of climate change) that must also be taken onto consideration as contributors to climate change (Speight, 2020).

The atmosphere is the source of oxygen for respiration and carbon dioxide for photosynthesis. Nitrogen is often (erroneously) considered to be a non-relevant gas in the atmosphere, but the nitrogen is employed by nitrogen-fixing bacteria and ammonia-manufacturing plants to produce chemically-bound nitrogen, necessary in many life-supporting molecules. As a basic part of the water cycle (Chapter 4), the atmosphere transports water from the oceans to land.

Unfortunately, the atmosphere also has been the recipient of many chemical pollutants (ranging from nitrogen oxides and sulfur oxides to more complex hydrocarbon-fluorocarbon-chlorofluorocarbon refrigerants) and other gases such as radon (Newman, 1993; Dewling et al., 2006; Ferrand, 2006) and detection of the pollutants involves a series of standard tests, many of which are applicable to the emissions testing of stacks and flues (ASTM, 2020). Such chemical intrusions into the atmosphere result in damage to life and materials (through corrosion) as well as altering the characteristics of the atmosphere.

In a protective role, the atmosphere absorbs most of the cosmic rays from outer space and also absorbs most of the electromagnetic radiation from the Sun are absorbed by the atmosphere. However, sufficient transmission of radiation in the regions of 300 to 2,500 nanometers (nm) (1 nanometer = 1 meter × 10^{-9} (near-UV, visible, and near-infrared radiation) and 0.01–40 m (radio waves) is allowed. Absorption of electromagnetic radiation at wavelengths lower than 300 nanometers (nm) removes the damaging UV radiation. The overall effect is that the atmosphere stabilizes the temperature of the land surfaces, thereby preventing the potential for extremes of temperature.

Atmospheric pressure decreases as a function of altitude. Thus:

$$P_h = P_o e^{mgh/RT}$$

In this equation, P_h is the pressure at any given height, P_o is the pressure at zero altitude (sea level); m is the average gram molecular mass of air (28.97 g/mole in the troposphere); g is the acceleration of gravity (981 cm × sec^{-1} at sea level); h is the altitude (in cm or meters or kilometers), and R is the gas constant (8.314 × 107 erg × deg^{-1} × $mole^{-1}$), and T is the absolute temperature. Furthermore:

$$\log P_h = \log P_o - (mgh \times 10^5)/2.303RT$$

At sea level where the pressure is 1 atmosphere:

$$\log P_h = (mgh \times 10^5)/2.303RT$$

The characteristics of the atmosphere also vary with location (latitude) as well as with solar activity. At a high altitude normally reactive species, such as atomic oxygen, are more persistent. At such altitudes, the pressure is low and the distance traveled by a reactive species before it collides with a potential reactant (the mean free path) is high.

3.2.1 OXYGEN AND OZONE

Oxygen is present (approximately 21% v/v) in the atmosphere of the Earth (Table 3.2) and makes up approximately 50% v/v of the crust of the Earth in the form of non-metal oxides (such as silicon dioxide, SiO_2) and metal oxides (such as ferrous oxide, FeO, and haematite, Fe_2O_3) provides the energy released in combustion and aerobic respiration. Oxygen is continuously replenished by photosynthesis in which the energy from sunlight produces oxygen from water and carbon dioxide. Atmospheric oxygen is also utilized by aerobic organisms in the degradation (decay and transformation) of organic material. Some oxidative weathering processes consume oxygen but the reverse reaction also occurs in which oxygen is returned to the atmosphere through photosynthetic processes.

Atmospheric oxygen takes part in energy-producing reactions, such as during the combustion of fossil fuels and products derived from fossil fuels. Using natural gas as the example:

$$CH_4 \text{ (natural gas)} + 2O_2 \rightarrow CO_2 + 2H_2O$$

As a result of the extremely rarefied atmosphere and the effects of ionizing radiation, elemental oxygen in the upper atmosphere can exist in other forms such as oxygen atoms ($O\cdot$), excited oxygen molecules (O_2^*), and ozone (O_3).

Ozone (sometimes referred to as trioxygen, O_3) is an inorganic molecule that is an allotrope of oxygen that is much less stable than the diatomic allotrope (O_2).

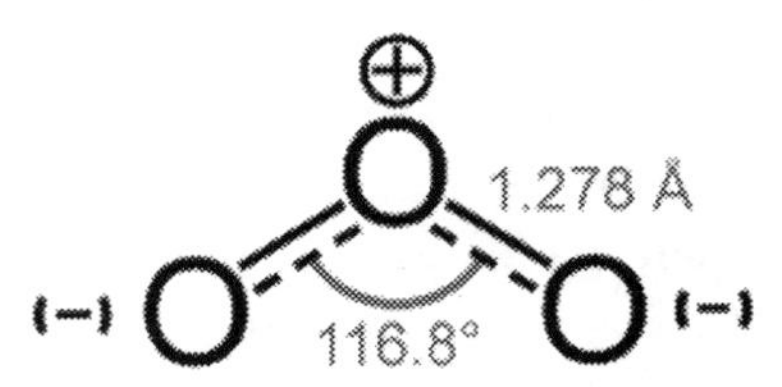

By way of definition, an allotrope is the property of some chemical elements to exist in two or more different forms.

Ozone is formed from dioxygen by the action of UV light and electrical discharges in the atmosphere. The highest concentration of ozone occurs in a layer in the lower stratosphere between the altitudes of 9 miles and 18 miles and absorbs most of the UV radiation from the Sun. This ozone results from the dissociation of molecular oxygen in the upper atmosphere and nitrogen dioxide in the lower atmosphere.

Ozone is a powerful oxidant which can cause damage to the mucous and respiratory tissues in faunal species (including humans) and also damage to tissues in plants when the concentration of ozone in the air is on the order of 0.1 ppm v/v and higher. While this makes ozone a potent respiratory hazard and pollutant near ground level, a higher concentration in the ozone layer (from two to eight ppm) is beneficial, preventing damaging UV light from reaching the surface of the Earth.

Although present in only trace quantities, atmospheric ozone plays a critical role for the biosphere by absorbing the UV radiation with a wavelength (λ) in the range 240 to 320 nanometers (1 nanometer = 1 meter × 10^{-9}), which would otherwise be transmitted to the land surface. This type of radiation is lethal to unicellular organisms (algae, bacteria, protozoa) and to many flora and fauna (plants and animals). Ozone can also cause damage to the genetic material of human cells (deoxyribonucleic acid, DNA) and is responsible for human skin condition known as sunburn, with the potential for the onset of skin cancer.

Ozone also plays a role in heating the upper atmosphere by absorbing solar UV and visible radiation. In consequence, the temperature increases steadily from approximately –50°C (–60°F) at the tropopause (5 to 10 miles altitude) to approximately 7°C (45°F) at the stratopause (30 miles altitude). This ozone heating provides the major energy source for driving the circulation of the upper stratosphere and mesosphere.

At this point, it is worthy of note that atmospheric circulation plays a major role in determining the climate. The energy and circulation of the atmosphere is manifested in the regular wind patterns such as the trade winds but, overall, the circulation of the air (in the form of the mass movement of air) can appear as two extremes: (i) as a gentle breeze; or (ii) as a tornado with other options that lie between these two extremes. The wind motion also transfers water that has evaporated from the oceans to the continents, thereby providing precipitation that is critical for the survival of terrestrial ecosystems. The occurrence of moist air rising near the equator and the dry air sinking in the subtropical regions (the Hadley Circulation) plays a major role in the general circulation of the atmosphere and assists in the transport of heat from the equatorial regions to higher latitudes. In fact, because of the Hadley Circulation, the tropical regions are hot and wet while the subtropical regions are warm and dry. As a result of the atmospheric circulation patterns, the mid-latitude regions experience seasonal contrasts in temperature variations and in rainfall while the Polar Regions are typically cold and dry.

The mesosphere forms the top-most layer of the homosphere and is the third layer of the atmosphere, directly above the stratosphere and directly below the thermosphere and immediately above the stratosphere results in a further temperature decrease to approximately –92°C (–134°F) at an altitude ca. 53 miles (ca. 85 km). The upper regions of the mesosphere define a region (the exosphere) from which molecules and ions can completely escape the atmosphere. The thermosphere is the region that extends to the far outer reaches of the atmosphere in which the highly rarified gas reaches temperatures as high as 1,200°C (2,200°F) by the absorption of energetic radiation of wavelengths less than approximately 500 nanometers (nm) by gas species in this region.

Other regions that exist in the atmosphere include: (i) the ionosphere in which most of the constituents are in ionized form and temperature increase with height; and (ii) the ozonosphere, which is contained within the stratosphere and is located in the lower portion of the stratosphere (approximately 49,000 to 115,000 feet) which does contain approximately 90% v/v of the ozone in the atmosphere.

Approximately 50% of the solar radiation entering the atmosphere reaches the land surface. The remainder is either reflected away from the Earth or absorbed into the atmosphere after which the energy is radiated away from the Earth. In either case, the effect may appear to be the same. Most of the solar energy reaching that is transmitted through the atmosphere must, at some time, be returned to space in order to maintain a heat balance. In addition, <l% of the total energy interacting with the atmosphere reaches the land surface by convection and conduction processes, and this must be returned to space.

Thus, energy transport, which is crucial to the disposition of energy within the atmosphere (as well as within the land and water systems) involves three major physical mechanisms which are: (i) conduction; (ii) convection; and (iii) radiation.

The conduction of energy occurs through the interaction of adjacent atoms or molecules while convection involves the movement of whole masses of air, which may be either relatively warm or cold. For example, convection is the cause of temperature variations that occur when large masses of air move across a land surface or across an ocean. Radiation, in the current context, is the mechanism by which heat is transported away from the Earth, usually after conduction and convection effects have transported the heat to atmosphere. In fact, horizontally moving air (wind) and vertically moving air (air currents) are very much involved and the prevailing wind direction is an important factor in determining the areas most affected by a source of chemical pollutants.

Atmospheric water, present as vapor, liquid, or ice, provides the humidity and the relative humidity, which is expressed as a percentage and is an indication of the amount of water vapor in the air as a ratio of the maximum amount of water that the air can hold at the designated temperature. Thus:

Relative humidity, % = (actual vapor density)/(saturated vapor density) × 100

Typically, clouds form when the air can no longer hold water vapor and the water forms small aerosol droplets. Clouds are important absorbers and reflectors of radiation (heat), and their formation is affected by human activities, especially pollution from PM and the emission of gases such as sulfur dioxide (SO_2) and hydrogen chloride (HCl). In fact, the concern over the emissions of potentially toxic substances from all sources and their possible effects in the environment has led to the introduction of legislative controls in several countries (Kyte, 1991), especially in the United States where the passage of the Amendments to the Clean Air Act (CAA) introduced stricter controls for present and future years (Chapter 10).

Condensation of water vapor to form clouds occurs prior to the formation of precipitation in the form of rain or snow when the air is cooled to a temperature below the dew point. The dew point is the temperature to which air must be cooled to become saturated with water vapor and when cooled further, the water vapor will condense to form liquid water (dew). At this time, it is appropriate to define and use the terms: (i) reservoir; (ii) burden; and (iii) flux.

A reservoir is a domain, such as the atmosphere, where a pollutant may reside for an indeterminate time. The amount of a specific pollutant in a reservoir is known as the burden, which is usually expressed in units of 106 metric tons (10^{12} g called tera-grams, Tg). The rate of transfer of a pollutant from one sphere or domain to another (the flux) is frequently expressed in units of tera-grams per year. The chemical reactions that occur in the atmosphere (Birks et al., 1993) primarily involve, on paper, a series of simple chemical reactions. Indeed, the chemistry of the solar system appears to involve simple chemical reaction. But the simple paper chemistry belies the fact that both are extremely complex systems. In addition, atmospheric chemistry also suffers, to a greater extent than the solar system, from the influence of human activities.

In terms of the nonhuman effects on atmospheric chemistry, an issue that is least understood is the role of the absorption of light by the chemical species which brings photochemical reactions. These reactions would not otherwise occur under atmospheric (particularly temperature) conditions in the absence of light.

Nitrogen dioxide (NO_2) is an essential participant in the smog-formation process because of the ability of this chemical species to absorb photon energy (hv) to produce the molecule in an excited state:

$$NO_2 + hv \rightarrow NO_2^*$$

These highly reactive species are strongly involved in chemical reactions in the atmosphere. The other two species are atoms or molecular fragments with unshared electrons (free radicals) and ions consisting of ionized atoms or molecular fragments. The free radicals are generated when the absorption of light promotes an orbital electron (usually paired in the stable molecular state) to a vacant orbital of higher energy. This interaction gives rise to an excited singlet state in which the promoted electron has a spin opposite to that of its former orbital partner. If the spin of the promoted electron is reversed, such that it has the same spin as its former partner, the result is an excited triplet state.

Above approximately 19 miles (30 km), oxygen is dissociated during the daytime by photo energy after which the oxygen atoms then form ozone. Thus:

$$O_2 + hv \rightarrow O + O$$
$$O + O_2 + M \rightarrow O_3 + M$$

In this equation, M is an arbitrary third-body molecule required to conserve the energy of the reaction. Ozone has a short lifetime during the day because of photodissociation:

$$O_3 + hv \rightarrow O_2 + O\cdot$$

Above 54 miles, where oxygen is a minor component of the atmosphere, there is no net destruction of ozone, and the oxygen is almost exclusively converted back to ozone.

The gases nitric oxide (NO) and nitrogen dioxide (NO_2) tend to promote ozone removal:

$$NO + O_3 \rightarrow NO_2 + O_2$$
$$NO_2 + O \rightarrow NO + O_2$$

Ozone destruction also involves chlorine atoms ($Cl\cdot$) and hypochlorite species ($ClO\cdot$) produced by decomposition in the stratosphere of chlorine species from various sources (e.g., $CFCl_3$, CF_2Cl_2, CH_3CCl_3, $CHCl_3$, and CCl_4):

$$Cl + O_3 \rightarrow ClO + O_2$$
$$ClO + O \rightarrow Cl + O_2$$
$$ClO + hv \rightarrow Cl + O$$
$$Cl + CH_4 \rightarrow HCl + CH_3Cl$$
$$OH + HCl \rightarrow Cl + H_2O$$

$$ClO + NO_2 + M \rightarrow ClNO_3 + M$$
$$ClNO_3 + hv \rightarrow ClO + NO_2$$

In addition to these chemical reactions, approximately 1% v/v of the atomic oxygen is removed by downward transport of ozone into the troposphere, where it is destroyed at or near the ground.

One of the objections to the use of supersonic aircraft, designed to fly in the lower stratosphere (such as the Anglo-French Concorde and Russian Tupolev-144), was related to the production of nitric oxide (NO) and nitrogen dioxide (NO_2) by the thermal decomposition of air (N_2 and O_2) in the engines of these aircraft. It was projected that a fleet of such aircraft would inject the nitrogen oxides into the stratosphere at a rate some three times greater than that from the present source. The ensuing perturbation to the atmospheric nitrogen cycle would cause a depletion of the stratospheric ozone. Although, there is the alternate view that the supersonic aircraft have a very much smaller effect on the stratospheric ozone than originally predicted. Later plans for such aircraft need to consider the potential effects of the aircraft on the atmosphere and the attempts to mitigate these effects; the time mitigation of the effects of the aircraft led to the prevention of their use.

The adage that for every reaction, there is an equal and opposite reaction is no different in any (albeit hypothetical) calculation of the potential changes to the atmosphere and there are obvious hazards associated with such predictions. In fact, predictions may be considered relatively easy, as long as they do not involve predictions related to the future! There has also been concern that industrial production of the chlorofluoromethanes Freon 11 (trichlorofluoromethane, CCl_3F) and Freon 12 (dichlorodifluoromethane, CCl_2F_2) may cause substantial changes to the natural chlorine cycle.

The Freon chemicals have been used principally used as refrigerants and as aerosol-can propellants. Once they are released into the atmosphere, their only presently recognized removal mechanism involves photodissociation in the stratosphere:

$$CFCl_3 + hv \rightarrow CFCl_2 + Cl$$
$$CF_2Cl_2 + hv \rightarrow CF_2Cl + Cl$$

A number of other potential ozone-altering processes with anthropogenic origins have also been identified. For example, increases in the levels of carbon dioxide in the atmosphere due to the combustion of carbon-based (fossil) fuel can lead to stratospheric cooling with the ensuing decrease in the ozone destruction rate. Atmospheric testing of nuclear weapons produces nitric oxide and nitrogen dioxide which passes into the stratosphere. Or to be more general, any process which produces chemicals that have the potential to react directly (or indirectly) to increase/decrease the level of ozone in the atmosphere has the potential to cause environmental effects.

Ozone is produced during the oxidation of unburned hydrocarbon derivatives (C_nH_{2n+2}, RH):

$$RH + OH \rightarrow R + H_2O$$
$$R + O_2 \rightarrow RO_2$$
$$RO_2 + NO \rightarrow RO + NO_2$$

$$NO2 + hv \rightarrow NO + O$$
$$O + O_2 \rightarrow O_3$$

Ozone is also generated by the oxidation of carbon monoxide. Thus:

$$CO + OH \rightarrow CO_2 + H$$
$$H + O_2 \rightarrow HO_2$$
$$HO_2 + NO \rightarrow OH + NO_2$$
$$NO_2 + hv \rightarrow NO + O$$
$$O + O_2 \rightarrow O_3$$

Some of the ozone in the troposphere is derived by injection of ozone from the stratosphere. Tropospheric ozone is involved in the formation of photochemical smog, in the production of acid rain from sulfur oxides and from nitrogen oxides, and in reactions with chemical pollutants (Stensland, 2006). Acid rain (or, as more generally named, acid deposition) is a form of pollution depletion in which pollutants are transferred from the atmosphere to soil or water. The acidic constituents of the rain are produced by the interaction of nitrogen oxides and sulfur oxides with water in the atmosphere:

$$2NO + H_2O \rightarrow 2HNO_2$$
Nitrous acid

$$2NO + O_2 \rightarrow 2NO_2$$
$$NO_2 + H_2O \rightarrow HNO_3$$
Nitric acid

$$SO_2 + H_2O \rightarrow H_2SO_3$$
Sulfurous acid

$$2SO_2 + O_2 \rightarrow 2SO_3$$
$$SO_3 + H_2O \rightarrow H_2SO_4$$
Sulfuric acid

It is possible to measure the acidity of the deposition by and electrometric methods (ASTM D5015). Acid rain is, in fact, precipitation that incorporates anthropogenic acids and acidic materials. However, the deposition of acidic materials on the land surface occurs as rain, snow, fog, dry particles, and gases. Acid precipitation, strictly defined, contains a greater concentration of hydrogen (H+) than of hydroxyl (OH) ions, resulting in a solution pH less than 7.

3.2.2 NITROGEN

The 78% v/v nitrogen contained in the atmosphere exists either in the elemental form (N_2) or in combination with other elements. The nitrogen cycle and nitrogen fixation by microorganisms are essential atmosphere-biosphere interactions. A small amount of nitrogen is thought to be fixed in the atmosphere by lightning, and some is also fixed by combustion processes, as in the internal combustion engine and industrial furnaces.

Molecular nitrogen, unlike molecular oxygen, which is almost completely dissociated to the monatomic form in higher regions of the thermosphere, is not readily dissociated by UV radiation. However, at altitudes on the order of 62.5 miles above sea level, atomic nitrogen is produced by photochemical reactions:

$$N_2 + h\nu \rightarrow N\cdot + N\cdot$$

3.2.3 OTHER GASES

Although only approximately 0.035% v/v (350 ppm) of air consists of carbon dioxide, and it is a species of some concern. Carbon dioxide is primarily responsible for the absorption of infrared energy re-emitted by the land such that some of this energy is radiated back to the land surface. Current evidence suggests that change in the atmospheric carbon dioxide level will substantially alter the climate through the greenhouse effect (Pickering and Owen, 1994; Speight, 2020).

The often-cited factor contributing to increased atmospheric carbon dioxide is consumption of carbon-containing fossil fuels but this is not the sole cause (Speight, 2020). In addition, release of carbon dioxide from the biodegradation of biomass and the uptake of carbon dioxide by photosynthesis are important factors in determining overall carbon dioxide balance in the atmosphere.

Forests have a much greater influence on the carbon dioxide balance than other types of vegetation because forests store enough readily oxidizable carbon in the form of wood and humus (Stevenson, 1994) to have a marked influence on atmospheric carbon dioxide content. Thus, during the summer months, forests are responsible for reducing the atmospheric carbon dioxide by photosynthesis content markedly. During the winter, metabolisms of biota, such as bacterial decay of humus, releases a significant amount of carbon dioxide. Therefore, the current worldwide trend toward destruction of forests and conversion of forest lands to agricultural uses will contribute substantially to a greater overall increase in atmospheric carbon dioxide levels.

Water circulates through the atmosphere in the hydrologic cycle (Chapter 4). Typically, the content of water vapor in the troposphere is within a range of 1 to 3% v/v with a global average of approximately 1% v/v. However, air can contain as little as 0.1% v/v or as much as 5% v/v water. Furthermore, water vapor absorbs infrared radiation even more strongly than does carbon dioxide, thereby influencing the heat balance of the Earth. Clouds formed from water vapor reflect light from the Sun and have a temperature-lowering effect. On the other hand, water vapor in the atmosphere acts as a blanket insofar as it retains heat from the land surface by absorption of infrared radiation.

Water vapor in the upper atmosphere is involved in the formation of hydroxyl ($HO\cdot$) radicals and hydroperoxyl ($HOO\cdot$) radicals. Condensed water vapor in the form of small droplets is of considerable concern in atmospheric chemistry (Anderson, 2006). The harmful effects of some air pollutants (such as nitrogen oxides and sulfur oxides) require the presence of water to form acid rain. In addition, atmospheric water vapor has an important influence upon pollution-induced fog formation. Water vapor interacting with pollutant PM in the atmosphere may reduce visibility to undesirable levels.

When ice particles in the atmosphere change to liquid droplets or when these droplets evaporate, heat is absorbed from the surrounding air. Reversal of these processes results in heat release to the air (as latent heat). This may occur many miles from the place where heat was absorbed and is a major mode of energy transport in the atmosphere. It is the predominant type of energy transition involved in thunderstorms, hurricanes, and tornadoes.

On a global basis, rivers drain approximately one-third of the precipitation that falls on the continents. This means that two-thirds of the precipitation is lost as combined evaporation and transpiration. During the summer, losses by evaporation and transpiration may exceed precipitation because of the large quantities of water stored in the root zone of the soil. In some cases, the evaporation and transpiration furnish atmospheric water vapor necessary for cloud formation and precipitation.

3.2.4 PARTICULATE MATTER

PM is Particles are common significant components of the atmosphere (ASTM D1704) and is injurious to the respiratory tract. Most aerosols (colloidal-sized PM) from natural sources have a diameter on the order of <0.1 micron. These particles originate in nature from sea sprays, smokes, dusts, and the evaporation of organic materials from vegetation. Other typical particles of natural origin in the atmosphere are bacteria, fog, pollen grains, and volcanic ash.

Many important atmospheric phenomena involve aerosol particles including electrification phenomena, cloud formation, and fog formation. Particles help determine the heat balance of the atmosphere by reflecting light. Probably the most important function of particles in the atmosphere is their action as nuclei for the formation of ice crystals and water droplets.

Particles are involved in many chemical reactions in the atmosphere (Figure 3.1) (Speight, 1996). Neutralization reactions, which occur readily in solution, may take place in water droplets suspended in the atmosphere. Particles may also participate in oxidation reactions induced by light-in fact, particles of metal oxides and carbon particles can have a catalytic effect on oxidation reactions.

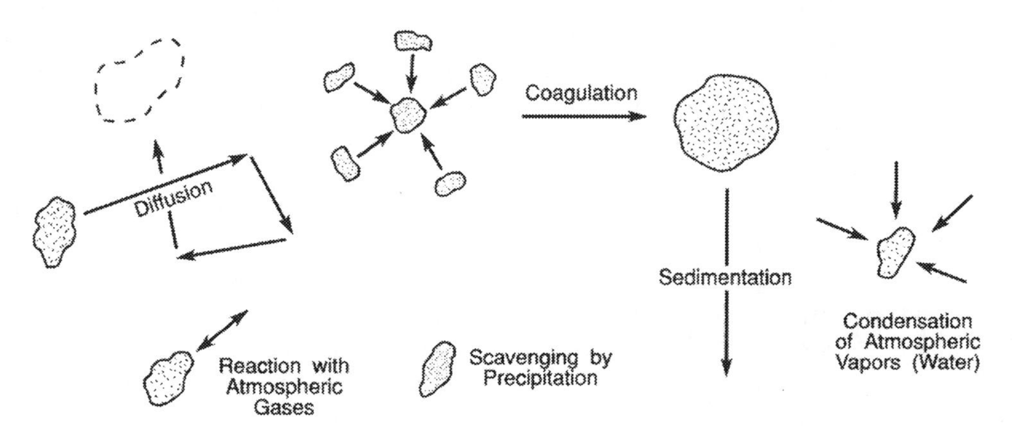

FIGURE 3.1 Particulate behavior in the atmosphere.
Source: Reproduced with permission from: Speight (1996). © Taylor and Francis.

3.3 STRUCTURE

In spite of many popular perceptions of the atmosphere, there are no exact fine-line boundaries within the atmosphere. More correctly, the atmosphere is divided into layers (stratified) based on temperature-density relationships resulting from physical and photochemical (light-induced chemical phenomena) processes in air. However, general features of the atmosphere can be identified to the point where space (the last frontier!) begins (Table 3.3).

TABLE 3.3 General Features of the Atmosphere

- Can be divided into four thermal layers: (i) the troposphere; (ii) the stratosphere; (iii) the mesosphere; and (iv) the thermosphere.
- Starts to blend with space at an altitude of approximately 50 to 100 miles above sea level*.
- The troposphere contains the weather systems.
- Air temperatures decrease upward in the troposphere.
- Temperatures increase with altitude in the stratosphere as ozone absorbs incoming solar radiation.
- Temperatures decline in the mesosphere but increase in the thermosphere.

*Other statements indicate the space commences at the top of the thermosphere (approximately 400 miles above sea level).

Excluding the exosphere, the atmosphere has four primary layers for the lowest to the highest which are: (i) the troposphere, which extends from 0 to 7 miles high; (ii) the stratosphere, which extends from 7 to 31 miles high; (iii) the mesosphere, which extends from 31 to 50 miles high; and (iv) the thermosphere, the thermosphere, which extends from, which extends from 440 to 6,200 miles high, and which extends to the layer known as the exosphere, which extends from 6,200 mile high and into space.

More generally and by way of further definition, the atmosphere of the Earth has been defined as having two major zones or segments: (i) the homosphere, which is the lowest part of the atmosphere of the Earth and lies between the heterosphere and the surface of the Earth; and (ii) the heterosphere. The homosphere is the lower segment of the two-part division of the atmosphere and consists of three regions namely: (i) the troposphere; (ii) the stratosphere; and (iii) the mesosphere. While all three sub-regions have the same composition of air, the concentration of the air decreases significantly with an increase in the altitude increases.

On the other hand, the heterosphere, which extends from the turbopause to the edge of the atmosphere and lies directly above the homosphere, is the layer of an atmosphere where the gases are separated out by molecular diffusion with increasing altitude such that the lighter (lower molecular weight and less dense) species become more abundant relative to the heavier (higher molecular weight and more dense) gases. Also, the so-called boundaries between the different molecular and atomic species in the atmosphere vary according to temperature and solar activity.

Of particular interest in the current context of gaseous emissions into the atmosphere (Table 3.4), the turbopause is the area of the atmosphere below which turbulent mixing dominates and the homosphere (the region below the turbopause) is the region in which the chemical species are well mixed. In fact, the heterosphere is the region where molecular diffusion dominates and the chemical composition of the atmosphere varies according to chemical species.

TABLE 3.4 Common Gaseous Pollutants of the Atmosphere*

Pollutant	Description
Ammonia (NH_3)	Ammonia: A gas with a characteristic pungent odor.
–	Emitted from agricultural processes.
–	Contributes significantly to the nutritional needs of terrestrial organisms.
–	Serves as a precursor to food and fertilizers.
–	Reacts with oxides of nitrogen and sulfur in the atmosphere to form secondary particles.
Carbon dioxide	Content in air has increased during the last century.
–	May cause greenhouse effect.
Carbon monoxide	Produced by incomplete combustion of fuel.
–	Combines with hemoglobin and reduces the oxygen-carrying capacity.
CFCs	Chlorofluorocarbons.
–	Rise into the stratosphere.
–	Cause damage to the ozone layer.
Fluorides	Release an extremely toxic gas (hydrogen fluoride, HF) on heating.
Hydrocarbons	Unburned discharges from incomplete combustion of fuels.
Oxides of nitrogen	Include NO and NO, which are released by automobiles and chemical industries as waste gases and also by burning of materials.
Oxides of sulfur	Produced by burning of coal and crude oil products.
–	Both oxides (SO_2 and SO_3) react with water to form sulfurous acid (H_2SO_3) and sulfuric acid (H_2SO_4)
–	May precipitate as rain or snow-acid rain.
Photochemical oxidants	Nitrogen oxides in the presence of sunlight react with unburned hydrocarbons to form peroxyacetyl nitrate (PAN), ozone, aldehydes, and other complex organic compounds.
VOCs	Volatile organic compounds.
–	Categorized as either methane (CH_4) or non-methane volatile organic compounds (NMVOCs).
–	Methane: a greenhouse gas which contributes to global warming.
–	The NMVOCs benzene (C_6H_6), toluene ($C_6H_5CH_3$), and xylene isomers [$C_6H_4(CH_3)_2$] are suspected carcinogens.

3.3.1 THE TROPOSPHERE

The troposphere (Figure 3.2) (Speight, 1996) is the lowest layer of the homosphere and is the layer of the atmosphere where life can exist without artificial life-support systems. It is also the region known as the weather layer of the Earth as it has all weather conditions. While temperatures at the bottom of the troposphere are hospitable for life, temperatures at the top of the troposphere are on the order of –50°C (–60°F).

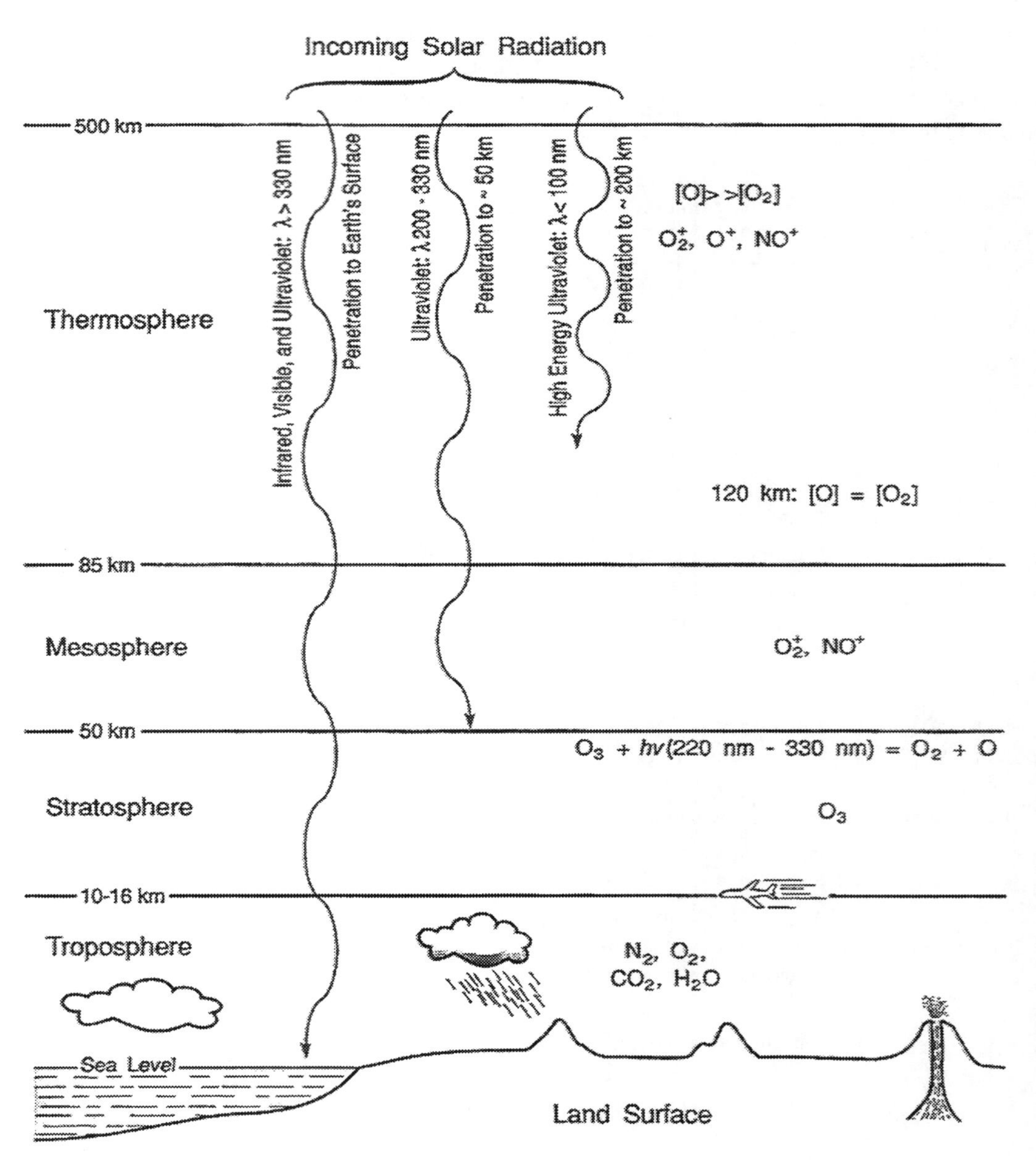

FIGURE 3.2 Structure of the atmosphere.

Source: Reproduced with permission from: Speight (1996). © Taylor and Francis.

Thus the troposphere is characterized by a generally homogeneous composition of major gases and a decrease in temperature with an increase in altitude. The upper limit of the troposphere, which has a temperature minimum of approximately –56°C (–69°F), varies in altitude by approximately a half-mile. In contrast to the gaseous constituents, the water vapor content of the troposphere is extremely variable because of cloud formation, precipitation, and evaporation of water from terrestrial water bodies.

3.3.2 THE STRATOSPHERE

The stratosphere forms the middle layer of the homosphere and lies directly above the troposphere in which the temperature rises to a maximum on the order of –2°C (28°F) with increasing altitude. The heating effect is caused by the absorption of UV radiation energy by ozone and temperature profile can become even more complicated due to the formation of additional ozone in the gaseous plumes emitted from urban centers (Colbeck and MacKenzie, 1994).

The stratosphere has a thickness in excess of 25 miles and contains the ozone layer. The temperature increases with altitude in the stratosphere because the ozone-which is concentrated in the upper two-thirds of the layer-absorbs UV solar radiation and the decreasing density of the air in the layer allows greater agitation of the molecular and atomic species. The maximum temperature in the region approaches 0°C (32°F) at the stratopause-the region that separates the stratosphere and the overlying mesosphere.

3.3.3 THE THERMOSPHERE

The outermost layer of the atmosphere, the thermosphere, prevents a variety of harmful cosmic radiation including X-rays, gamma rays, and some UV radiation from reaching the Earth. Isolated gas molecules in the thermosphere are converted positively charged and negatively charged atomic and molecular species) through the effects of the solar radiation. As a result, the visible effects (such as auroras, for example, the Aurora Borealis in northern latitudes and the Aurora Australis in southern latitudes) occur when electrons and protons interact in the ionosphere.

At these high altitudes, the residual atmospheric gases segregate into strata according to molecular size. The temperature in the thermosphere increases with altitude due to absorption of highly energy solar radiation and, in fact, the temperature is highly dependent on solar activity.

3.3.4 THE EXOSPHERE

In the exosphere, beginning at approximately (375 miles above sea level, the atmosphere turns into space (the final frontier!), although by the criteria set for the definition of the Karman line (the altitude-approximately 62 miles high)-generally recognized as the altitude where space begins).

The highly attenuated gas in this layer can reach temperatures but, despite the high temperature, an observer or object will experience cold temperatures in the thermosphere, because the extremely low occurrent of gaseous species (practically a high vacuum is insufficient for the molecules to conduct heat. In the anacoustic zone (at an elevation above 99 miles), the occurrence of the various gaseous species is sufficiently low and molecular interactions are too infrequent to permit the transmission of sound.

3.3.5 OTHER REGIONS

Within the principal layers that are above, several secondary layers may be distinguished by properties other than temperature and are: (i) the ozone layer; (ii) the ionosphere; (iii) the homosphere and the heterosphere; and (iv) the planetary boundary layer.

The ozone layer is, as discussed above, is contained within the stratosphere. In this layer, the ozone concentration is on the order of 2 to 8 parts per million, which is a higher concentration than in the lower atmosphere but the concentration of the ozone is low compared to the main components of the atmosphere.

The ionosphere is a region in which, due to the effect of solar radiation, there is a prevalence of ionic species. UV light is the primary producer of ions in the ionosphere but during periods of darkness, there is a tendency for the positive ions slowly recombine with free electrons. In the lower regions of the ionosphere where the concentration of species is relatively high, the process is relatively rapid and, thus, at night, IoT is possible to transmit radio waves over much greater distances.

3.4 HUMIDITY

Humidity is a measure of the moisture contained in air. The absolute humidity is a measure of the amount of water in air, while the relative humidity is a measure of the amount of water in the air relative to the maximum amount of water that the air can hold. More specifically, the absolute humidity is a measure of the mass of water (for example, grams) in a volume of air (for example, cubic meters)-thus, the absolute humidity is expressed as g/m^3 (grams per cubic meter) whereas the relative humidity is expressed as a percentage.

The distribution of water on Earth is dependent upon the interactions that occur between the surface of the Earth and the atmosphere. The circular path of the hydrologic cycle illustrates phenomena such as evaporation, condensation, runoff, infiltration, percolation, and transpiration (Figure 3.3) (Speight, 1996) which can cause the physical state of water (vapor, liquid, or solid) to change as it moves between different elements of Earth system. However, more than 97% v/v of the water of the Earth is present in the oceans, with the majority of the remaining water (3% v/v)on the continents (the geosphere) in the form of groundwater, streams, lakes, and ice. However, less than 0.001% v/v of all of the water on the Earth is in the atmosphere.

Evaporation (sometimes referred to as vaporization) occurs from the surface of a liquid as the liquid changes from the liquid phase to the gas phase When evaporation occurs, the energy removed from the vaporized liquid reduces the temperature of the liquid which results in evaporative cooling. On the other hand, condensation occurs when the air becomes saturated with moisture (relative humidity = 100%) and as the temperature falls, the relative humidity of the air rises. Condensation typically occurs on preexisting surfaces such as dust particles and aerosols to form droplets which are kept airborne by air turbulence. As the droplets collide and coalesce, they form larger droplets which eventually form a rain-drop and, with decreasing temperature (less than –10°C, 14°F) the water droplets form ice crystals.

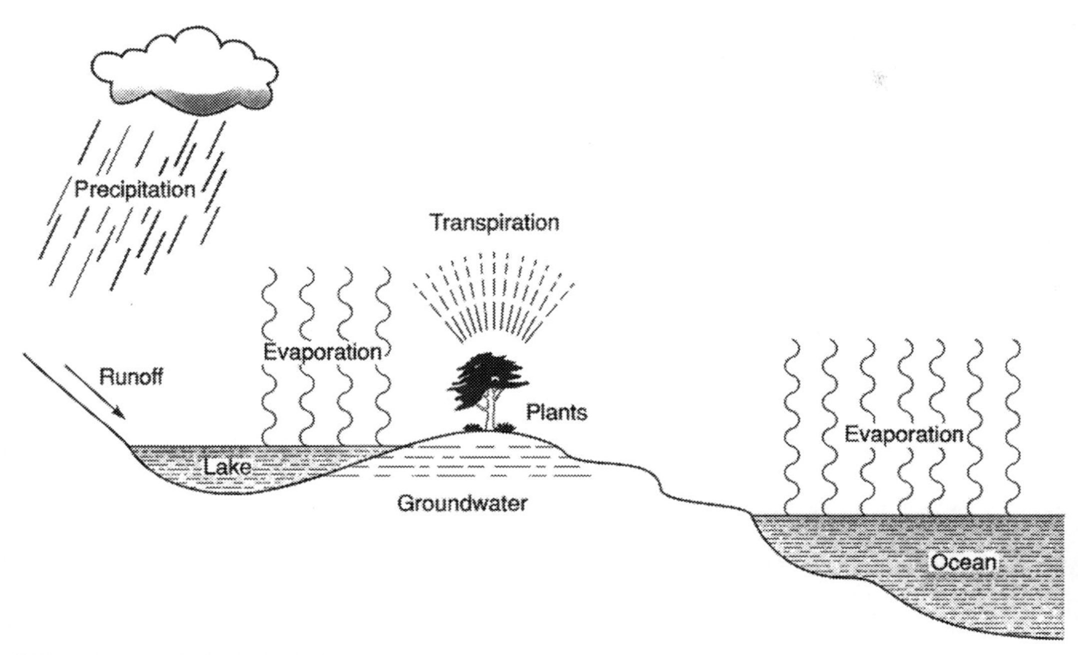

FIGURE 3.3 The hydrological cycle.
Source: Reproduced with permission from: Speight (1996). © Taylor and Francis.

3.5 DISPERSION OF POLLUTANTS

Air pollution is an unwanted change in the quality of the atmosphere of the Earth that is caused by the emission of gases and solid PM or liquid PM. Furthermore, air pollution is considered to be one of the major causes of global climate change and depletion of the ozone layer, which may have (or will have) serious consequences for all flora and fauna. Polluted air is carried everywhere by winds and air currents and is not confined by national boundaries. Therefore, air pollution is a concern for everybody irrespective of the type and location of the source.

Gaseous pollutants can be divided into two main categories: (i) primary; and (ii) secondary pollutants (Table 3.5). Primary pollutants are typically chemicals that are emitted directly from the source-typical examples include sulfur dioxide emissions from combustion sources and organic compound emissions (volatile organic compounds (VOCs)) from other surface facilities. On the other hand, secondary pollutants are gaseous and vapor phase compounds that form due to chemical (sometimes physical) interactions between primary pollutants in the atmosphere or between primary pollutants and naturally occurring compounds in the atmosphere.

The dispersion of chemical pollutants in the atmosphere is typically determined by atmospheric turbulence or by wind. Turbulence results from such factors as the friction of the land surface, physical obstacles to wind flow, and the vertical temperature profile of the lower atmosphere. Stability refers to the degree of turbulence in the atmosphere. For air quality purposes, stability usually refers to the lower layers of the atmosphere where pollutants are emitted. The idea of discrete stability classes (as in the following list) is a simplification of the complex nature of the atmosphere, but has proved useful in predictive studies.

TABLE 3.5 Examples of Primary and Secondary Gaseous Pollutants

Pollutant Type	Example*
Primary	Ammonia
	Carbon monoxide
	Carbon dioxide
	Hydrogen chloride
	Hydrogen fluoride
	Hydrogen sulfide
	Mercaptide derivatives
	Nitrogen oxide
	Nitrogen dioxide
	Particulate matter
	Sulfide derivatives
	Sulfur dioxide
Secondary	Nitric acid
	Nitrogen dioxide
	Nitrous acid
	Ozone
	Particulate matter
	Sulfuric acid vapor
	Sulfurous acid vapor
	Volatile organic compounds (VOCs)

*Listed alphabetically and not by order of occurrence or relative amounts produced.

A stable atmosphere is marked by air that is cooler at the ground than aloft, by low wind speeds, and consequently, by a low degree of turbulence. A pollutant plume released into a stable lower layer of the atmosphere can remain relatively intact. On the other hand, an unstable atmosphere is marked by a high degree of turbulence. An intermediate turbulence class between stable and unstable conditions is the "neutral" stability class. A visible plume released into a neutral stability condition may display a cone-like appearance as the edges of the plume spread out in a V-shape.

The term inversion refers to a layer in the atmosphere where temperature increases with height rather than decreases, as is usually the case. This inversion layer serves as a cap and restricts chemical pollutants from any further upward dispersal. As a result, pollutant levels will increase below the cap and an extended period of such a pollutant buildup is usually referred to as a stagnation episode during which smog (Chapter 10) can occur.

In order to reduce the emissions at a given locale, there has been the tendency to build larger (taller) flue gas (or smoke) stacks. Pollutants emitted from a source with a tall stack tend to (but not always) produce lower a concentration of a pollutant at the ground level than the same concentration of the same pollutant emitted from a pollutant source with a short flue gas stack. However, it must be recognized that sources with the same flue gas stack height can produce a different impact depending on the plume rise above the stack (Figure 3.4) (Speight, 1996)-the impact depends on: (i) the exit velocity of the gas stream; (ii) the temperature of the emissions; and (iii) the atmospheric conditions. The combination of the physical stack height and plume rise above the stack is referred to as the effective stack height.

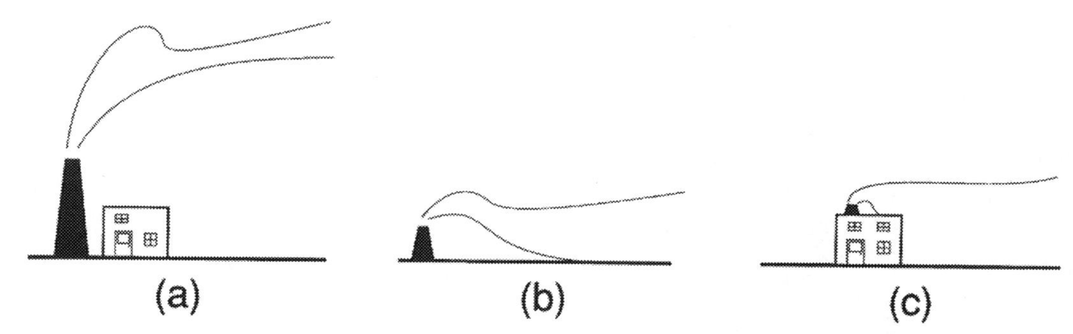

FIGURE 3.4 Structure of a smokestack plume depending upon the location of the stack and wind conditions. *Source:* Reproduced with permission from: Speight (1996). © Taylor and Francis.

Pollutants emitted into the atmosphere or formed in the atmosphere are eventually depleted. In addition to chemical transformation which acts as a depletion mechanism for precursors, two other common depletion mechanisms are: (i) dry deposition; and (ii) washout. Typically, the dispersion of pollutant chemical in the atmosphere are often referred to as a plume which disperses in the form of a chemical plume that spreads in various directions that are dictated by the atmospheric conditions. The plume in the form of vapor or smoke are of considerable importance in the atmospheric dispersion modeling of air pollution. There are three primary types of air pollution emission plumes which are: (i) buoyant plumes; (ii) dense gas plumes; and (iii) passive plumes, often referred to as neutral plumes.

A buoyant plume is a plume which is lighter (have a lower density than the air) and are often at a higher temperature than the ambient air which surrounds the plume or the plume may be at a lower temperature or at approximately the same temperature as the ambient air but have a lower molecular weight constituents than the constituents of the ambient air. For example, the emission from the flue gas stack of an industrial furnace is usually emitted as a buoyant because the gases (that constitute the plume) are considerably warmer and, therefore, less dense than the ambient air. Another example is the emission plume of methane gas (from a gas well or from the flatulence of cattle) at ambient air temperatures is buoyant because methane has a lower molecular weight (16) than the molecular weight of the ambient air (molecular weight of nitrogen = 28, molecular weight of oxygen = 32).

A dense gas plume is heavier than air because the constituents of the plume have a higher density than the constituents of the surrounding ambient air. This type of plume may have a higher density than air because it has a higher molecular weight than air (for example, a plume of carbon dioxide, molecular weight = 44 whereas air has oxygen (molecular weight = 32) and nitrogen (molecular weight = 28). The carbon dioxide plume may also have a higher density than air if the plume is at a much lower temperature than the air. For example, a plume of evaporated gaseous methane from an accidental release of natural gas (predominantly methane) from a gas well or, in the case of a release from liquefied natural gas (LNG), the methane may be at a temperature on the order of –161°C (–258°F). Finally, and by definition, a passive or neutral plume is typically neither lighter nor heavier than air.

In the case of an inversion, the air near the surface of the Earth surface is warmer than the air above it because the atmosphere is heated from below as solar radiation warms the surface of the Earth, which in turn warms the layer of the atmosphere directly above it. In such as case, the atmospheric temperature normally decreases with increasing altitude. However, under certain meteorological conditions, atmospheric layers may form in which the temperature increases with increasing altitude rather than the converse to form inversion layers. Any rising air within the inversion soon expands, thereby undergoing adiabatic cooling to a lower temperature than the surrounding air and the air stops rising. Any sinking air soon compresses adiabatically to a higher temperature than the surrounding air and the parcel stops sinking.

Two other phenomena that influence the dispersion of chemical in the atmosphere are: (i) dry deposition; and (ii) wet deposition.

Dry deposition refers to the removal of both PM and gases as they come into contact with the land surface. Also often included in the category of dry deposition, is the phenomenon known as gravitational sedimentation which occurs during periods without precipitation. These particles include: (i) aerosols; (ii) sea salts; (iii) particulate matter; and (iv) adsorbed/reacted gases captured by vegetation.

On the other hand, wet deposition involves the formation of droplets followed by droplet precipitation (such as by rain or snow) or impaction of the droplets on the surface of the Earth (as, for example, in the case of fog). Of serious concern is the wet deposition of radionuclides from a pollution plume by rain, often leading to the formation of so-called hot spots of radioactivity on the surface of the Earth.

On the other hand, wet deposition is the removal of pollution plume components by the action of rain. Also, a pollutant may travel more rapidly through the atmosphere as a result of a chemical change. Dry deposition is approximately equivalent to wet deposition for sulfur (as sulfur dioxide) and nitrate (as nitric acid) but dry deposition in the form of PM is the dominant source of ammonium salts (NH_4^+). The dominant forms of gaseous nitrogen are nitrogen dioxide (NO_2) and nitric acid (HNO_3) vapor, and trace amounts of ammonia (NH_3). The conversion of nitrogen dioxide (NO_2) to nitric acid (HNO_3) in the atmosphere is much more rapid than the conversion of sulfur dioxide (SO_2) to sulfuric acid (H_2SO_4) resulting in the longer-distance transport of sulfur dioxide as a pollutant in the atmosphere.

Washouts refer to the uptake of gases and PM by water droplets or by snow and their removal from the atmosphere. The actual removal of pollutants from the atmosphere by wet deposition requires the formation of precipitation within the clouds. Without cloud elements greater than approximately 100 microns in diameter, the pollutant mass remains in the air, largely in association with the relatively small cloud drops, which have negligible rates of descent.

The consequence of upshot of all of the aforementioned effects is that the atmosphere can be seriously polluted by any one or more anthropogenic effects or natural effects. Determination of the pollutants in the atmosphere, through strict sampling protocols, is a major issue where accuracy and reliability are a necessity (Lodge, 1988).

KEYWORDS

- **atmosphere**
- **climatology**
- **geosphere**
- **liquefied natural gas**
- **nitrogen dioxide**

REFERENCES

Anderson, L. G., (2006). Atmospheric chemistry. In: Pfafflin, J. R., & Ziegler, E. N., (eds.), *Environmental Science and Engineering* (5th edn., Vol. 1. pp. 118–136). CRC Press, Taylor & Francis Group, Boca Raton, Florida.

ASTM D1704, (2020). *Amount of Particulate Matter in the Atmosphere*. Annual Book of Standards. ASTM International, West Conshohocken, Pennsylvania.

ASTM D5015, (1995). *PH of Atmospheric Wet Deposition Samples.* Annual Book of Standards. ASTM International, West Conshohocken, Pennsylvania.

ASTM, (2020). *Annual Book of Standards*. ASTM International. West Conshohocken, Pennsylvania.

Birks, J. W., Calvert, J. G., & Sievers, R. W., (1993). *The Chemistry of the Atmosphere: Its Impact on Global Change.* American Chemical Society, Washington, D.C.

Dewling, R. T., Deieso, D. A., & Nicholls, G. P., (2006). Radon. In: Pfafflin, J. R., & Ziegler, E. N., (eds.), *Environmental Science and Engineering* (5th edn., Vol. 2. pp. 1047–1057). CRC Press, Taylor & Francis Group, Boca Raton, Florida.

Ferrand, E. F., (2006). Air Pollutant Effects. In: Pfafflin, J. R., & Ziegler, E. N., (eds.), *Environmental Science and Engineering* (5th edn., Vol. 1, pp. 29–43). CRC Press, Taylor & Francis Group, Boca Raton, Florida.

Hidy, G. M., Mueller, P. K., Grosjean, D., Appel, B. R., & Wesolowski, J. J., (1979*). The Character and Origins of Smog Aerosols*. John Wiley & Sons Inc., New York.

Jazib, J., (2018). *Basics of Environmental Sciences*. Iqra Publishers, New Delhi, India.

Kyte, W. S., (1991). *Desulphurization 2: Technologies and Strategies for Reducing Sulphur Emissions.* Institute of Chemical Engineers, Rugby, Warwickshire, United Kingdom.

Lodge, J. P. Jr., (1988). Methods of Air Sampling and Analysis (3rd edn.). Lewis Publishers Inc., Chelsea, Michigan.

Newman, L., (1993). *Measurement Challenges in Atmospheric Chemistry*. Advances in Chemistry Series No. 232. American Chemical Society, Washington, DC.

Pickering, K. T., & Owen, L. A., (1994). *Global Environmental Issues*. Routledge Publishers, New York.

Prinn, R. G., (1987). In: Parker, S. P., (ed.). *McGraw-Hill Encyclopedia of Science and Technology* (Vol. 2. p. 171 & 185). McGraw-Hill, New York.
Speight, J. G., (1996). *Environmental Technology Handbook*. Taylor & Francis, Washington, DC.
Speight, J. G., (2020). *Global Climate Change Demystified*. Scrivener Publishing, Beverly, Massachusetts.
Stensland, G. J., (2006). Acid Rain: In: Pfafflin, J. R., & Ziegler, E. N., (eds.), *Environmental Science and Engineering* (5th edn., Vol. 1. pp. 1–14). CRC Press, Taylor & Francis Group, Boca Raton, Florida.
Stevenson, F. J., (1994). Humus Chemistry. John Wiley & Sons Inc., New York.
Westberg, H., & Zimmerman, P., (1993). In: Newman, L., (ed.), *Measurement Challenges in Atmospheric Chemistry*. Advances in Chemistry Series No. 232. American Chemical Society, Washington, DC. Chapter 10.

CHAPTER 4

The Aquasphere

4.1 INTRODUCTION

Following the definitions presented in Chapter 1, the term Earth system refers to the three interacting physical, chemical, and biological systems in which the Earth acts as a livable planet in which life-supporting processes occur and interact. The systems consist of: (i) the atmosphere; (ii) the aquasphere, which consists of the oceans, the rivers, the lakes, and all other water systems; and (iii) the geosphere, which consists of the landmasses. Life is an integral part of the Earth system, and the Earth system now includes human society and, in many cases, the human activities are causing changes to the Earth system.

On occasion, a fourth component (the biosphere) and a fifth component (the cryosphere) are added (Table 4.1) (Chapter 3). Also, the ice in the polar regions may also be referred to as the cryosphere which comprises snow, river ice, lake ice, sea ice, glaciers, and ice sheets play a major role in the climate of the Earth system through the impact on the water cycle, surface gas, exchange, and sea level. Since all of the components of the cryosphere are sensitive to temperature change, the cryosphere is a natural determinant of the variability of the climate (Vaughan et al., 2013). These systems are interconnected, but for the purposes of this book, the three predominant systems are presented as: (i) the atmosphere; (ii) the aquasphere; and (iii) the geosphere.

By way of introduction, water (H_2O) is a polar inorganic compound that is (at room temperature) a tasteless and odorless liquid which is almost colorless with a gentle suspicion of blueness. It is often considered to be a universal solvent because of the ability to dissolve many substances and is, therefore also considered to be the solvent of life. In fact, natural water (i.e., water as it exists in nature invariably includes various dissolved substances, and special process steps are required to obtain pure water.

Water—a liquid under the standard conditions of temperature and pressure—is the only common substance to exist as a gas (water vapor or steam), solid (ice), and liquid in normal terrestrial conditions. The application of heat or the absence heat can cause phase transitions such as freezing to a solid (liquid water to ice), melting (ice to liquid water), vaporization (liquid water to vapor), condensation (vapor to liquid water), sublimation (ice to vapor), and deposition (vapor to ice). In addition, water has the unique property of becoming less dense as it freezes. At a pressure of 1 atmosphere (14.7 psi), water reaches a maximum density of 1,000 kg/m^3 (62.43 lb/ft^3) at 3.98°C (39.16°F)—the density of ice is

The Science and Technology of the Environment. James G. Speight (Author)

917 kg/m^3 (57.25 lb/ft^3) which represents an expansion on the order of 9% than can exert pressure on pipes. In a lake or ocean, water at 4°C (39°F) is the lower phase and ice floats on the liquid water. This ice insulates the water phase thereby acting as an insulator and, prevents the water from any further freezing (to the solid phase-ice) which assists in the survival of aquatic organisms.

TABLE 4.1 The Main Components of the Earth System

Component	Description
Atmosphere	The gaseous layer surrounding the Earth.
	Held to the surface by gravity.
	Receives energy from solar radiation, which warms the surface of the Earth.
	Absorbs water from the surface of the Earth via evaporation.
	Redistributes heat and moisture across the surface of the Earth.
	Contains substances that are essential for life, including carbon, nitrogen, oxygen, and hydrogen.
Aquasphere	The parts of the earth system composed of water in its liquid, gaseous (vapor) and solid (ice) phases.
	Includes oceans, seas, ice sheets, sea ice, glaciers, lakes, rivers, streams, atmospheric moisture, ice crystals, and areas of permafrost.
	Also includes the moisture in the soil (soil water) and within rocks (groundwater).
	Can be sub-divided into the fluid water systems (the hydrosphere) and solid water (ice) systems (the cryosphere).
Geosphere	Composed of rocks and minerals.
	Includes the solid crust, the molten mantle, and the liquid and the core of the Earth.
	Includes the soil in which nutrients become available to living organisms.
	Provides an important ecological habitat and the basis of life forms.
	Subject to the processes of: (i) erosion; (ii) weathering; and (iii) transport, as well as tectonic forces and volcanic activity.

Overall, water is a unique chemical that is a necessary component of all living organisms and is a major driver in many of the chemical and physical properties (Eisenberg and Kauzmann, 1969; Gerstein and Levitt, 1998; Sharp, 2001). Some of the properties of water are essential for life, while other properties affect on the size and shape of living organisms and the constraints within which living organisms must operate. Moreover, many of the basic physical properties of water can be explained in molecular and structural terms, which is relevant to the behavior of water as the properties pertaining to the aquasphere. In particular, it is the readiness of water to participate in hydrogen bonding interactions (Eisenberg and Kauzmann, 1969; Gerstein and Levitt, 1998; Sharp, 2001).

4.2 COMPONENTS

The aquasphere consists of a variety of water systems (Table 4.1) that are essential to life and all of the systems are interrelated—the many interactions between the systems of the Earth are complex and occur constantly and simultaneously. Some examples of Earth systems interacting include volcanic eruptions and ocean tsunamis, but there are also slow changes that change: (i) the chemistry of the oceans; (ii) the content of the atmosphere; and (iii) the microbial biodiversity in soil. Each part of the Earth—from the inner molten core to the outer reaches of the atmosphere—has a role in making Earth suitable for the existence of billions of lifeforms (1 billion = 1×10^9).

Within the aquasphere, the phenomenon known as: (i) precipitation; (ii) evaporation; (iii) freezing; (iv) melting; and (v) condensation are part of the hydrological cycle (also known as the water cycle), which is a continuous process of water circulation throughout the Earth systems. Figure 4.1 shows the hydrological cycle—a representation of the means by which water evaporates from the surface of the Earth, rises into the atmosphere, cools, condenses to form clouds, and falls again to the surface as precipitation. The major physical components of the global water cycle include: (i) the evaporation from the ocean and land surfaces; (ii) the transport of water vapor by the atmosphere, precipitation onto the ocean and land surfaces; (iii) the net atmospheric transport of water from land areas to ocean; and (iv) the return flow of freshwater from the land back into the ocean.

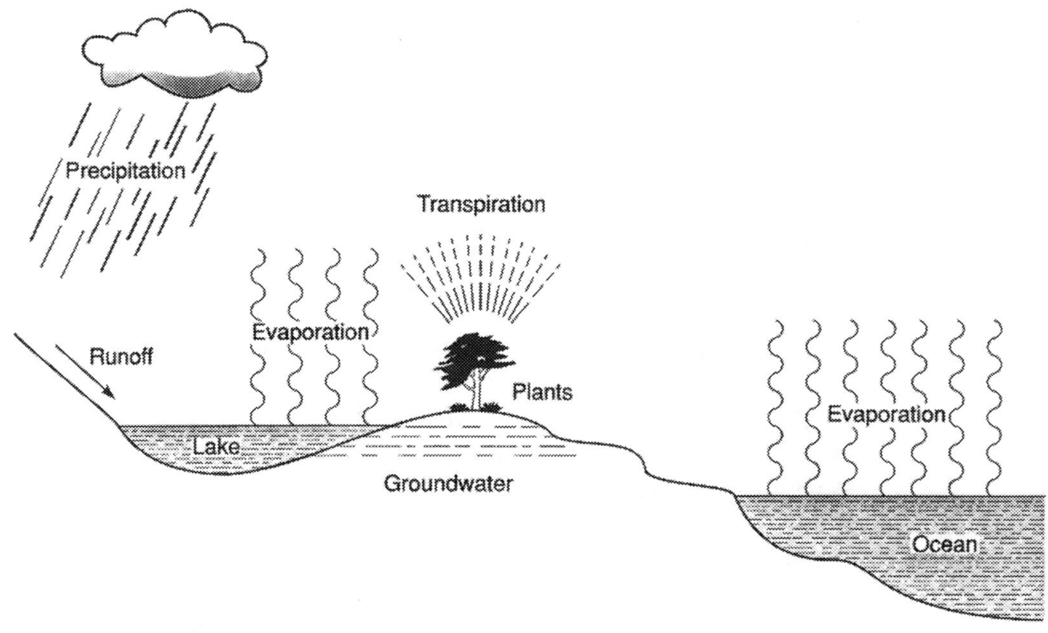

FIGURE 4.1 The hydrological cycle.

Source: Reproduced with permission from: Speight (1996). © Taylor and Francis.

This system of the cycling of water is intimately linked with the interchange of energy between the atmosphere, the ocean, and the land and plays a role in determining the variability of the climate. For example, approximately 75% of the energy (or heat) in the atmosphere of the Earth is transferred through the evaporation of water from the surface of the Earth. On land, water evaporates from the ground, mainly from soil, plants (through transpiration), lakes, and streams. In fact, approximately 15% v/v of the water entering the atmosphere is from evaporation from the land surfaces and evapotranspiration from plants which: (i) cools the surface of the Earth; (ii) cools the lower atmosphere; and (iii) provides water to the atmosphere to form clouds.

On land the water cycle includes: (i) the deposition of rain and snow; (ii) water flow in runoff; (iii) infiltration of water into the soil and groundwater; (iv) storage of water in soil, lakes, and streams, and groundwater; (v) polar and glacial ice; and (vi) the use of water in vegetation and human activities (Figure 4.1).

A water system comprises a drainage basin and the various physical, chemical, and biological constituents, including: (i) water networks; (ii) ecosystems; (iii) the oceanic and atmospheric systems that govern evaporation and precipitation in the basin; and (iv) the water systems, such as terminal lakes or seas-into which the water flows. In whichever system water is located, water is a complex system of chemical species that has received considerable study (Eglinton, 1975; Jenne, 1979; Melchior and Bassett, 1990). Indeed, the public has been introduced to the issues of water pollution for the past several years, not the least of which is the pollution of Boston Harbor by tea being imported from England, which played a significant role in the formation of the United States (Easterbrook, 1995).

Briefly, when rain falls and seeps deep into the Earth, filling the porous spaces of an aquifer (an underground formation in which water can be stored) thereby becoming groundwater which is one of the least visible but most important natural sources of water—in some rural areas, groundwater is the only freshwater source. However, groundwater can be polluted when contaminants (such as pesticides, fertilizers as well as waste leached from landfills and septic systems) pass into an aquifer, thereby rendering the aquifer water unsafe for human consumption and use. Moreover, removal of groundwater contaminants can vary from difficult to impossible. Once polluted, an aquifer (as a source of freshwater) may be unusable for decades (or even years), and the contamination can spread far from the original polluting source by passing into lakes, streams, rivers, and oceans.

On the other hand, surface water covers approximately 70% of the surface of the Earth in the form of oceans, lakes, rivers, and streams. In terms of non-oceanic water (Table 4.2), surface water from freshwater sources is in danger from various pollutants (Jazib, 2018). In fact, pollution by nutrients, which includes nitrate-based and phosphate-based chemical, is the leading type of contamination in freshwater sources. While various floral and faunal species do need these nutrients to grow, these chemicals have become a major type of pollutant due to the runoff of farm waste and fertilizer. Municipal and industrial waste discharges into freshwater sources also contribute toxins along with the waste from domestic sources and also from and industrial sources, although many countries have laws in place to cease such discharges. Furthermore, water is uniquely vulnerable to pollution because it can dissolve more substances than any other liquid on the Earth. Toxic substances from farms, towns, and factories readily dissolve into and mix with water leading to water pollution.

In addition, the rates of evaporation and precipitation, humidity, and available soil moisture are factors governing water availability for various life forms. Precipitation varies in relation to the position and movement of air masses and weather system, location relative to mountain ranges (rain shadow effect), and altitude. Seasonal distribution of rainfall is as important as the total amount; rainfall evenly distributed throughout the year usually results in greater availability.

TABLE 4.2 Sources of Non-Oceanic Water and Water Systems

Source	Description
Surface water	A water system such as a river, lake, or freshwater wetland.
–	Naturally replenished by precipitation.
–	Naturally lost by discharge to the oceans, evaporation, evapotranspiration, and groundwater recharge.
Groundwater	Freshwater located in the subsurface pore space of soil and rocks. Also, water that is flowing in aquifers below the water table.
Aquifer water	Fresh water in a layer of sediment or rock (such as sand or gravel) that is highly permeable.
Unconfined aquifer	An aquifer that is overlaid by permeable materials.
–	Recharged by water seeping down from above in the form of rainfall and snowmelt.
Confined aquifer	An aquifer which is sandwiched between two impermeable layers of rock or sediments.
–	Recharged only in those areas where the aquifer intersects the land surface.

The water that humans use is primarily fresh surface water and groundwater, but in arid regions and in some semi-arid regions, a portion of the water supply is taken from the ocean, which is a source that is likely to become more important as the demand for water outstrips the supply. Saline or brackish groundwater may also be utilized in some areas. Throughout history, the quality and quantity of water available to humans have been vital factors in determining not only the quality of life but also the existence of life Indeed, whole civilizations have disappeared not only because of water shortages resulting from changes in the climate but also because of water-borne diseases, such as cholera and typhoid (Cartwright and Biddiss, 1972). In fact, one can wonder whether or not the castle moat actually protected the inmates from the attackers or whether the disease ridden moat was fatal to both sides. Also, serious epidemics of waterborne diseases such as cholera, dysentery, and typhoid fever were caused by underground seepage from privy vaults into town wells (James and Thorpe, 1994). Such direct bacterial infections through water systems can be traced back for several centuries, even though the germ or bacterium as the cause of disease was not proved for nearly another century.

In terms of pollution of the oceans (also referred to as marine pollution), a considerable amount of the pollution originates on land-whether it occurs along the coast or inland- and is not always from ships at sea. Contaminants are carried from farms, factories, and cities by streams and rivers into bays and estuaries from which the contaminants travel out to sea.

By definition, an estuary is the place where freshwater (from a land-based water system such as a river) mingles with the saltwater of the ocean and within which seawater is measurably diluted with freshwater from river. Estuaries differ in size, shape, and volume of water flow and are influenced by the geology of the region in which they occur. Also, as the river water reaches the encroaching seawater, the particulate matter (PM) carried by the river is deposited in the estuary and accumulates to form deltas in the upper reaches of the mouth of the river.

Also, marine debris—particularly plastic materials—that is carried by the wind or washed in via storm drains and sewers. The oceans are also subject to spills and leaks from ships carrying a variety of products from the points of origin to the point of sales and use. Moreover, the indirect reuse of wastewater that has been discharged into rivers is a common occurrence, and the practice is acceptable as long as the discharges are treated and contamination from any form of chemical waste or physical waste is avoided (Lacy, 1983). Less acceptable is the direct reuse of wastewater for potable use, even after a high level of treatment. Of most concern here is the possibility of an outbreak of disease.

Water pollutants can be divided among some general classifications (Table 4.3), and water pollution control is closely allied with the water supplies of communities and industries because both generally share the same water resources (Noyes, 1991). There is great similarity in the pipe systems that bring water to each home or business property, and the systems of sewers or drains that subsequently collect the wastewater and conduct it to a treatment facility. Treatment should prepare the flow for return to the environment so that the receiving watercourse will be suitable for beneficial uses such as general recreation, and safe for subsequent use by downstream communities or industries. In considering water pollution, it is useful to keep in mind an overall picture of possible pollutant cycles (Chapter 1) leading to pollutant interchange among the biotic, terrestrial, atmospheric, and aquatic environments.

Water supply can be considered to be one part of a hydrologic cycle (Figure 4.1) (Speight, 1996) in which a major portion of the water occurs in the oceans (Pickering and Owen, 1994). Water (as water vapor) is also present in the atmosphere and other water is contained as ice and snow in snowpack, glaciers, and the polar ice caps. Surface water occurs in lakes, rivers, streams, and reservoirs. Groundwater is located in aquifers underground. However, for the present purposes, water supply is generally considered to occur in four accessible locations: (i) groundwater; (ii) river water; (iii) lake water; and (iv) ocean water.

Groundwater is a vital resource that plays a crucial role in geochemical processes and the quality and mobility of groundwater are dependent upon the geology of the area (i.e., the rock formations in which the water is held). Groundwater is the part of the aquasphere that is most vulnerable to damage from chemical waste. Although surface water supplies are subject to contamination, groundwater can become almost irreversibly contaminated by the improper land disposal of chemicals. Once there is penetration into aquatic systems, chemical species are subject to a number of chemical and biochemical processes. These include acid-base, oxidation-reduction, precipitation-dissolution, and hydrolysis reactions, as well as biodegradation. Under many circumstances, biochemical processes largely determine the fates of chemical species in water, particularly the oxidation of biodegradable organic waste in water. Microorganisms such as bacteria produce organic acids and chelating agents, such as citrate, which have the effect of solubilizing heavy metal ions.

Physically, an important characteristic of such formations is their porosity, which determines the percentage of rock volume available to contain water. A second important physical characteristic is permeability, which describes the ease of flow of the water through the rock. High permeability is usually associated with high porosity. However, clays tend to have low permeability even when a large percentage of the volume is filled with water.

TABLE 4.3 Examples of Water Pollutants and the Effects

Class of Pollutant*	Effect
Acidity, alkalinity, salinity (in excess)	Water quality
–	Aquatic life
Algal nutrients	Eutrophication
Asbestos	Human health
Biochemical oxygen demand	Water quality
–	Oxygen levels
Chemical carcinogens	Incidence of cancer
Crude oil, products, and wastes	Effect of wildlife, health effects
Detergents	Eutrophication, wildlife, health effects
Inorganic pollutants	Toxicity, aquatic biota
Metal-organic combinations	Metal transport
Organic pollutants	Toxicity
Pathogens	Health effects
Pesticides	Toxicity, aquatic biota, wildlife
Polychlorinated biphenyls	Biological effects
Radionuclides	Toxicity
Sediments	Water quality, aquatic biota, wildlife
Sewage	Water quality, oxygen levels
Taste, odor, and color	Health effects, esthetics
Trace elements	Health, aquatic biota

*Listed alphabetically.

The water table is crucial in explaining and predicting the flow of wells and springs and the levels of streams and lakes. It is also an important factor in determining the extent to which pollutant and hazardous chemicals are likely to be transported by water. The water table tends to follow the general contours of the surface topography and varies with differences in permeability and water infiltration. The water table is at surface level in the vicinity of swamps and frequently above the surface where lakes and streams are encountered. The water level in such bodies may be maintained by the water table. In flowing streams or reservoirs are located above the water table; they lose water to the underlying aquifer and cause an upward bulge in the water table beneath the surface water. Furthermore, the flow of the groundwater is an important consideration in determining the accessibility of the water for use and transport of pollutants from underground waste sites. Various parts of a body of groundwater are in hydraulic contact so that a change in pressure at one point will tend to affect the pressure and level at another point. For example, infiltration from a heavy, localized rainfall may affect water level at a point remote from the infiltration. Groundwater flow occurs as the result of the natural tendency of the water table to assume even levels by the action of gravity.

An aquifer, which is a subsurface zone that yields economically important amounts of water to wells, is a water-bearing formation and maybe in the form of porous rock, unconsolidated gravel, fractured rock, or cavernous limestone.

Aquifers are important reservoirs storing large amounts of water relatively free from evaporation loss or pollution. If the annual withdrawal from an aquifer regularly exceeds the replenishment from rainfall or seepage from streams, the water stored in the aquifer will be depleted. Lowering the pressure in an aquifer by over-pumping may cause the aquifer and confining layers of silt or clay to be compressed under the weight of the overburden. The resulting subsidence of the ground surface may cause structural damage to the aquifer and to surface buildings, damage to wells, and other problems. By contrast, an aquiclude is a rock formation that is too impermeable to yield groundwater. Impervious rock in the unsaturated zone may retain water infiltrating from the surface to produce a perched water table that is above the main water table and from which water may be extracted.

A river is a natural, freshwater surface system that has considerable volume compared with its smaller tributaries (often referred to as streams as well as brooks, creeks, branches, or forks. Rivers are usually the main stems and larger tributaries of the drainage systems that convey surface runoff from the land. Rivers flowing to the ocean drain approximately two-thirds of the land systems.

4.3 PROPERTIES

The study of water (hydrology) is divided into a number of subcategories. OPn the other hand, limnology is the branch of science that deals with the characteristics of freshwater, such as the chemical properties, the physical properties, and the biological properties as well as chemical and physical properties. Oceanography is the science of the ocean and its physical and chemical characteristics.

Water, the pure state, is an inorganic chemical substance that is transparent, tasteless, and odorless (Forsberg et al., 2006). Water is a major component of all living things and it is anomalous in many of the physical and chemical properties. Some of the properties—such as cohesion, adhesion, capillary action, surface tension, the ability to dissolve many substances, and high specific heat—are essential for life while other properties have significant (often detrimental) effects on the size and shape of living organisms, how the organisms work, and the constraints within which the organisms must operate. Many of unusual properties of water can be assigned to a causative forces which arise because water is a polar substance that leads to attraction between the water molecules.

In terms of the structural chemistry of water, the molecule is made up of two hydrogen atoms bonded to an oxygen atom (H_2O) and is often referred to as a bent molecule (in a tetrahedral configuration) in which each end of a water molecule has a slight electric charge giving the molecule polarity.

O

H 104.45° H

In some scientific circles. Because of the tetrahedral configuration of the molecule, water has been jokingly referred to as tetrahedral soup.,

This uneven distribution of charges results in two ends the water molecule has a fractional positive change ($H^{\delta+}$) while the oxygen atom has a fractional negative charge ($O^{\delta-}$) and this influences the polarity of the water molecule. The fractionally positive hydrogen atoms attract the fractionally negative oxygen atoms of nearby water molecules, causing the molecules to associate and this attraction causes water molecules to form temporary bonds (hydrogen bonds) that can break easily. Also, water is an amphoteric molecule insofar as it can exhibit the properties of an acid or a base, depending on the pH of the aqueous solution. Water readily produces both hydrogen ions (H^+) and hydroxyl ions (OH^-) and because of the amphoteric character nature water undergoes self-ionization. The product of the concentrations of hydrogen ions and hydroxyl ions is a constant, so the respective concentrations of these ions are inversely proportional to each other.

Because of the polarity, water molecules can attract each other as well as exhibit the tendency to associate with other chemical substances through adhesion. Both cohesion and adhesion allow water to move in one continuous column from the roots to the leaves of a plant. This upward movement (capillary action) the force that causes a liquid to climb upward against the force of gravity—is the combined force of attraction between water molecules and between water with the molecules of surrounding materials. Thus, capillary action allows water to move through porous materials and also causes water molecules to cling to the fibers of other materials such as paper and cloth.

Water is often referred to as the universal solvent because of the ability to dissolve more substances than any other known solvent. Water can dissolve substances such as sugars (carbohydrates, $C_6H_{12}O_6$), bleach (sodium hypochlorite. NaOCl), and salt (NaCl). Water can also dissolve gases such as carbon dioxide (CO_2) and oxygen (O_2). Water also dissolves nutrients but does not dissolve nonpolar substances such as (organic) oil and (organic) wax which are water-repellent (hydrophobic). Thus, water has a number of unique properties (Table 4.4) that make it suitable for living organisms and a solvent for many materials. It is the basic transport medium for nutrients and waste products() but a point that must not be missed, there is also the very real potential for water to transport toxins (Hall and Strichartz, 1990; Stollenwerk and Kipp, 1990).

TABLE 4.4 Properties of Water

Property	Effects
High dielectric constant	High solubility of ionic substances.
High heat capacity	Stabilization of temperatures of organisms.
High heat of evaporation	Determines transfer of heat between the atmosphere and water systems.
High latent heat of fusion	Temperature stabilized at the freezing point of water.
Maximum density at 4°C	Allows ice to float.
–	Vertical circulation restricted in stratified bodies of water.
Solvent	Transport of nutrients
–	Transport of waste products.

The solvent properties of water are influenced insofar as most ionic materials are dissociate (and dissolve) in water. In addition, water has a high heat capacity. Thus, a relatively large amount of her is required to change appreciably the temperature of a mass of water. Thus, water can have a stabilizing effect upon the temperature of nearby geologic formations because the high heat capacity of water prevents sudden changes of temperature in bodies of water and tends to mitigate any abrupt temperature variations thereby protecting aquatic organisms from the shock of the sudden change in temperature. The high heat of vaporization of water (585 cal/g at 20°C, 68°F) also stabilizes the temperature of bodies of water and influences the transfer of heat and water vapor between bodies of water and the atmosphere.

Water has a maximum density at 4°C (39°F), a temperature above its freezing point (0°C, 32°F) which causes the lower density ice to float on the surface of water and, as a result, few large bodies of water ever freeze solid. Furthermore, the occurrence of the vertical circulation of water in a lake is a determining factor in the chemistry and biology of water.

Surface water occurs primarily in rivers, streams, lakes, and reservoirs. Lakes may be classified as oligotrophic, eutrophic, or dystrophic (Table 4.5) (Dresnack, 2006; Jazib, 2018). Oligotrophic lakes are deep, generally clear, deficient in nutrients, and without much biological activity. Eutrophic lakes have more nutrients, support more life and are more turbid. Dystrophic lakes are shallow with abundant plant life and typically contain colored water with a low pH.

TABLE 4.5 General Types of Lakes

Types of Lakes	Description
	Based on Nutrients
Oligotrophic lake	Low nutrient concentrations.
Eutrophic lake	Over supply by nutrients such as nitrogen and phosphorus; usually a result of agricultural run-off or discharge of municipal sewage.
Dystrophic lake	Low pH, high humic acid content and brown waters.
	Based on Origin
Volcanic lake	Receives water from magma after volcanic eruptions; have highly restricted biota.
Endemic lake	Very ancient, deep, and have endemic (native to that area) fauna.
Artificial lake	Created due to construction of a dam.
	Based on of Salinity
Freshwater lake	Low concentration of salts.
Saltwater lake	High concentration of salts.
Meromictic lake	Rich in salts; permanently stratified.
Desert salt lake	Occur in arid regions; high salt concentrations due to high evaporation.

Wetlands are flooded areas in which the water is sufficiently shallow that growth of bottom-rooted plants is possible (Dennison and Berry, 1993; Easterbrook, 1995). Wet flatlands are areas where mesophytic vegetation is more important than open water and which are commonly developed in filled lakes, glacial pits, and potholes, or in poorly drained coastal plains or flood plains. The term swamp is usually applied to a wetland where trees and shrubs are able to grow whereas the term bog is a water-filled area that lacks a solid foundation. In fact, some bogs consist of a thick covering of vegetation that floats on the water.

Some constructed reservoirs are similar to lakes, while others differ a great deal from them. Reservoirs with a large volume relative to their inflow and outflow are called storage reservoirs. Reservoirs with a large rate of into the reservoir and out of the reservoir (flow-through) compared to the reservoir volume are run-of-the-river reservoirs. The physical, chemical, and biological properties of water in the two types of reservoirs may vary appreciably. Also, water in storage reservoirs more closely resembles lake water, whereas water in run-of-the-river reservoirs is much like river water.

Impounding water in reservoirs may have some profound effects upon water quality and these changes result from factors such as different velocities, changed detention time, and altered surface-to-volume ratios relative to the streams that were impounded. Some resulting beneficial changes due to impoundment are a decrease in the level of organic matter, a reduction in turbidity, and a decrease in hardness dues to a decrease in (or low content of) the amount of calcium ions (Ca^{2+}) and magnesium ions (Mg^{2+}) in the water content).

Some detrimental changes are lower oxygen levels due to decreased re aeration, decreased mixing, accumulation of pollutants, lack of a bottom scour produced by flowing water scrubbing a stream bottom, and increased growth of algae. Algal growth may be enhanced when suspended solids settle from impounded water, causing increased exposure of the algae to sunlight. Stagnant water in the bottom of a reservoir may be of low quality. Oxygen levels frequently go to almost zero near the bottom, and hydrogen sulfide is produced by the reduction of sulfur compounds in the low oxygen environment. Insoluble iron (Fe^{3+}) species are reduced to soluble iron (Fe^{2+}) species which must be removed prior to use of the water.

The temperature–density relationship that exits in water results in the formation of distinct layers (stratification) within non-flowing bodies of water (Figure 4.2) (Speight, 1996). During the summer, a surface layer (epilimnion) is heated by solar radiation and, because of its lower density, floats upon the bottom layer (hypolimnion). When an appreciable temperature difference exists between the two layers, mixing does not occur and the behavior of each layer is independent of the other, and the layers exhibit different chemical and biological properties.

The epilimnion, which is exposed to light, may have a heavy growth of algae because of the exposure of this upper-most later to the atmosphere and (during daylight hours) because of the photosynthetic activity of algae. In the hypolimnion, the biodegradable organic material may cause the water to become anaerobic. A The shear-plane, or layer between epilimnion and hypolimnion (the thermocline) and during the autumn when the epilimnion cools and a point is reached at which the temperatures of the epilimnion and hypolimnion are equal (Cheremisinoff, 1995).

The chemistry and biology of the oceans are unique because of the high salt content and the depth. The environmental problems of the oceans have increased greatly in recent years because of the dumping (inadvertent discharge and deliberate discharge) of pollutants and, moreover, many of the pollutants have a considerable lifetime in the ocean (Figure 4.3) (Speight, 1996). Discharge of pollutants into the oceans also emanate from large conduits that carry both sanitary and storm wastes.

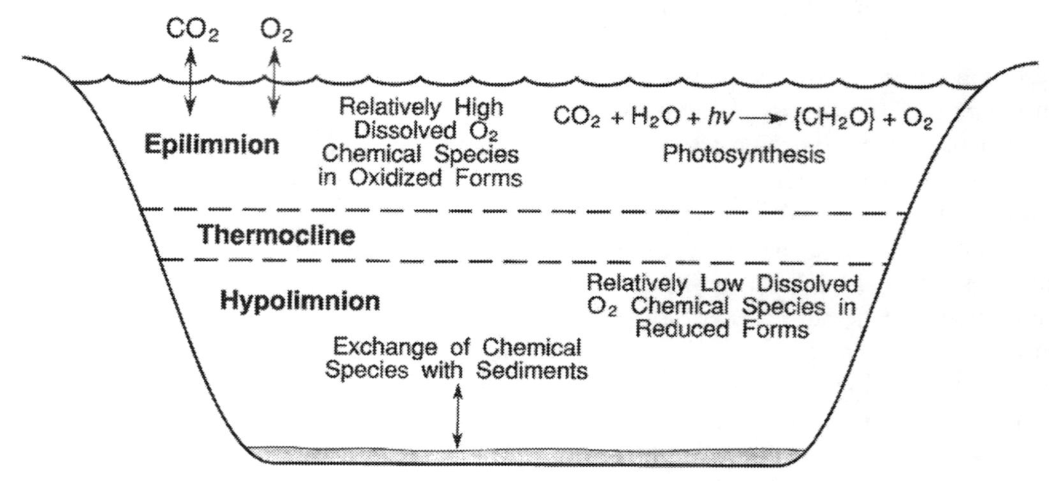

FIGURE 4.2 Representation of the stratification of a lake.
Source: Reproduced with permission from: Speight (1996). © Taylor and Francis.

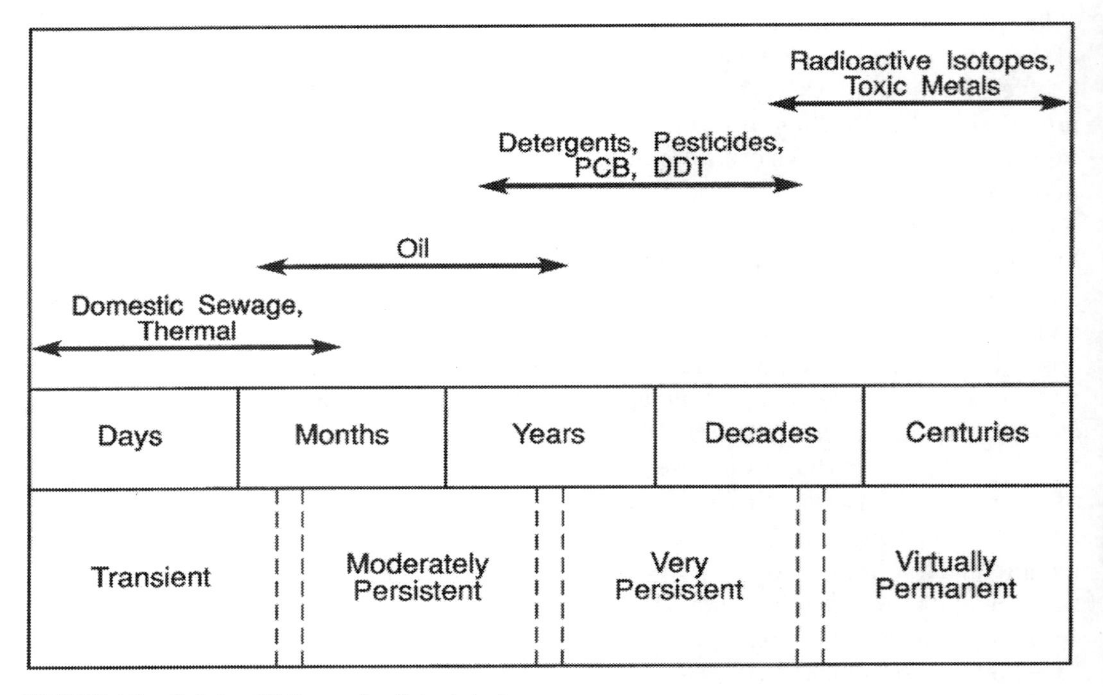

FIGURE 4.3 Relative lifetimes of pollutants in the oceans.
Source: Reproduced with permission from: Speight (1996). © Taylor and Francis.

4.3.1 GASES

Generally, the solubility of gases in water decreases as temperature increases. However, dissolved gases (oxygen for fish and carbon dioxide for photosynthetic algae) are crucial to the welfare of living species in water. Dissolved oxygen frequently is the key substance in determining the extent and types of life in water because oxygen-deficient water is fatal to many aquatic animals, but the presence of an overabundance of oxygen can be equally fatal to many kinds of anaerobic bacteria. In addition, carbon dioxide, bicarbonate ion (HCO_3^-), and carbonate ion (CO_3^{2-}) have an extremely important influence upon the chemistry of water and also of wastewater (Benefield et al., 1982; Schwarzenbach et al., 1993; Huang et al., 1995). Many minerals are deposited as salts of the carbonate ion and algae utilize dissolved carbon dioxide in the synthesis of biomass.

Carbon dioxide exists in air on the order of 0.035% v/v of the air and as a consequence of the low level of atmospheric carbon dioxide, water totally lacking in alkalinity in equilibrium with the atmosphere contains only a low level of carbon dioxide. However, the formation of bicarbonate HCO_3^-)and carbonate (CO_3^{2-}) greatly increases the solubility of carbon dioxide. High concentrations of free carbon dioxide in water may adversely affect respiration and the gas exchange of aquatic animals. Because of the presence of carbon dioxide in air and its production from the microbial decay of organic matter, dissolved carbon dioxide is present in virtually all natural water systems.

4.3.2 ACIDITY AND ALKALINITY

The acidity of water is the capacity of the water to neutralize hydroxyl ions (OH^-) and the alkalinity of water is the capacity of water to neutralize the hydrogen ion (H^+). Acidity can be caused by weak organic acids, such as acetic and tannic acids, and strong mineral acids including sulfuric and hydrochloric acids. However, the most common source of acidity in unpolluted water is carbon dioxide in the form of carbonic acid (H_2CO_3).

The alkalinity of water refers to the capability of water to neutralize acid in which the water has a buffering capacity—a buffer is a solution to which an acid can be added without changing the concentration of available hydrogen (H^+) ions (without changing the pH) appreciably. In a surface water body, such as a lake, the alkalinity of the water is caused predominantly by the adjoining from the rock formations. For example, when precipitation (rain or snow) falls in the area around (or leading into) the lake and if the area (the rock formation and the land) contain formation or rocks such as limestone, the runoff dissolves minerals such as calcium carbonate ($CaCO_3$), which raises the pH and alkalinity of the water.

On the other hand, acidic water is not frequently encountered in nature, except in cases of severe pollution. When acidity does exist in natural water systems, it generally results from the presence of weak acids such as phosphoric acid carbon dioxide, hydrogen sulfide, proteins, fatty acids, and acidic metal ions, particularly ferric ions (Fe^{3+}). However, acids—such as sulfuric acid (H_2SO_4) and hydrochloric acid (hydrogen chloride, HCl)—are the pollutants that contribute to the acidity of water systems.

The alkalinity of natural water systems is important for fish and other aquatic life because the alkalinity of the system protects (buffers) the water system against rapid pH changes. Typically, aquatic life species function best in a pH range on the order of 7.0 to 9.0. Moreover, the higher alkalinity of surface water will protect the system against acid rain (as well as against the discharge of acid waste) and prevent any changes to the pH changes that are harmful to aquatic life. If an increasing amount of acid material is added to a body of water, the buffering capacity (protecting ability) of the system is consumed.

4.3.3 CHEMICAL SPECIES

Natural water systems have a broad range of chemical species—especially the species known as total dissolved solids (TDS)—which vary with the source of the water. Some chemical substances, particularly redox-sensitive trace metals (such as iron, manganese, and lead as well as other metals species), are more soluble when natural waters are depleted in dissolved oxygen.

Metal ions in aqueous solution tend to reach a state of maximum stability through chemical reactions. Acid-base interactions, precipitation, complex formation, and oxidation-reduction reactions provide the interactions by which metal ions in water are transformed to more stable chemical forms. Hydrated metal ions (for example illustrated as $M^{+}.H_2O$), particularly those with a high positive charge tend to lose protons in aqueous solution. Hydrated trivalent metal ions, such as iron (Fe^{3+}), generally are minus at least one hydrogen ion at neutral pH (= 7.0) or higher. For tetravalent metal ions, the completely protonated forms are rare, even at low pH (i.e., the acidic range).

Of the cations found in most freshwater systems, calcium (Ca^{2+}) generally has the highest concentration. Calcium is a key element in many geochemical processes, and the different mineral forms (polymorphs) of calcium carbonate are among the primary contributing minerals: gypsum ($CaSO_4.2H_2O$), anhydrite ($CaSO_4$), dolomite ($CaCO_3.MgCO_3$), and calcite ($CaCO_3$) and aragonite ($CaCO_3$).

Water that contains a high level of carbon dioxide will dissolve calcium ions (Ca^{2+}) from the various carbonate minerals which along with ions of magnesium (Mg^{2+}) and iron (Fe^{2+}) account for the hardness of water. A number of other chemical species are present naturally in water (Table 4.6) and some of these species can also be pollutants. In particular, chelating agents are common potential water pollutants and occur in sewage effluent and industrial wastewater such as metal plating wastewater.

Chelating agents are complex-forming agents and have the ability to solubilize heavy metals (i.e., metals with a high atomic number and high density) as a chelate derivative. The formation of soluble complexes increases the leachability of these metals from waste disposal sites and reduces the efficiency with which the metals can be removed from sludge during biological waste treatment. In addition, there are chelating agents of biological origin such as ferrichrome. This chelating agent (a cyclic hexapeptide)is composed of three glycine and three modified ornithine residues with hydroxamate groups –N(OH)C(=O)C-] is synthesized by, and extracted from fungi, which forms extremely stable chelates with iron (Fe^{3+}).

TABLE 4.6 Chemical Species Occurring in Water

Substance*	Source
Aluminum	Aluminum-containing minerals
Chloride	Minerals, pollution
Fluoride	Minerals, water additive
Iron	Minerals, mine water
Magnesium	Minerals such as dolomite ($CaCO_3.MgCO_3$)
Manganese	Minerals, decay of nitrogenous organic matter, pollution
Potassium	Mineral matter, fertilizer runoff, forest fire runoff
Phosphorus	Minerals, fertilizer runoff, domestic waste (from detergents)
Silicon	Minerals, such as sodium feldspar albite, $NaAlSi_3O_8$, pollutants
Sulfur	Minerals, pollutants, acid mine water, acid rain
Sodium	Minerals, pollution

*Listed alphabetically.

Many organic compounds interact with suspended material in a water system, and settling of the suspended material (which may also contain containing adsorbed organic matter) can carry any adsorbed organic compounds into the sediment of a stream or lake. Furthermore, suspended PM affects the mobility of organic compounds adsorbed on to particles which undergo chemical degradation and biodegradation at different rates and by different pathways compared to the organic matter in solution.

The most common types of sediments for the ability to adsorbed and bind with organic matter are sediment which contain or are composed of clay minerals (Chapter 3), organic humic substances, and clay-humic complexes. Both clay minerals and humic substances act as cation exchangers and, therefore, these materials adsorb cationic organic compounds through ion exchange. When adsorbed by clay minerals, cationic organic compounds are generally held between the layers of the clay mineral structure where their biological activity is minimal, if not zero.

4.3.4 AQUATIC ORGANISMS

Aquatic organisms can be classified according to: (i) the various biological characteristics; (ii) the habitat; and (iii) the adaptations of the organism. For example, microorganisms (algae, bacteria, and fungi) are living catalysts for a variety of chemical processes to occur in water and in soil.

Furthermore, the living organisms (biota) in an aquatic ecosystem are classified as either: (i) autotrophic organisms; or (ii) heterotrophic organisms. The former category (the autotrophic organisms) utilize solar or chemical energy to convert fix elements from simple, nonliving inorganic material into complex molecules that occur in living organisms. As an example, algae are typical autotrophic aquatic organisms and carbon dioxide (CO_2), nitrate (NO_3^-), and phosphate derivatives (PO_4^{3-}) are sources of carbon, nitrogen, and phosphorus, respectively, for autotrophic organisms.

Heterotrophic organisms utilize the organic substances produced by autotrophic organisms as energy sources and as the raw materials for the generation of their own biomass. The decomposers (decomposing organisms, also called or reducers or reducing organisms) are a subclass of the heterotrophic organisms and consist of chiefly bacteria and fungi, which ultimately break down material of biological origin to the simple compounds originally used (or needed) by the autotrophic organisms.

Macrophytes are individual aquatic plants that can be observed without a microscope (or other similar device) and are categorized based on where they grow and how they grow. For example, rooted macrophytes are rooted in the riverbed or lake bed and are restricted to areas where flow is low enough to permit sediments to accumulate. The rooted macrophytes may have leaves and are: (i) entirely submerged, i.e., under the water; (ii) floating on the surface of the water; or (iii) emergent above the surface of the water. In turbid water, the lack of light restricts photosynthetic processes little light penetrates and photosynthesis is restricted, and only plants with floating leaves or emergent leaves can survive. On the other hand, floating **aquatic macrophytes** are rootless plants that persist only in backwater areas where the flow is minimal—if there is any flow at all; otherwise these macrophytes are carried downstream. Because the photosynthetic surfaces are above the surface of the water, these plants can grow in deep, turbid water and in places where rooting sites are sparse.

A large population of aquatic macrophytes can have a negative effect on aquatic ecosystems and, in some cases, they can form dense mats that cover the water surface. An extreme situation occurs when the macrophytes cause the underlying water to become deoxygenated which can be disruptive to natural aquatic ecosystems.

Many groups of invertebrates (which include all animals without a backbone and are more diverse and abundant than vertebrates) are found in aquatic systems. Invertebrates living on or in aquatic sediments (benthic invertebrates) and these benthic invertebrate communities are often used as indicators of aquatic ecosystem health through measurements of population abundance and diversity.

Zooplankton—because they cannot swim against currents—are more important in non-running water (such as in lakes) than in running water. Running water typically carries the zooplankton downstream but can be abundant in large slow-flowing rivers. The zooplankton are heterotrophic and are significant sources of energy and nutrients to carnivorous invertebrates and also to some vertebrates.

Insects are the most diverse group of animals on the Earth—most insects are terrestrial, while some (such as dragonflies and mosquitoes) have life stages that are aquatic. While most aquatic insects live on or near the bottom of water systems, some are able to engage in a form of swimming and many need water with dissolved oxygen to survive while other insects, such as mosquito larvae, breathe through the surface film of still (non-moving or non-changing) water in the water system. Stream and river insects are crucial in the processing (biodegradation) of organic matter. For example, some of the insects degrade the biofilm, other insects shred larger leaves into smaller particles, while another category of insects filter or collect these smaller particles.

Finally in the group, fish display every major feeding type and there are: (i) herbivorous fish, which feed on periphyton or macrophytes, or may even filter phytoplankton from the water; (ii) carnivorous fish, which feed on mollusks, worms, insects, zooplankton, and other fish; and (iii) omnivorous fish, which may feed on specific types of prey, or in the interest of survival may feed indiscriminately on any of the flora and fauna that they can consume. Due to this diversity in feeding habits, different fish can occupy different places in a food chain—the shark might be cited as the most prominent member of the food chain, with humans as their only enemy.

Fish typically, like other faunal species (humans are the possible exception), occupy specific habitats while others exist in a wide variety of lakes and rivers. Furthermore, there are a number of items that influence the distribution of fish and these are: (i) oxygen concentration in the water; (ii) temperature of the water; (iii) the presence of macrophytes in the water; (iv) the availability of suitable substrate for spawning; (v) the current speed in streams and rivers; and (vi) the salinity or the acidity-alkalinity of the water. Furthermore, changes in fish habitat (such as reduction of flooding due to damming as well as the opposite effect in which flooding can introduce foreign-non-typical-minerals into the water) can favor some types of fish, and disadvantage others.

4.3.5 ALGAE

Algae (such as kelp or phytoplankton) are a diverse group of aquatic organisms that have the ability to engage photosynthetic activities and can be considered to be microscopic organisms that use inorganic nutrients to exist. The general nutrients that algae use are carbon (from carbon dioxide, CO_2, or from bicarbonate, HCO_3^-), nitrogen (generally as nitrate, NO_3^-), phosphorus (as some form of orthophosphate, PO_4^{3-}), sulfur (as sulfate, SO_4^{2-}), and trace elements such as sodium (Na), potassium (K), calcium (Ca), magnesium (Mg), iron (Fe), cobalt (Co), and molybdenum (Mo).

Algae have, because of their diversity, the ability to exist in freshwater lakes or in saltwater oceans and also can endure a range of temperature, oxygen concentration, carbon dioxide concentration, and turbidity. Generally, algae live independently in the various forms (single cells or colonies) and can also exist in symbiotic relationships with a variety of non-photosynthetic organisms, including sponges, mollusks, and fungi (as lichens).

Algae are capable of photosynthesis and produce their nourishment by using photoenergy (from the sun) and carbon dioxide to generate carbohydrate derivatives ($C_6H_{12}O_6$) and oxygen. In fact, most algae are autotrophs (more specifically, they are photoautotrophs) insofar as they use of light energy to generate nutrients for survival. However, there are algae that need to obtain nutrition from outside sources, and they are heterotrophic insofar as they can apply a variety of heterotrophic strategies to acquire nutrients from organic materials (carbon containing compounds such as carbohydrate derivatives, protein derivatives and fats).

The term algal bloom refers to the rapid growth (and expansion) of an microalgal colony which leads to the production of toxins, disruption of the natural aquatic ecosystems, and requires careful water treatment. In freshwaters, cyanobacteria (blue-green algae) are the main toxin producers, though some eukaryotic algae (algae that have a nucleus enclosed within a nuclear envelope) use the toxins to protect themselves from being eaten by small animals.

The main cause of an algal bloom is nutrient pollution, in which an excess of nitrogen and phosphorus stimulate the algae to uncontrollable growth. The phenomenon is caused by events such as the runoff of a variety of agricultural fertilizers and animal manure that are rich in nitrogen as well as wastewater that has both a high content of nitrogen and phosphorus.

4.3.6 BACTERIA

Bacteria (pl; singular: bacterium) live in soil, the ocean and inside the human digestive system either individually or as groups ranging up to millions of individual cells Most bacteria are in the size range of 0.3 to 50 microns (1 micron = 1 meter $\times$ 10^{-6}).

The metabolic activity of bacteria is greatly influenced by their small size-the surface-to-volume ratio is high and, thus, the inside of a bacterial cell is readily accessible to a chemical substance in the surrounding medium. As a result, bacteria can cause chemical reactions to be rapid compared to the chemical reactions caused or mediated by larger organisms. In addition, bacteria excrete enzymes that can exist outside of the cell (exoenzymes) and these enzymes break down solid food material to soluble components which can penetrate bacterial cell walls, where the digestion process is completed.

Bacteria are essential participants in many important elemental cycles in nature, including: (i) the nitrogen cycle; (ii) the carbon cycle; and (iii) the sulfur cycle. Bacteria are also responsible for the formation of mineral deposits, such as mineral deposits containing iron (Fe) or manganese (Mn).

4.3.7 FUNGI

Fungi (pl: singular: fungus) are aerobic (oxygen-requiring) eukaryotic organisms (i.e., oxygen-requiring organisms that have a nucleus enclosed within a nuclear envelope) that includes microorganisms such as yeasts and molds, as well as the more familiar mushrooms. They are non-photosynthetic organisms that frequently possess a filamentous structure and, generally, are much larger than bacteria (typically: 5 to 10 microns in width. Fungi can generally thrive in more acidic media than can bacteria and fungi are also have (relative to bacteria) more tolerant to higher concentrations of heavy metal ions (ions of the high density high atomic weight metals) than are bacteria.

An extremely important function of fungi in the environment is the ability of the fungi cause the breakdown of cellulose. In this process, fungal cells secrete a bio-catalyst (enzyme)—the enzyme cellulase, which converts the insoluble cellulose to soluble carbohydrate derivatives that can be absorbed by the fungal cell.

Although fungi do not survive readily in aqueous systems, they do play an important role in determining the composition of natural water systems and in wastewater systems because of the decomposition products arising from fungal-based conversions that enter water. An example of such a product is humic material (organic material that consists of the organic compounds that are important components of humus, the major organic fraction of soil and peat), which interacts with hydrogen ions and metal ions.

4.4 BIODEGRADATION PROCESSES

Biodegradation (transformation of a chemical by microorganisms) is the decay or breakdown of chemicals that occurs when microorganisms use an organic substance as a source of carbon and energy. Biodegradation processes are necessary for waste recycling and the processes are processes which involve bacteria and fungi. Temperature is an important parameter and typically occurs at temperatures between 10 and 35°C (50 and 95°F)—water is essential for the biodegradation process. The process can be divided into three stages, which are: (i) biodeterioration; (ii) bio-fragmentation; and (iii) assimilation.

The first stage (biodeterioration) is a surface-level degradation that modifies the chemical properties and/or the physical properties and/or the mechanical properties of the contaminant which occurs when the material is exposed to abiotic factors and allows for further degradation by changing the structure of the contaminant to a more amenable (reactive) structure. Some abiotic factors that influence these initial changes of the process are compression (through buildup of the pressure exerted by the overburden or depth, light, temperature, and the presence of other chemicals in the environment.

Although biodeterioration is often cited as the first stage of biodegradation, the process can, in some cases, occur in parallel (simultaneously) to bio-fragmentation which is the conversion of the spilled chemical to lower molecular weight fragments (or products) that are more amenable to further chemical or physical reaction and, hence, removal from the environment (or the ecosystem). Assimilation occurs when the fragment (or fragments) are assimilated into the environment (or ecosystem) without any deleterious effect on the system.

On the other hand, the resulting products from bio-fragmentation can be assimilated into microbial cells (the assimilation stage) by membrane carriers. However, other products of the bio-fragmentation stage may still have to undergo biotransformation reactions to yield products that can then be transported to the inside of the cell. Once inside the cell, the products enter catabolic pathways.

By way of explanation, the catabolic pathways are those metabolic pathways that break down molecules into smaller units that are either oxidized to release energy or used in other reactions. Catabolism breaks down larger molecules into smaller units and is, in fact, the molecular breaking-down aspect of metabolism, whereas anabolic pathways is the building-up aspect.

The biodegradation process is, in general, an important process for the removal of chemical compounds (especially organic chemicals) from the environment. As an example, the biodegradation of phosphorus compounds is important in the environment for two reasons which are: (i) the process provides a source of algal nutrient orthophosphate from the hydrolysis of polyphosphates; and (ii) the process deactivates highly toxic organophosphate compounds, such as the organophosphate-based insecticides. As another example, sulfur-containing compounds are common in water and sulfate ions (SO/) occur in varying concentration in natural water systems. The biodegradation of organic sulfur compounds, both those of natural origin and sulfur-containing anthropogenic pollutants, is an important microbial process. Sometimes the degradation products, such as the odorous and toxic hydrogen sulfide H_2S), can have a detrimental effect on water quality.

One consequence of bacterial action on metal compounds is the occurrence of drainage of acidic aqueous solutions from active mines or from abandoned mines (Hall, 2006). In fact, acid mine drainage is a common and damaging problem in the waters flowing from coal mines and draining from the spoil piles (mine tip page, gob piles).

A gob pile is composed of the accumulated spoil or the waste rock removed during mining. This waste material is typically composed of shale, as well as smaller quantities of carboniferous shale and various other residues. The washings are highly acidic and have the ability to sterilize the surrounding land and water systems with the ensuing serious (often fatal) effects on the flora and fauna.

Acidic mine water results from the presence of sulfuric acid (H_2SO_4) produced by the oxidation of pyrite (FeS_2). Thus:

$$4FeS_2 + 15O_2 + 14H_2O \rightarrow 4Fe(OH)_3 + 8H_2SO_4$$

Pyrite oxygen water + Sulfuric acid

The prevention of the effects of acid mine water on the environment is a major challenge. In addition, selenium, which may also occur in the mine water or in the mine tip page or gob piles and give rise to selenium-containing runoff, is also subject to bacterial oxidation and reduction. These transitions are important because selenium is a crucial element in nutrition, particularly of livestock. Microorganisms are involved with the selenium cycle-a soil-dwelling strain of *Bacillus megaterium* is capable of oxidizing elemental selenium to selenite (SeO).

4.5 SOLUBILITY OF POLLUTANTS

Chemicals are the most common type of contaminant in the aquasphere and the amount and dispersion of chemical pollutants in the aquasphere is dependent upon the solubility of the chemical in the water or the miscibility of the chemical with water. In fact, many pollutants pass from the source into surface water systems because of: (i) terrestrial runoff; (ii) direct discharge; (iii) inadvertent discharge; or (iv) atmospheric deposition in which the chemical descend to the water system on the surface of the Earth in rain or in snow as well as in fog. In turn, rivers carry many of these pollutants to the sea.

Aquatic microorganisms thrive on biodegradable substances and when many of these chemicals are discharged into a waterway, the number of microorganisms increases. When the available oxygen in the water has been consumed, the depletion of oxygen leads to the death of aerobic microorganisms but promotes the growth of anaerobic organisms (eutrophication). In addition, anaerobic microorganisms can cause water contamination by producing toxins such as sulfide derivatives and ammonia. On the other hand, there are chemical contaminants that are insoluble in water and exist as a suspension of the chemical (or PM) in the water, which can form a layer on the surface of the water, thereby preventing oxygen penetration leading to oxygen depletion of the water system. However, the more course PM may settle at the bottom of the lake, ocean, or river and have an effect on the life that exists at the base of the water system; in some cases, the material can comprise of harmful toxins.

In fact, the ability of a body of water to produce living material (the productivity of the water) results from a combination of chemical and physical factors. For example, water that has a low productivity generally is desirable for water supply or for swimming while relatively high productivity is required for the support of fish. In addition, the excessive productivity of a water system can result in choking by weeds and can cause odor problems in the system. For example, the growth of algae may be high when excessive content of nutrients in a lake or other body of water, frequently due to runoff from the land—the waters are said to be highly productive—with the consequence is that the decomposition of the dead can algae reduce the oxygen level in the water to a low value which can result in the death of flora and fauna due to the lack of oxygen (eutrophication).

Moreover, when a population increase necessitates an expanded utilization of lakes and streams, cultural eutrophication becomes a major water resource problem. This form of eutrophication (i.e., cultural eutrophication) is reflected in changes in species composition, population sizes, and productivity in groups of organisms throughout the aquatic ecosystem.

Chemicals and heavy metals from industrial and municipal wastewater also contaminate waterways. These contaminants are toxic to aquatic life and often reducing the life span of an organism as well as the ability to reproduce, thereby accumulating in the food chain as predator eats prey. An example is the accumulation of mercury within various species fish.

The three main physical properties that affect life in a water system are: (i) temperature of the water; (ii) transparency of the water; and (iii) turbulence in the water. Low water temperatures result in slow biological processes, whereas high temperatures are fatal to most organisms, and a temperature of only a few degrees can produce significant differences in the types of organisms present in the water system. For example, the thermal discharge of hot water (i.e., the discharge of water from the water-cooling system) from a power plants can kill temperature-sensitive fish while increasing the growth of fish and other species that are adapted to higher temperatures.

The transparency of the water in the system affects the growth of algae and turbulence is a factor in mixing and transport processes in water. The turbulence of the water is largely responsible for the transport of nutrients to living organisms and of waste products away from the living organisms.

Biochemical oxygen demand, which refers to the amount of oxygen utilized when the organic matter in water system undergoes biodegradation, is another important water-quality parameter. A body of water with a high biochemical oxygen demand, and no means of rapidly replenishing the oxygen, obviously cannot sustain organisms that require oxygen. The degree of oxygen consumption by microbial oxidation of contaminants in water is called the biochemical oxygen demand (or biological oxygen demand).

The levels of nutrients in water frequently determine its productivity. Aquatic plant life requires an adequate supply of carbon (carbon dioxide, CO_2), nitrogen (nitrate, NO_3), phosphorus (orthophosphate, HPO_4^{2-}), and trace elements such as iron. In many cases, phosphorus is the limiting nutrient and is generally controlled in attempts to limit excess productivity. The salinity and acidity-alkalinity of a water system also have an effect on the types of life forms present in the water. In fact, water runoff from land irrigation may contain harmful levels of salt as well as harmful levels of acidic chemical or alkaline chemicals that cannot be tolerated by aquatic life forms.

Finally, the majority of the chemical reactions that occur in water systems-particularly the reactions that involving organic matter and oxidation-reduction processes-occur through the agency of bacterial intermediates. In fact, various types of microorganisms are responsible for the formation of many sediments and mineral deposits, and they (the microorganisms) also play a major role in the treatment of waste.

KEYWORDS

- **aquasphere**
- **atmosphere**
- **earth system**
- **hydrological cycle**
- **molybdenum**
- **total dissolved solids**

REFERENCES

ASTM, (2020). *Annual Book of Standards*. ASTM International, West Conshohocken, Pennsylvania.

Benefield, L. D., Judkins, J. F., & Weand, B. L., (1982). *Process Chemistry for Water and Wastewater Treatment.* Prentice-Hall, Englewood Cliffs, New Jersey.

Cartwright, F. F., & Biddiss, M. D., (1972). *Disease and History*. Dorset Press, New York.

Cheremisinoff, P., (1995). *Handbook of Water and Wastewater Treatment Technology*. Marcel Dekker Inc., New York.

Dennison, M. S., & Berry, J. F., (1993). *Wetlands: Guide to Science, Law, and Technology.* Noyes Data Corp., Park Ridge, New Jersey.

Dresnack, R., (2006). Eutrophication. In: Pfafflin, J. R., & Ziegler, E. N., (eds.), *Environmental Science and Engineering* (5th edn., Vol. 1. pp. 389–401). CRC Press, Taylor & Francis Group, Boca Raton, Florida.

Easterbrook, G., (1995). *A Moment on the Earth: The Coming Age of Environmental Optimism.* Viking Press, New York.

Eglinton, G., (1975). *Environmental Chemistry* (Vol. 1). Specialist Periodical Reports. The Chemical Society, London, United Kingdom.

Eisenberg, D., & Kauzmann, W., (1969). *The Structure and Properties of Water*. Oxford University Press, Oxford University, Oxford, United Kingdom.

Forsberg, M., Gherini, S., & Stumm, W., (2006). Water: Properties, structure, and occurrence in nature. In: Pfafflin, J. R., & Ziegler, E. N., (eds.), *Environmental Science and Engineering* (5th edn., Vol. 2. pp. 1289–1306). CRC Press, Taylor & Francis Group, Boca Raton, Florida.

Gerstein, M., & Levitt, M., (1998). Simulating Water and the Molecules of Life. Scientific American, 279: 100–105.

Hall, E. P., (2006). Mine Drainage. In: Pfafflin, J. R., & Ziegler, E. N., (eds.), *Environmental Science and Engineering* (5th edn., Vol. 2. pp. 1016–1021). CRC Press, Taylor & Francis Group, Boca Raton, Florida.

Hall, S., & Strichartz, G., (1990). *Marine Toxins: Origin, Structure, and Molecular Pharmacology*. Symposium Series No. 418. American Chemical Society, Washington, D.C.

https://www.ipcc.ch/site/assets/uploads/2018/02/WG1AR5_Chapter04_FINAL.pdf (accessed on 20 November 2021).

Huang, C. P., O'Melia, C. R., & Morgan, J. J., (1995). *Aquatic Chemistry: Interfacial and Interspecies Processes*. Advances in Chemistry Series No. 244. American Chemical Society, Washington, D.C.

James, P., & Thorpe, N., (1994). *Ancient Inventions*. Ballantine Books, New York.

Jazib, J., (2018). *Basics of Environmental Sciences*. Iqra Publishers, New Delhi, India.

Jenne, E. D., (1979). *Chemical Modeling in Aqueous Systems*. Symposium Series No. 93. American Chemical Society, Washington, D.C.

Lacy, W. J., (1983). In: Kent, J. A., (ed.), *Riegel's Handbook of Industrial Chemistry* (p. 14). Van Nostrand Reinhold, New York.

Melchior, D. C., & Bassett, R. L., (1990). *Chemical Modeling of Aqueous Systems II.* Symposium Series No. 416. American Chemical Society, Washington, D.C.

Noyes, R., (1991). *Handbook of Pollution Control Processes*. Noyes Data Corp., Park Ridge, New Jersey.

Pickering, K. T., & Owen, L. A., (1994). *Global Environmental Issues*. Routledge Publishers Inc., New York.

Schwarzenbach, R. P., Gschwend, P. M., & Imboden, D. M., (1993). *Environmental Organic Chemistry*. John Wiley & Sons Inc., New York.

Sharp, K. A., (2001). *Water.* Encyclopedia of Life Sciences. John Wiley & Sons Inc., Hoboken, New Jersey.

Speight, J. G., (1996). *Environmental Technology Handbook.* Taylor & Francis, Washington, DC.

Stollenwerk, K. G., & Kipp, K. L., (1990). In: Melchior, D.C., & Bassett, R. L., (eds.), *Chemical Modeling of Aqueous Systems II.* Symposium Series No. 416. American Chemical Society, Washington, D.C. Chapter 19.

Vaughan, D. G., Comiso, J. C., Allison, I., Carrasco, J., Kaser, G., Kwok, R., Mote, P., et al., (2013). Observations: Cryosphere. In: Stocker, T. F., Qin, D., Plattner, G. K., Tignor, M., Allen, S. K., Boschung, J., Nauels, A., et al., (eds.), *Climate Change 2013: The Physical Science Basis.* Contribution of Working Group I to the Fifth Assessment Report of the Intergovernmental Panel on Climate Change. Cambridge University Press, Cambridge, University, Cambridge, United Kingdom. Chapter 4.

CHAPTER 5

The Geosphere

5.1 INTRODUCTION

The Earth is an integrated system that consists of three major physical components: (i) the atmosphere; (ii) the aquasphere; and (iii) the geosphere (Table 5.1). On occasion, a fourth component and a fifth component are added which are (iv) the biosphere, which is a term that includes all of the living organisms on the Earth and (v) the cryosphere, which is an all-encompassing term for those portions of surface of the Earth where water is in solid form ice, which includes snow cover, glaciers, ice caps, ice sheets, sea ice, lake ice, river ice, and frozen ground. Each of these five components are also sub-systems and they are tightly interconnected (Jazib, 2018).

TABLE 5.1 The Main Components of the Earth System

Component	Description
Atmosphere	The gaseous layer surrounding the Earth.
	Held to the surface by gravity.
	Receives energy from solar radiation, which warms the surface of the Earth.
	Absorbs water from the surface of the Earth via evaporation.
	Redistributes heat and moisture across the surface of the Earth.
	Contains substances that are essential for life, including carbon, nitrogen, oxygen, and hydrogen.
Aquasphere	The parts of the earth system composed of water in its liquid, gaseous (vapor) and solid (ice) phases.
	Includes oceans, seas, ice sheets, sea ice, glaciers, lakes, rivers, streams, atmospheric moisture, ice crystals, and areas of permafrost.
	Also includes the moisture in the soil (soil water) and within rocks (groundwater).
	Can be sub-divided into the fluid water systems (the hydrosphere) and solid water (ice) systems (the cryosphere).
Geosphere	Composed of rocks and minerals.
	Includes the solid crust, the molten mantle, and the liquid and the core of the Earth.
	Includes the soil in which nutrients become available to living organisms.
	Provides an important ecological habitat and the basis of life forms.
	Subject to the processes of: (i) erosion; (ii) weathering; and (iii) transport, as well as to tectonic forces and volcanic activity.

The Science and Technology of the Environment. James G. Speight (Author)

The geosphere (also referred to as land systems and sometimes as the lithosphere) are those parts of the Earth which are not permanently covered by water (sometimes referred to as dry land) and upon which the human species, and many other animal species, live, and from which they extract most of their food and energy. Throughout history, the majority of human activity has occurred in land areas that support habitat and agriculture as well as various natural resources and varieties of flora and fauna (Table 5.2). A continuous area of land (landmass or landmass) surrounded by ocean may be designated as a continent or an island.

TABLE 5.2 Examples of Names for Various Areas of Land

Area	**Definition***
Acreage	An area of land measured in acres.
Common land	Land that everyone has a right to use.
Conservancy	A conservation area.
Conservation area	Land that is protected from being damaged.
Domain	Land owned and controlled by a particular person.
Enclosure	Land surrounded by a fence or wall.
Exclusion zone:	Land where a specified type of activity is not allowed.
Flatland	Level land without mountains, hills, or valleys.
Footprint	The amount of land that a person uses in order to exist.
Lowland	Land that is low and flat.
Permafrost	Ground that remains permanently frozen all year.
Terrain	An area of land, usually with a particular physical feature.
Territory	Land that belongs to or is used by someone.
Wasteland	An area of land that is empty or cannot be used.
Wilderness area	An area where the government has decided that roads or buildings cannot be built so that area can be enjoyed for the natural beauty habitat for animals.

*Can vary by region and by country.

In this respect, the land systems may be expected to be the ecosystems that receive the highest degree of anthropogenic stress. This is true, in part, but the water systems and the atmosphere are also subject to pollution and to single out any system as the major recipient of pollutants would be misleading, even erroneous.

The term land systems refers to all activities that occur on the land as well the human use of land and the benefits gained from land (Crossman et al., 2013; Verburg et al., 2013). Humans alter and modify the quantity and quality of the available land. In fact, changes to land systems are the direct result of human decision making and the impact of many local land system changes has far reaching consequences for the Earth system.

Areas where land meets large bodies of water (oceans) are referred to as the coastal zones and the demarcation line between land and water can vary by local jurisdiction. For example, a maritime boundary is a political demarcation (which can vary due to tides and weather). However, there are natural boundaries that can be used to define where water meets land, but, in general, there is no clear line of demarcation at which the land ends and a body of water begins.

Moreover, at times past and even in the present (although the perception is changing), land systems were believed to contain unlimited resources and would be able to sustain human activities and survive development beyond imaginable time. As resource development continued, it became evident that this was not the case. In fact, the original premise of the unchanging ecosystems is changing to a general consensus that accepts the original consensus to be false. Land systems are now recognized as being extremely fragile and have suffered irreparable damage as a result of human activities. As a result, the ability of the land systems to survive development by humans beyond imaginable time is now unacceptable.

However, there are two atmospheric phenomena that need to be given consideration: (i) the greenhouse effect; and (ii) the tendency for the deposition of acid rain. Both of these phenomena have the potential to cause changes in the flora and fauna that inhabit the land, as well as affect the life forms that are dependent upon the water in acidified lakes that can also result from such deposition. Moreover, this phenomena illustrate many of the complex relationships that can, and do, exist between the atmosphere (Chapter 3), the water systems (Chapter 4), and the land systems. A very general rule is that if one of the three systems is subject to pollution, at least one of the other two will also be polluted the others will also be polluted.

One of the greater impacts of humans upon land systems is the creation of desert areas (barren areas of land where there is little-to-no precipitation and, consequently, living conditions are hostile for floral and faunal life) by misuse/abuse of land where the yearly water deposition (rainfall or snow) borders on the arid/semi-arid allocation. The result will be a noticeable decrease in groundwater availability, a reduction in the availability of surface water, soil erosion, and not only a decrease in the types of vegetation but also a decrease in the total vegetation. The problem is severe in many parts of the world insofar as large, arid, and semi-arid areas of the world are experiencing the need for water to prevent desert formation.

Thus, with increasing population and industrialization, one of the more important aspects of the use of the land systems is concerned with the protection of water resources. Mining waste, agricultural waste, chemical waste, and radioactive waste are different types of contaminants, and there are limitations to the tolerance of the Earth for continual increases in the human population. Even though environmental issues related to the land systems may be discussed in generalities, there are some specific items that need to be understood.

Land systems are composed predominantly of four subsystems: (i) minerals; (ii) sedimentary strata; (iii) clays; and (iv) soil. Each subsystem is, in effect, a different physical and chemical system. Therefore, each subsystem can be expected to react differently to pollutants and/or contaminants. Thus, some knowledge of each of these subsystems is essential if the different modes of reaction of chemical waste are to be understood.

All of these systems are intertwined which can lead to harmful consequences for the land system. One specific example of interaction between all the spheres is human fossil fuel consumption and the combustion byproducts, such as carbon dioxide, end up in the atmosphere and they contribute to global warming, although there is relevant questions related to the precise exact contribution of fossil fuel-related carbon dioxide to global warming and climate change since the natural effect that also make a contribution to climate change are often ignored (Speight, 2020).

The many interactions between systems of the Earth are complex and are occurring constantly Extremely dramatic examples of the natural systems of the Earth-such as volcanic eruptions and tsunamis that are rarely taken into account when climate change is discussed-but there are also slow, nearly undetectable changes that alter ocean chemistry, the content of the atmosphere, and the microbial biodiversity in soil.

For the purposes of this book, the Earth is presented as an integrated system of three predominant sub-systems which are: (i) the atmosphere; (ii) the aquasphere; and (iii) the geosphere. Using a volcano as the example, the reactions between the Earth systems are complex and substantial (Table 5.3).

TABLE 5.3 Example of the Complex Interactions that Occur Within the Spheres of the Earth as a result of a Volcanic Eruption

Sphere	Results
Atmosphere	Volcanic emissions can contribute to blocking of the rays of the Sun.
	If photosynthesis is reduced, atmospheric concentrations of carbon dioxide can build up and stimulate global warming.
Aquasphere	Atmospheric sulfur dioxide and sulfur trioxide can combine with water to form sulfurous acid (H_2SO_3) and sulfuric acid (H_2SO_4).
Geosphere	Rain may bring the sulfurous and sulfuric acid into this sphere, thereby acidifying the soil.
Aquasphere	Rain may bring the sulfurous and sulfuric acid into this sphere, thereby acidifying the lakes and rivers and other water systems.
	Acid rain reduces the pH (to <7.0) of the water in lakes, rivers, and streams which results in a decrease in the growth of phytoplankton (microscopic marine algae that are the base of several aquatic food systems) and zooplankton (microscopic animals such as krill, sea snails, and pelagic worms, the young of larger invertebrates and fish, as well as weak swimmers such as jellyfish).

5.2 MINERALS

The land systems are divided into strata (layers) including: (i) the solid iron-rich inner core; (ii) the molten outer core; (iii) the mantle; and (iv) the crust. The latter (i.e., the crust) is the outer skin of the Earth and ranges in depth from 2 to 25 miles of the land systems that is accessible to humans. The crust consists of rocks-a rock is a mass of a pure mineral or an aggregate of two or more minerals.

A mineral is a substance that occurs naturally and is of elements and have a definite chemical composition and structure. Typically, using definitions from the geological sciences and mineralogical science as the guideline, a mineral or mineral species is a solid chemical compound that occurs naturally in pure form and has a well-defined chemical composition and a specific crystal structure.

The definition has been clarified by the International Mineralogical Association to classify a mineral that must: (i) be a naturally occurring substance formed by natural geological processes; (ii) be a solid substance in its natural occurrence; (iii) have a well-defined crystallographic structure; and (iv) have a well-defined chemical composition. A major exception to this rule is naturally-occurring mercury (Hg°) which is still classified as a mineral by the International Mineralogical Association, even though crystallizes only below –39°C (–38°F), it was included before the current rules were established,

The exceptions to this definition are the fossil fuels resources that are often classed (in the United States and Canada) among the mineral resources of the country, the state, or the province. However, the United States and Canada are not alone in this expansion of the definition of minerals; many other countries also class the fossil fuel as mineral resources.

Using the geological and mineralogical definition gives rise to more than 3,000 known minerals of which some are rare and precious such as gold and diamond, while others are more abundant, such as quartz (silica, SiO_2). Approximately 99% of the minerals that occur in the crust of the Earth are composed of two or more elements from a group of eight elements, most of which occur in combination with other elements (Tables 5.4 and 5.5).

TABLE 5.4 Elements in the Crust of the Earth*

Element	Symbol	% w/w in the Crust
Aluminum	Al	8
Calcium	Ca	3.5
Iron	Fe	5
Magnesium	Mg	2
Oxygen	O	47*
Potassium	K	2.5
Silicon	Si	28
Sodium	Na	3
All other elements		1

*Listed alphabetically rather than by occurrence.

**Not free oxygen but typically in rock-forming minerals made up of carbonates, oxides, and silicate groups such as the orthosilicate derivatives (SiO_4^{4-}) (see Table 5.5).

Minerals are naturally-occurring inorganic solids with well-defined crystalline structures resulting from definite internal structures and composition (Wenk and Bulakh, 2004; Mills et al., 2009; Rafferty, 2011). Minerals are present within many rocks. In fact, rocks may consist of one type of mineral or maybe an aggregate of two or more different types of minerals By far the most abundant, silicate minerals comprise approximately 90% of the crust of the Earth. Other important mineral groups include (alphabetically and not in order of abundance): carbonate derivatives ($–CO_3$), halide derivatives –Cl), native elements such as gold and silver, oxide derivatives (-O-), phosphate derivatives ($–PO_3$), sulfate derivatives ($–SO_4$), and sulfide derivatives (-S-).

TABLE 5.5 Names and Chemical Composition of Common Minerals*

Mineral Group	Example	Formula
Carbonates	Calcite	$CaCO_3$
	Dolomite	$CaCO_3.MgCO_3$ or $CaMg(CO_3)_2$
Halides	Fluorite	CaF_2
	Halite	NaCl
Oxides	Bauxite	Al_2O_3
	Haematite	Fe_2O_3
	Magnetite	Fe_3O_4
	Rutile	TiO_2
Silicates	Quartz	SiO_2
	Olivine	$(Mg.Fe_2)SiO_4$
	Feldspar (potassium)	$K.AlSi_3O_8$
Sulfides	Chalcocite	$CuFeS_2$
	Galena	PbS
	Gypsum	$CaSO_4.2H_2O$
	Pyrite	FeS_2
Native elements	Copper	Cu
	Gold	Au
	Silver	Ag
	Sulfur	S

*Listed alphabetically and not by occurrence.

A mineral is formed by a natural process-anthropogenic compounds are excluded from the definition of a mineral. In addition, a mineral is typically abiogenic and does not result from the activity of living organisms. Biogenic substances are chemical compounds produced by biological processes without a geological component and, therefore, are not typically classed as minerals.

The mineral is stable or metastable (i.e., in a state of equilibrium provided it is subjected to no more than small disturbances. Generally, a mineral can be represented by a chemical formula. Many mineral groups and species are composed of a solid solution-a solid solution is not commonly found in nature because of contamination or chemical substitution. For example, the olivine group is described by the variable formula Mg_2SiO_4 and Fe_2SiO_4 [sometimes written as $(MgFe)_2(SiO_4)_2$ which is a solid solution of two molecular species. Also, in a mineral, there must be an ordered atomic arrangement which is generally interpreted to mean crystalline. Thus, a more formal definition of a mineral is an element or chemical compound that is normally crystalline and that has been formed as a result of geological processes (Nickel, 1995).

The actual chemical composition of a mineral species may vary by the inclusion of small amounts of impurities. In fact some mineral species can have variable amounts of two or more chemical elements that occupy equivalent positions in the structure of the mineral. In addition, to the chemical composition and crystal structure of the mineral, the description of a mineral species usually includes common physical properties such as habit, hardness, color, parting, specific gravity, radioactivity, and the reaction of the mineral to acid.

Color is characteristic of many minerals and can vary widely due to the presence of impurities. Most minerals are silicates such as quartz (SiO_2) or potassium aluminum silicate (orthoclase, $KAlSi_3O_8$). Thus, oxygen, and silicon make up approximately 47% and 28% w/w, respectively, of the land systems. Other prominent elements are aluminum (8%), iron (5%), calcium (4%), sodium (3%), potassium (3%), and magnesium (2%) and titanium (0.4%). These nine elements make up approximately 99% w/w of the crust of the Earth.

The hardness-an important property of a-mineral is expressed using the Moh scale which is a criterion for comparing the resistance of the mineral to crushing. The scale ranges from 1 to 10 and is based upon minerals that vary from talc (hardness = 1), to diamond (hardness = 10). Another property is cleavage which, in the mineral sense, denotes the manner in which a mineral will break along planes and the angles in which these planes intersect. Most minerals fracture irregularly, although some minerals fracture along smooth curved surfaces or into fibers or splinters. Specific gravity, density relative to that of water, is another important physical characteristic of minerals.

Evaporites are soluble salts that precipitate from solution as, for example, the result of the evaporation of seawater. The most common evaporite is halite (sodium chloride, NaCl). Other simple evaporite minerals are sylvite (potassium chloride, KCl), the nardite (sodium sulfate, Na_2SO_4), and anhydrite (calcium sulfate, $CaSO_4$). Many evaporites are hydrates, including bischofite (also spelled bischofite, which is magnesium chloride hexahydrate, $MgCl_2.6H_2O$), gypsum (calcium sulfate dihydrate, $CaSO_4.2H_2O$), kieserite (magnesium sulfate monohydrate, $MgSO_4.\ H_2O$), and epsomite (magnesium sulfate heptahydrate, $MgSO_4.7H_2O$). Double salts, such as carnallite (potassium magnesium chloride hexahydrate, $KCl.\ MgCl_2.6H_2O$), polyhalite (potassium magnesium calcium sulfate dihydrate, $K_2SO_4.\ MgSO_4.2CaSO_4.2H_2O$), and loeweite (sodium magnesium sulfate pentadecahydrate, $6Na_2SO_4.7MgSO_4.15H_2O$) are common in evaporites. The solidification of molten rock (magma) produces igneous rock. Common igneous rocks are granite, basalt, quartz (SiO_2), magnetite (Fe_3O_4 or $FeO.\ Fe_2O_3$), and the feldspar minerals ($KAlSi_3O_8$, $NaAlSi_3O_8$, and $CaAl_2Si_2O_8$). Igneous rocks are formed under chemically-reducing conditions of high temperature and high pressure (Table 5.6).

TABLE 5.6 The Three Types of Rocks that Occur on the Surface of the Earth

Rock Type	Description
Igneous	Formed by the cooling and solidification of the molten rock material (magma).
Sedimentary	Developed as a result of gradual accumulation and hardening of mineral particles brought together by wind, water, or other effects. Forms distinct layers during the process of formation.
Metamorphic	Formed by the metamorphosis of igneous and sedimentary rocks which occurs under high pressure and extreme heat.

Exposed igneous rocks are not in chemical equilibrium with their surroundings and disintegrate by weathering. The extent of the weathering of a mineral depends upon: (i) time; (ii) chemical conditions such as exposure of the mineral to air, carbon dioxide, and water; as well as (iii) physical conditions, such as temperature and mixing with water and air. Weathering is generally a slow process, especially in the case of igneous rocks which are often nonporous, and of low reactivity. Erosion from wind, water, or glaciers can convert fragments from weathered rocks to sedimentary rock and soil, which are often porous and chemically reactive.

Carbonate minerals of calcium and magnesium (limestone, $CaCO_3$, or dolomite, $CaCO_3$. $MgCO_3$) are especially abundant in sedimentary rocks. Reactive and soluble minerals such as carbonates, gypsum, olivine, feldspars, and iron-rich substances cannot survive prolonged weathering. These substances are subject to advanced-stage weathering and also subject to leaching by freshwater, low pH, aluminum hydroxy oxides, and silica.

When sedimentary rocks contain organic material (such as the residues of plant and animal remains/decay) they are usually designated organic sedimentary rocks. Crude oil may be found in such sediments (Speight, 2014).

5.3 CLAY MINERALS

The term clay mineral refers to a type of fine-grained natural material that contains hydrous aluminum phyllosilicate derivatives (which are the actual clay minerals). Geologic clay deposits are mostly composed of phyllosilicate minerals containing variable amounts of water trapped in the mineral structure. Phyllosilicate minerals (sometimes referred to as sheet silicates, are an important group of minerals that includes the mica minerals, chlorite minerals, serpentine minerals, talc, and clay minerals.

Clay minerals (often simply referred to as clays) exhibit a form of plasticity due to particle size and geometry as well as water content but become hard, brittle, and non-plastic upon drying-especially when the water of crystallization in eliminated from the mineral, structure. Depending on the soil character in which the clay mineral exist, the clay mineral can appear in various colors which vary from near-white to dull gray or brown to deep orange-red. On the other hand, clay minerals (which some investigators would state incorrectly) are silicate minerals that usually contain some aluminum, are one of the most significant classes of secondary minerals. Although many naturally occurring deposits include both silts and clay minerals, the latter (the clay minerals) are distinguished from other fine-grained soils by differences in size and mineralogy (Table 5.7). Silts are fine-grained soils that typically have a larger particle size distribution than the clay minerals (Table 5.8).

More specifically, clay minerals often contain variable amounts of (alphabetically rather than by abundance) alkaline earth metals (i.e., any one or more of the six chemical elements that comprise Group IIa of the Periodic Table of the Elements, notably beryllium (Be), magnesium (Mg), calcium (Ca), strontium (Sr), barium (Ba), and radium (Ra), as well as alkali metals (such as sodium and potassium), iron, and magnesium (Table 5.9).

TABLE 5.7 General Groups of Clay Minerals

Group	Description
Kaolin	Includes kaolinite, halloysite minerals and nacrite.
	Polymorphs (different chemical structures) of $Al_2Si_2O_5(OH)$.
Smectite	Includes dioctahedral smectites such as montmorillonite, nontronite, beidellite, and trioctahedral smectites such as saponite.
Illite	Includes the clay-micas minerals.
	Illite is the only common mineral.
Chlorite	Includes a wide variety of similar minerals with considerable chemical variation.
Others	Includes palygorskite (also known as attapulgite and sometimes referred to as Attapulgus clay) and sepiolite.
	Clay minerals with long water channels internal to the structure.

TABLE 5.8 Approximate Sizes of the Various Rock Types

Name	Diameter (Millimeters)
Boulder	>256
Cobble	64 to 256
Pebble	2 to 64
Sand	1/6 to 2
Silt	1/256 to 1/6
Clay	<1/256

TABLE 5.9 Periodic Table of the Elements

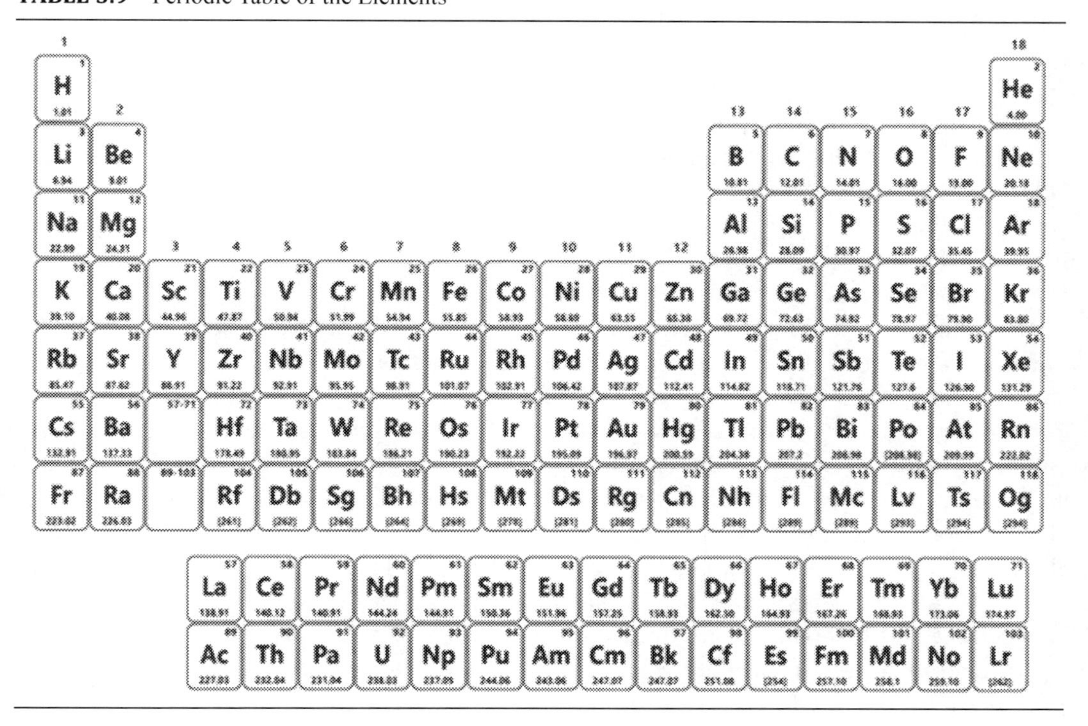

The weathering process-the primary process by which clay minerals are formed-involves the physical disintegration as well as the chemical decomposition of rocks. The weathering process is an uneven process and factors governing rock weathering and soil formation include: (i) the initial type of rock; (ii) the ratio of water to rock; (iii) the temperature; (iv) the presence of organisms and organic material; and (v) the time. The types of clay minerals found in weathering rocks strongly control the behavior of the weathered rock under various climatic conditions (such as humid-tropical, dry-tropical, and temperate conditions).

In terms of the time involved in the formation of clay minerals, these minerals typically form over long periods of (geologic) time as a result of the weathering of rocks (usually silicate-bearing rocks) by low concentrations of, for example, carbonic acid (H_2CO_3). This acidic chemical usually migrates through the rock after leaching through upper weathered layers. In addition to the weathering process, some clay minerals are formed through hydrothermal activity (geothermal activity in the presence of water).

In addition, there are two types of clay deposits: which are described as: (i) primary deposits; and (ii) secondary deposits. In the former types of deposits (i.e., primary deposits), the clay minerals (frequently referred to as primary clay minerals) were formed as residual deposits in the soil and remain at the site of formation. In the latter types of deposits (i.e., secondary deposits) the clay minerals (frequently referred to as secondary clay minerals) are clay minerals that have been transported from their original location by water erosion and deposited as a new sedimentary deposit.

Another process that is active in the formation of clay minerals is diagenesis, which is the in-place (in situ) alteration of a clay mineral (or any mineral) to a more stable form, excluding surficial alteration (such as weathering). The diagenetic process occurs, for example, when a mineral that is stable in one depositional environment is exposed to other depositional factors by burial and compaction.

Clay minerals are defined by size, as are other inorganic constituents of the Earth (Table 5.8). Secondary minerals are formed by the alteration of parent mineral matter. Olivine, augite, hornblende, and feldspar minerals all form clay minerals. These materials constitute the most important class of common minerals occurring as colloidal matter in water. Clay minerals consist largely of hydrated aluminum and silicon oxides (Figure 5.1) (Speight, 1996) and are secondary minerals.

The most abundant clay minerals are illite, montmorillonite, chlorite, and kaolinite. These clay minerals are distinguished from each other by complex chemical formulas, structure, as well as the chemical and physical properties. The three major groups of clay minerals are montmorillonite [$Al.\ Mg.\ Na.(OH)_2.\ Si_4O_{10}$], illite [$Al_3.\ K.\ Si_8O_{10}.(OH)_4$], and kaolinite [$Al_2Si_2O_5(OH)_4$]. Many clay minerals contain large amounts of sodium, potassium, magnesium, calcium, and iron, as well as trace quantities of other metals. Physically, clay minerals consist of sheet-like layered structures consisting of sheets of silicon oxide alternating with sheets of aluminum oxide units of two or three sheets make up a unit layer. Some clay minerals, particularly montmorillonite, may absorb large quantities of water between the unit layers that is accompanied by swelling of the clay.

Structurally, the silicon oxide sheet is composed of tetrahedra in which each silicon atom is surrounded by four oxygen atoms. Of the four oxygen atoms in each tetrahedron, three are shared with other silicon atoms that are components of other tetrahedra. On the other hand, the oxide is contained in an octahedral sheet in which each aluminum atom is surrounded by six oxygen atoms in an octahedral configuration. The overall structure is such that some (but not all) of the oxygen atoms are shared between aluminum atoms while the other oxygen atoms are shared within the tetrahedral sheet. In another aspect of the structure of the clay minerals, some of the minerals may be classified as either two-layer clay minerals in which oxygen atoms are shared between a tetrahedral sheet and an adjacent octahedral sheet, and three-layer clay minerals, in which the oxygen atoms in an octahedral sheet are shared with oxygen atoms in the tetrahedral sheets on either side of the octahedral sheet.

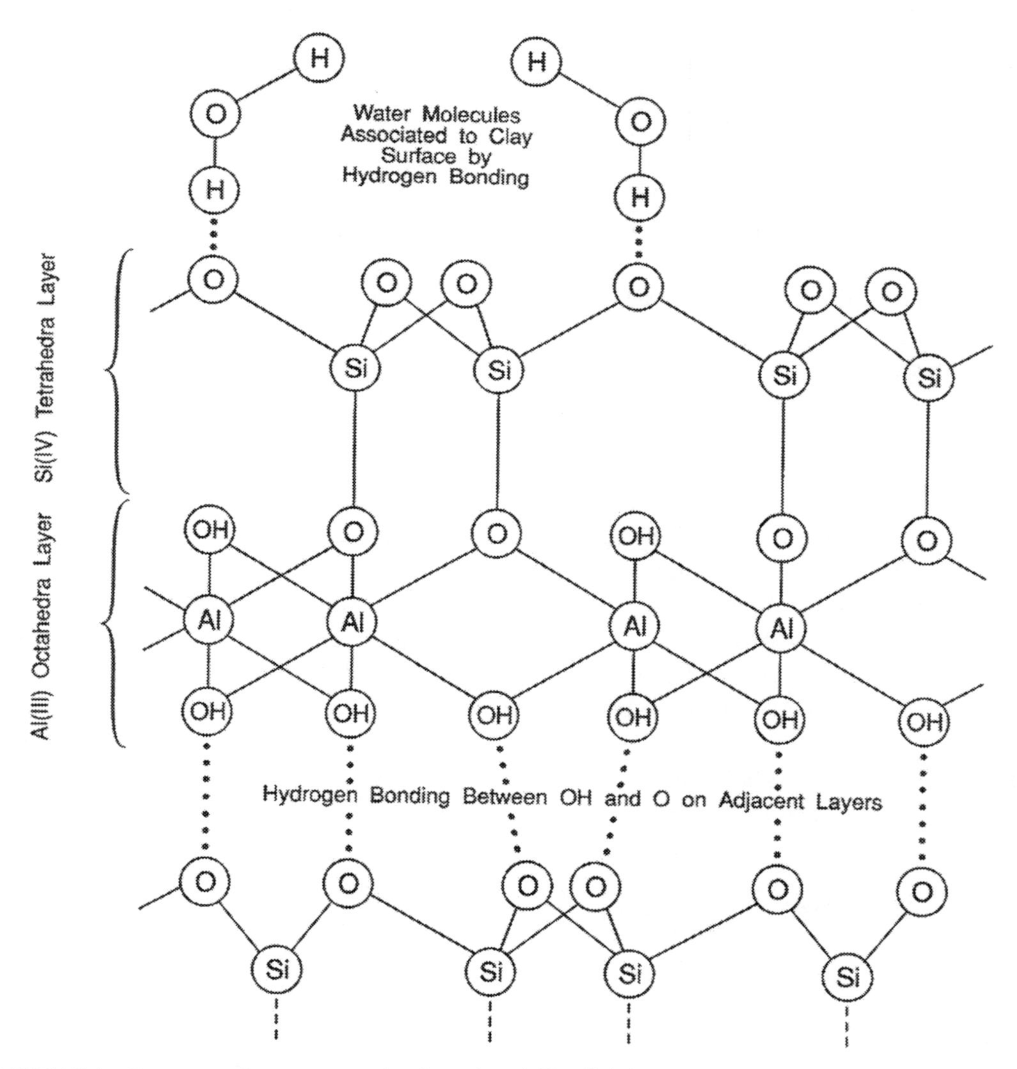

FIGURE 5.1 Representative structure of a clay mineral (Kaolinite).
Source: Reproduced with permission from: Speight (1996). © Taylor and Francis.

Because of the structure and high surface area per unit weight, clay minerals have a strong tendency to adsorb chemical species from water. Thus, because of their physical and chemical reactivity with organic and inorganic substrates (Theng, 1974; Anderson and Rubin, 1981), clay minerals play a major role in the transport and reactions of biological waste, organic chemicals, gases, and other pollutant species in water. In addition, clay minerals also may effectively immobilize dissolved chemicals in water and so exert a purifying action with the pollutant remaining adsorbed on the mineral thereby purifying the water or at least removing some of the pollutants from the water.

Clay minerals are typically nontoxic-although swallowing a large amount of a clay mineral is not advised without medical consultation! The turn of the 21st Century awakened a renewed interest in the utility of clay minerals mineral which are applicable in the area of environmental control. These minerals can be used in: (i) bioremediation processes; (ii) adsorption processes; and (iii) redox processes. Surface area, cation exchange capacity (CEC), surface acidity, and interlayer cations play a vital role in selection of clay minerals for any of the above remediation processes.

Therefore, clay minerals have a role to play in the environment (Aboudi et al., 2017) and, in fact, clay minerals bind cations such as calcium, magnesium, and ammonium (NH_4^+), which protects the cations from leaching by water and, hence, these cations remain in the soil as plant nutrients. Also, since many clay minerals can be suspended in water as colloidal particles, they may be leached from soil or carried to lower soil layers. In the current context of clay minerals in environmental issues, clay minerals exhibit plasticity when mixed with water in certain proportions. However, when dry, the clay mineral becomes firm, and when fired in a kiln, permanent physical and chemical changes occur which convert the clay into a ceramic material.

In summary, the characteristics that are common to all clay minerals derive from the chemical composition, layered structure, and size. The process by which some clay minerals swell when they take up water is reversible. Swelling clay expands or contracts in response to changes in environmental factors: namely: (i) wet and dry conditions; and (ii) temperature. Hydration and dehydration can vary the thickness of a single clay particle by almost 100%; for example, a 10Å-thick clay mineral can expand to 19.5Å in water (Velde, 1995). Another important property of clay minerals is the ability of the minerals to exchange ions because of the charged surface of the clay minerals. Ions can be attracted to the surface of a clay particle or taken up within the structure of these minerals. This property allows clay minerals to be an important vehicle for transporting (and widely dispersing contaminants) from one area to another.

5.4 SEDIMENTARY STRATA

A rock is a naturally-formed coherent aggregate mass of solid matter that constitutes part of the Earth and there are three families of rocks: (i) igneous rocks; which formed from the cooling and consolidation of magma or lava; (ii) sedimentary rocks, which are formed from either chemical precipitation of material or deposition of particles transported in suspension; and (iii) metamorphic, rocks which are formed from changing a rock as a result of high temperature, high pressure, or both. It is the second of the three types of rocks (the sedimentary rocks) to which this section is devoted.

Sedimentary strata (also generally referred to as sediments) typically consist of mixtures of clay minerals, silt, sand, organic matter, and various minerals. The strata may vary in composition from a stratum predominantly composed of mineral matter to a stratum composed predominantly organic matter-a notable example of this later type of stratum is coal. The properties and type of material in a stratum depends upon the origin and transport of the material that eventually form the stratum.

Water is the main vehicle of sediment transport, although wind can also be significant but various physical processes, chemical processes, and biological processes can result in the deposition of sediments in the base of a water system. Sedimentary material may be washed into a of water system by erosion or through sloughing (caving in) of the shore. Thus, many minerals, inorganic pollutants, organic pollutants, algae, and bacteria exist in the form of small particles suspended in water. These types of particles have the characteristics of soluble materials and larger particles in suspension, which range in diameter from approximately 0.001 micron (1 micron = 1 meter $\times$ 10^{-6}) to approximately 1 micron. Moreover, materials that from sediments may be carried by flowing water as: (i) dissolved material in solution; (ii) suspended material that is carried by the water in suspension; and (iii) bed material that is dragged along the bottom of the stream channel by the flowing water. In fact, bottom sediments are important sources of inorganic and organic matter in streams, freshwater impoundments, estuaries, and oceans. In fact, it is incorrect to consider bottom sediments simply as wet soil. Soil is in contact with the atmosphere and is aerobic, whereas generally, the environment around a bottom sediment is anaerobic in which case sediment is subject to reducing (non-oxidative) conditions. Also, bottom sediments undergo continuous leaching, whereas soil is not subject to such an effect.

Sedimentary rocks form as a stratum (or layer) that was deposited over time because of physical or chemical effect that caused the rock to form particles that then deposited in the layer. A sequence of strata (pl: layers; singular: stratum or layer) deposited without interruption is a conformable layer, but if there are breaks in the layer that represent times of non-deposition or erosion, the term to be applied is unconformity.

Thus, stratification is the layering that occurs in most sedimentary rocks and in at the surface of the Earth typically from fragmental deposits. The layers can range in thickness from fractions of an inch to several feet (or even to several yards) and vary in shape. The strata may also range in length and breadth from several feet that cover s few square feet to converge of many square miles; there are also thick lens-like bodies that extend (laterally) only a few feet.

Thus, stratigraphy is the branch of geology that focuses on the study of strata (rock layers) and layering (stratification). Stratigraphy has two related subfields: (i) lithostratigraphy (also referred to as lithologic stratigraphy) and (ii) biostratigraphy (also referred to as biologic stratigraphy). The former subfield (lithostratigraphy) focuses on the variation in rock units that displayed as visible layering and is due to physical contrasts in the rock type (differences in lithology) which can occur: (i) vertically as layering which is also called bedding; or (ii) laterally, and reflects changes in the environments of the deposition and is referred to as facies change.

The planes of parting (i.e., the separation planes or stratification planes) between the individual rock layers are horizontal where sediments are typically deposited as flat-lying layers but will exhibit inclination where the depositional site was a sloping surface. The bottom surface of a stratum generally conforms to irregularities of the underlying surface, but the stratification plane above the deposited material tends to be on a horizontal plane.

Key concepts in the study of stratigraphy involve an understanding of the means by which geometric relationships between rock layers arise and what the meaning of any such geometry in relation to the original depositional environment. The basic concept in stratigraphy (the law of superposition) requires that in an undeformed stratigraphic sequence, the oldest strata occur at the base of the sequence. Also, the term chemostratigraphy relates to any changes in the relative proportions of the elements (particularly the trace elements and the isotopes of the elements) within and between lithologic units. Another term, cyclostratigraphy, relates to the cyclic changes in: (i) the relative proportions of the minerals, particularly the carbonate minerals; (ii) the grain size of the minerals, (iii) the thickness of sediment layers, also called the vavres; (iv) any fossil diversity with time; (v) any seasonal or longer-term changes in the prevalent paleoclimate (Christopherson, 2008). All of these subcategories can assist in an understanding of the manner in which an environmental cleanup of a chemical spill is necessary.

5.4.1 INORGANIC MATTER

The inorganic component (mineral matter) of the soil is composed of a variety of minerals, any one of which can influence the properties of the soil. Moreover, the inorganic matter in soil is composed of rock particles that have been broken down into smaller particles that vary in size. For example: (i) sand particles are on the order of 0.1 to 2 mm in diameter; (ii) particles on the order of 0.002 and 0.1 mm in diameter are called silt; and (iii) clay particles that are on the order of 0.002 mm (or less) in diameter. However, some soils have no dominant particle size and contain a mixture of sand, silt, and humus; this type of soil is generally referred to as loam.

Finally, the most common elements in the crust of the Earth are oxygen (as mineral oxides), silicon (Si), aluminum (Al), iron (Fe), calcium (Ca), sodium (Na), potassium (K), and magnesium (Mg). Therefore, minerals composed of these elements (particularly silicon and oxygen) constitute most of the mineral fraction of the soil. Common soil mineral constituents are: (i) finely divided quartz (SiO_2); (ii) orthoclase ($KAl.\ Si_3O_8$); (iii) albite [$Na.\ Al_3Si_8)_8O$]; (iv) epidote [$4CaO_3(Al.\ Fe)_2O_3.6SiO_2.\ H_2O$]; (v) goethite [FeO(OH)]; (vi) magnetite (Fe_3O_4); (vii) calcium carbonate ($CaCO_3$); (viii) dolomite, the mixed calcium and magnesium carbonates ($CaCO_3.\ MgCO_3$); and (ix) the oxides of manganese (MnO) and titanium (TiO, titanium monoxide).

5.4.2 NUTRIENTS

An important function of soil is to provide the essential nutrients for plants (Table 3.5) which can be in the form of: (i) macronutrients which are the elements that occur in standard levels in plant materials or in fluids in the plant; and (ii) micronutrients which are the elements that are essential at lower levels and are generally required for the functioning of essential enzymes.

The elements that are generally recognized as essential macronutrients for plants are carbon, hydrogen, oxygen, nitrogen, phosphorus (P), potassium (K), calcium (Ca), magnesium (Mg), and sulfur (S). The former three elements (carbon, hydrogen, and oxygen) are obtained from the atmosphere while the other essential macronutrients are obtained from the soil—of these, nitrogen, phosphorus, and potassium are the most likely to be lacking and are commonly added to the soil as fertilizers.

In the uncommon likelihood that a soil is calcium-deficient, a process known as liming used to treat acid soils by the addition of lime (CaO), can provide an adequate supply of calcium for the plant life. However, calcium uptake by plants and leaching of the calcium by carbonic acid (H_2CO_3, formed from carbon dioxide and water) may produce a calcium deficiency in the soil. Acid soil ($pH < 7.0$) may still contain an appreciable level of calcium which is not available to plants, but treatment of an acid soil to restore the pH to near-neutrality ($pH = 7.0$) can mitigate the calcium deficiency. In a soil that is alkaline ($pH > 7.0$), the presence of high levels of sodium (Na), magnesium (Mg), and potassium (K) can produce a calcium deficiency because these ions (sodium, magnesium, and potassium) can compete with the calcium for availability to plants.

Although magnesium makes up approximately 2% w/w of the land system, most of the calcium is in the form of minerals. Generally, exchangeable magnesium is considered available to plants and is held by ion-exchanging organic matter or clay minerals. The availability of magnesium to plants depends upon the calcium/magnesium ratio in the soil. If this ratio is high in favor of calcium, magnesium may not be available to plants with the consequence of a magnesium deficiency results. Similarly, excessive levels of potassium or sodium may also cause deficiency of the magnesium that is available for plant life.

Sulfur is assimilated by plants as the sulfate ion (SO_4^-), but in areas where the atmosphere is contaminated with sulfur dioxide (SO_2), the leaves of plants may absorb the sulfur as sulfur dioxide. Soil that is deficient in sulfur does not support plant growth because sulfur is a component of some essential amino acids and of thiamin (thiamine, vitamin B1) and biotin (part of the vitamin B family, also known as vitamin H). Sulfate ions are generally present in the soil as insoluble and immobilized sulfate-containing minerals or as soluble salts, which are readily leached from the soil and lost as water runoff.

Nitrogen, phosphorus, and potassium are plant nutrients that are obtained from soil and are important for crop productivity. In most soils, more than 90% w/w of the nitrogen is organic and participates in the nitrogen cycle (Chapter 1), and it is this organic nitrogen that is primarily the product of the biodegradation of dead floral species and faunal species. The organic nitrogen in the soil is eventually hydrolyzed to ammonia (NH_3) which can be oxidized to nitrate ($-NO_3$) by the action of bacteria in the soil.

Nitrogen fixation is the process by which atmospheric nitrogen is converted to nitrogen compounds available to plants, i.e.:

$$N_2 \rightarrow NO_3^-$$

Artificial sources (such as human activities) account for 30 to 40% of all nitrogen fixed and these include: (i) chemical fertilizer manufacture; (ii) nitrogen available for fixation as a result of fuel combustion as well as the combustion of nitrogen-containing fuels; and (iii) the increased cultivation of nitrogen-fixing legumes. An issue that can arise as a result of the increased fixation of nitrogen is the possible effect upon the atmospheric ozone layer by nitrogen dioxide released during denitrification of fixed nitrogen.

Although the percentage of phosphorus in plant materials is relatively low, it is an essential component of plants. Phosphorus, like nitrogen, must be present in a simple inorganic form before it can be taken up by plants. In the case of phosphorus, the utilizable species is some form of orthophosphate (PO_4^{3-}) which is most available to plants at pH values near neutrality. It is believed that in relatively acidic soils, orthophosphate ions are precipitated or sorbed by species of aluminum (Al^{3+}) and iron (Fe^{3+}).

Potassium (K) is one of the most abundant elements and constituents and approximates 3% w/w of the crust of the Earth. However, much of this potassium is not easily available to plants-some silicate minerals contain bound potassium that is difficult to release but exchangeable potassium held by clay minerals is relatively more available to plants. Relatively-high levels of potassium are utilized by growing plants because of the activation (by potassium) of some enzymes as well as playing a role in the water balance in plants, and it is also essential for the transformation of carbohydrates.

Boron (B), chlorine (Cl), copper (Cu), iron (Fe), manganese (Mn), molybdenum (Mo, for nitrogen-fixation), sodium (Na), vanadium (V), and zinc (Zn) are essential plant micronutrients at low concentrations. However, these elements-like many chemicals-are frequently are toxic at higher concentrations levels. In addition, manganese, iron, chlorine, zinc, and vanadium may be involved in photosynthesis.

Iron and manganese occur in a number of soil minerals while other micronutrients and trace elements are found in primary (non-weathered) minerals that occur in soil. Boron is substituted for silicon in mica (any of a group of hydrous potassium, aluminum silicate minerals) and is present in the mineral tourmaline (a crystalline boron silicate mineral compounded with elements such as aluminum, iron, magnesium, sodium, lithium, or potassium and which is also classified as a semi-precious stone). Copper is present in feldspar, amphibole, olivine, pyroxene, and mica as well as occurring at trace levels in silicate minerals. Molybdenum occurs as molybdenite (molybdenum disulfide, MoS_2) while vanadium is substituted for iron or aluminum in oxides, pyroxenes, amphiboles, and mica. Zinc also occurs naturally as the result of isomorphic substitution for magnesium, iron, and manganese in oxides, amphiboles, olivines, and pyroxenes and as traces of zinc sulfide in silicates. Other trace level elements that occur as specific minerals, sulfide inclusions, or by isomorphic substitution for other elements in minerals are chromium, cobalt, arsenic, selenium, nickel, lead, and cadmium.

5.4.3 ORGANIC MATTER

The organic material in soil (humus) is made up of microorganisms (dead and alive), and dead animals and plants in varying stages of decay and improves soil structure, providing plants with water and minerals although some soils, such as peat soils, may contain as much as 95% w/w organic material. Though typically comprising approximately less than 5% w/w of a productive soil, the organic matter in the soil (Table 3.4) determines the productivity of the soil because it (the organic matter): (i) serves as a source of food for microorganisms; (ii) undergoes chemical reactions such as ion exchange; and (iii) influences the physical properties of soil. Some organic compounds even contribute to the weathering of mineral matter, the process by which soil is formed.

Humus, which is derived from lignin, is a generic term for the water-insoluble material that makes up the bulk of soil organic matter (Stevenson, 1994). Humus is composed of a base-soluble: fraction (humic and fulvic acids) and an insoluble fraction referred to as humin, which is the insoluble component of soil organic matter that remains after extraction of the other components of the soil organic matter that are soluble in aqueous base) and is the residue from the biodegradation (by bacteria and fungi) of plant material. The bulk of plant biomass consists of biodegradable cellulose and degradation-resistant lignin, a complex polymeric substance that is second only to carbohydrates in natural abundance (Sarkanen and Ludwig, 1971).

An increase in the atomic nitrogen/carbon ratio is a feature of the transformation of plant biomass to humus. During the process, microorganisms convert organic carbon to carbon dioxide to obtain energy. Simultaneously, the bacterial action incorporates bound nitrogen with the compounds produced by the decay processes. The result is a nitrogen/carbon ratio of approximately 1/10 upon completion of the humification process.

Humic materials in soil strongly adsorb many solutes in soil water and have a particular affinity for heavy polyvalent cations. Thus, water becomes depleted of cations (or purified) in passing through humic-rich soils. Humic substances in soils also have a strong affinity for organic compounds with low water-solubility such as the types of herbicides widely used to kill weeds.

In some cases, there is a strong interaction between the organic and inorganic portions of soil, and this is especially true of the strong complexes formed between clay minerals and humic (fulvic) acid compounds-fulvic acids are a family of organic acids that are components of the humus. In many-soils, the majority (50 to 100%) of the soil carbon is in the form of complexes with clay minerals which play a role in determining: (i) the physical properties of soil; (ii) the fertility of the soil; and (iii) the stabilization of soil organic matter. One of the mechanisms for the chemical binding between the colloidal particles of clay minerals and humic organic particles (which are probably of the flocculation type) in which anionic organic molecules interact with carboxylic acid functional groups and serve as bridges in combination with cations to bind clay colloidal particles together as a floc.

5.4.4 SOIL

Soil is an important aspect of any environmental studies because, on the land, the soil is often the major recipient of any spilled chemicals(s). Soil covers most of the land surfaces as a continuum and each soil grade into the rock materials and into other soils at its margins, where changes occur in groundwater, vegetation, types of rocks, or other factors which influence the development of soil. The organic and inorganic material in which soils form is the parent material and soil can be conveniently subdivided into various types, each with different composition and properties (Tables 5.10 and 5.11).

TABLE 5.10 Types of Soils Based on Composition

Classification	% w/w Clay	% W/W Silt	%w/w Sand
Clay soil	40–100	0–40	0–45
Loam soil	7–27	28–50	23–52
Sandy soil	1–10	1–15	85–100%

TABLE 5.11 Types of Soils Based on Properties

Soil Type*	Descriptions
Alluvial soil	Formed as a result of deposition by the rivers and streams; lacks a well-developed profile; supports luxuriant growth of vegetation.
Chernozemic soil	Occurs in humid to semi-arid temperate climatic conditions under grasses.
Desertic soil	Found in arid climatic conditions.
Latosolic soil	Develops in humid tropical or semitropical-forested regions.
Mountain soil	Occurs in hilly regions under colder climates.
Podzolic soil	Also called podzolonic soil.
	Found in humid temperate climate, under forest vegetation.
Saline soil	Occurs in the dry climates where rapid evaporation of water results in surface deposition.
Tundra soil	Develops in colder regions and under vegetation such as lichens, mosses, herbs, and shrubs.

*Listed alphabetically.

Soil is a complex mixture of minerals, organic matter, as well as water (Figure 5.2) (Speight, 1996) and is the final product of the physical, chemical, and biological processes on rocks. In engineering terms, soil is included in the broader concept of regolith (the layer of unconsolidated rocky material covering the bedrock).

The organic portion of soil consists of plant biomass in various stages of decay involving the populations of bacteria, fungi, and animals such as earthworms. In fact, the presence of living organisms (which can produce pores and crevices) greatly affects soil formation and structuralisms. Plant roots can penetrate into crevices to produce more fragmentation and plant secretions promote the development of microorganisms around the root in an area known as the rhizosphere-more specifically, the region of soil in the vicinity of plant roots in which the chemistry and microbiology is influenced by their growth, respiration, and nutrient exchange. Additionally, the decomposition of the leaves and other material that fall from plants contribute to soil composition.

The formation of soil involves the formation of unconsolidated materials by the weathering which leads to the phenomenon known as the soil profile. Soil is a product of several factors some of which are (i) the influence of climate, (ii) elevation, orientation, and slope of terrain, (iii) organisms, and (iv) the parent materials of the soil (i.e., the original rocks and minerals) interacting over time. The influence of the climate (weathering) involves the physical disintegration and chemical decomposition of the rocks and minerals contained within the rocks and minerals occurs. The physical disintegration process involves the breakdown of the rocks into smaller fragments and eventually into sand and silt particles that are commonly made up of individual minerals. Simultaneously, the decomposition of the minerals occurs and soluble materials are released and new minerals are often created. The new minerals form either by minor chemical alterations of the original minerals or by a complete chemical breakdown of the original minerals. Based on the location of soil mineral particles formation and deposition, the soils are classified as (i) residual soil and (ii) transported soil.

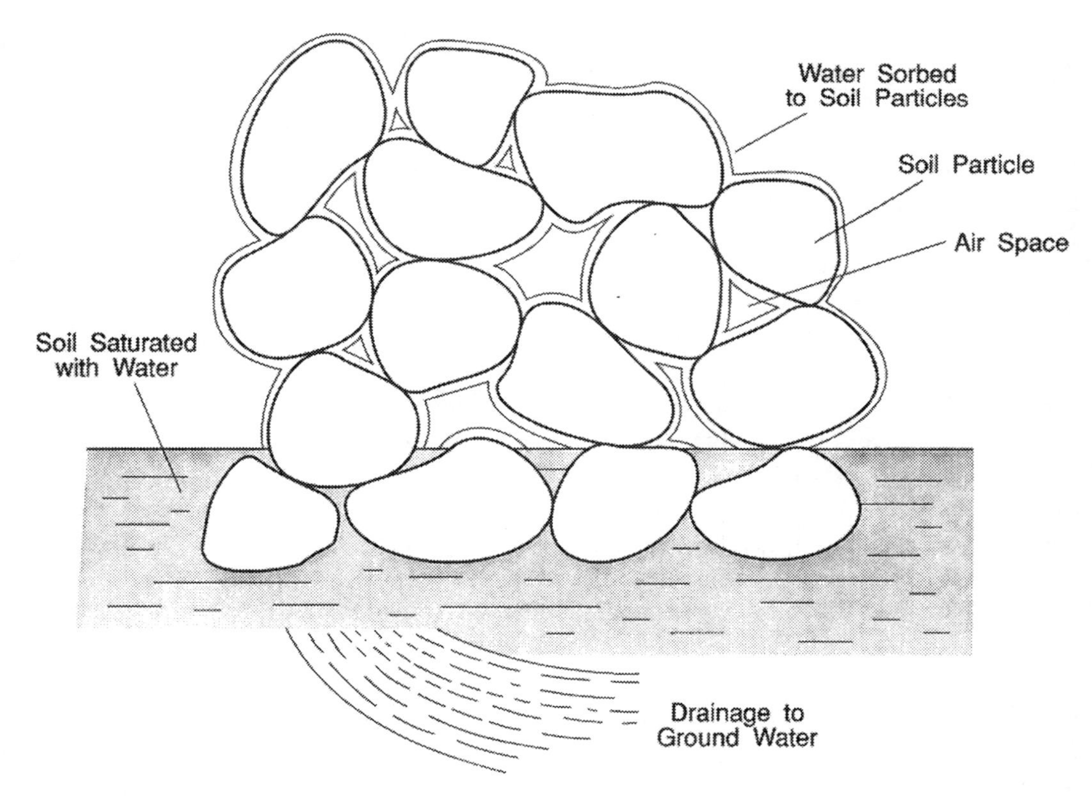

FIGURE 5.2 Representation of soil structure.

Source: Reproduced with permission from: Speight (1996). © Taylor and Francis.

Residual soil is the soil that has been formed in place while on the other hand, transported soil is the soil in which the mineral particles have been carried (transported) from some other location by wind, water, ice, or gravity. Furthermore, the transported soil can be classified (alphabetically) into (i) alluvium, which is soil that has been transported by the movement of water, (ii) colluvium, which is soil that has been transported by gravity, (iii) eolian soil, which is soil that has been transported by wind, and (iv) glacial soil, which is soil that has been transported by the movement of glaciers (Kumar and Mina, 2018).

Mineral soils are soils that are formed directly from the weathering of bedrock and these types of soils have a similar composition to the original rock. Other soils form in materials that came from elsewhere, such as sand and glacial drift. Materials located in the depth of the soil are relatively unchanged compared with the deposited material. Sediments in rivers may have different characteristics, depending on whether the stream moves quickly or slowly. A fast-moving river could have sediments of rocks and sand, whereas a slow-moving river could have fine-textured material, such as clay.

Thus, soil is a mixture of organic matter, minerals, gases, and liquids, and organisms that together support life on the Earth (Voroney and Heck, 2007; Chesworth, 2008; Jazib, 2018). The soil on the Earth (the pedosphere) has four important functions which are (i) a medium for plant growth, (ii) a medium for water storage, supply, and purification, (iii) a modifier of the atmosphere of the Earth, and (iv) a habitat for organisms. The soil is the most important part of any land system insofar as it is the medium upon which the majority of the terrestrial life forms depend upon the soil for their existence. In addition, the productivity of soil is affected by environmental conditions and pollutants. Also, the soil continually undergoes development by way of numerous physical, chemical, and biological processes, which include weathering and the associated erosion.

Most soils have a dry bulk density (the density of the soil that takes into account the voids in dry soil) on the order of 1.1 and 1.6 g/cm^3, while the soil particle density is much higher, typically in the range of 2.6 to 2.7 g/cm^3.

Soil contamination is due to the discharge of chemical waste from landfill leachate, lagoons, and other sources as well as to contaminants from domestic, industrial, and agricultural activities. More specifically, contamination of the soil involves two different processes which are: (i) the slow but steady degradation of soil quality, such as organic matter, nutrients, water-holding, capacity, porosity, purity as well as to contaminants such as domestic and industrial wastes or chemical waste from agricultural activities; and (ii) the concentrated pollution of smaller areas, mainly through dumping or leakage of wastes. In some cases, land farming of degradable organic wastes is practiced as a means of disposal and degradation in which microbial processes cause the degradation of the material, as for example by the application of sewage and fertilizer-rich sewage sludge to soil.

Physically, approximately 35% v/v of a typical soil is composed of air-filled pores. As a point of reference, the atmosphere at sea level contains 21% v/v oxygen and <0.1% v/v carbon dioxide but these amounts may be quite different in the air in the soil because of the decay of organic matter which produces carbon dioxide by consumption of oxygen. As a result, the oxygen content of air in soil may be as low as 15% v/v and the carbon dioxide content may be on the order of several percent.

To describe soil, it is preference; to use the phenomenon known as the soil profile. The description of the soil profile commences at the underlying rock which supplies the parent material for the formation of soil, a series of layers are developed. Each layer is referred to as soil horizon (layer), and there are five main types of horizons in soil which are denoted as the O, A, B, C, and R horizons. Each horizon is different from the others in terms of the properties, such as color, texture, structure, consistency, porosity, and reactivity.

The O-horizon is the topmost portion of soil and consists of organic matter and is commonly found in forest soils. It is further subdivided into: (i) the O_1 horizon, which is the organic horizon in which the original forms of plant and animals residues can be recognized by the naked eye; the O_2 horizon in which the original plant and animal forms cannot be distinguished by the naked eye; (ii) the A-horizon, which is the mineral horizon that lies at or near the surface and is recognized as the zone of maximum leaching or eluviation-which is the sideways or downward movement of dissolved or suspended material within soil caused by rainfall while illuviation is (geologically) the accumulation of suspended material and soluble compounds that have been leached from an overlying stratum; (iii) the B-horizon which is the layer below the A layers in which there is the accumulation of iron and aluminum oxide and silicate clay minerals the occurs and is sometimes also referred to as subsoil and is further divisible in B1, B2, and B3 horizons. The R Horizon represents the bedrock from which the other horizons may have originated (Sigel and Sigel, 1995).

Typically, the soil profile is shown as the A, B, C, and R horizons (Figure 5.3) (Speight, 1996) which are typically parallel (or close to parallel) to the surface of the Earth and maybe thick or thin. In general, the boundary of soils with the underlying rock or rock material occurs at depths ranging from 1 foot to 6 feet, though there are extremes that lie outside of this range.

One of the more important classes of productive soils is the podzol-type of soil (podzolonic soil) which is a soil that is formed in temperate-to-cold climates under relatively high rainfall conditions under coniferous or mixed forest or heath vegetation. These generally rich soils tend to be acidic (pH = 3.5 to 4.5) such that alkali and alkaline earth metals and, to a lesser extent, aluminum and iron, are leached from the A horizon, leaving kaolinite as the predominant clay mineral. At somewhat higher pH in the B horizon, hydrated iron oxides and clay minerals are redeposited.

The mechanical properties of soil are largely determined by particle size. The four major categories of soil particle sizes are: (i) gravel; 2 to 60 mm; (ii) sand; 0.06 to 2 mm; (iii) silt, 0.06 to 0.006 mm; and (iv) clay minerals, <0.002 mm, which represent a size fraction rather than a specific class of mineral matter. Soil receives large quantities of contaminants as a result of a variety of activities. Pesticides (Chapter 6) are a particular contaminant because of their application to crops (Bourke et al., 1992).

5.4.5 WATER

Although not usually recognized as a nutrient, water (Chapter 4) is important for the condition of the soil and the maturation of the flora and fauna (Kramer and Boyer, 1995). Therefore, in the present context, water can also be classed as a nutrient (Chapter 3).

Typically, because of the small size of soil particles and the presence of small capillaries and pores in the soil, the water phase is not totally independent of the solid matter that comprises the soil. In fact, the availability of water to the plants is governed by gradients arising from capillary action and from gravitational forces. The availability of nutrient solutes in water depends upon concentration gradients and electrical potential gradients.

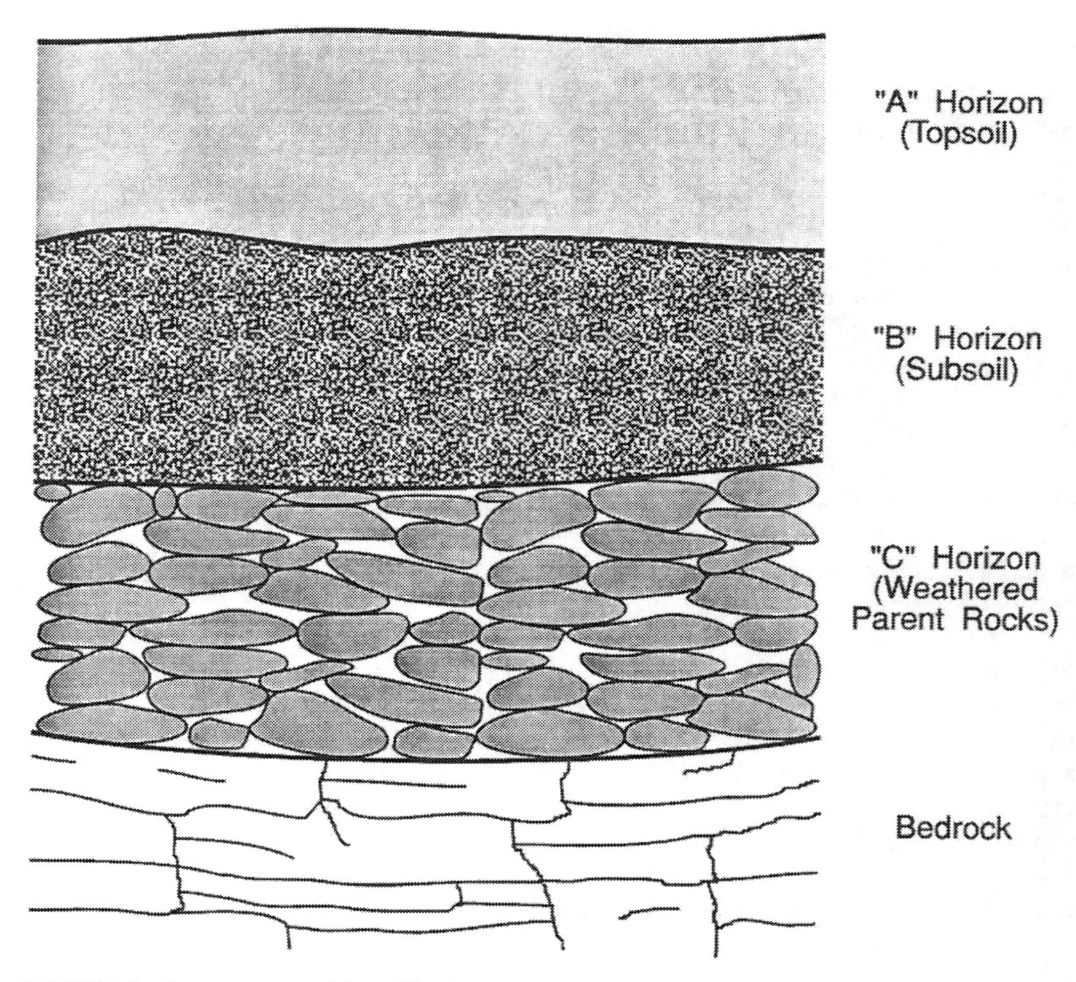

FIGURE 5.3 Representation of the soil horizons.

Source: Reproduced with permission from: Speight (1996). © Taylor and Francis.

There is a strong interaction between water and clay minerals in soil. Water is absorbed on the surfaces of clay particles and because of the high surface/volume ratio of colloidal clay particles, a substantial volume of water may be bound in this manner. Water is also held between the unit layers of the expanding clay minerals, such as the montmorillonite minerals.

As soil becomes waterlogged (water-saturated) it undergoes drastic changes in the physical, chemical, and biological properties. In water-saturated soil, the bonds holding soil colloidal particles together are broken, thereby causing disruption of soil structure. Thus, the excess water in such soils is detrimental to plant growth, and the soil does not contain the air required by most plant roots. In fact, the decay of organic matter in soil increases the equilibrium level of dissolved carbon dioxide in groundwater which lowers the pH and contributes to weathering of carbonate minerals, particularly calcium carbonate ($CaCO_3$).

5.5 ADSORPTION ON POLLUTANTS

The types of non-clay minerals and clay minerals in the soil give the soil adsorption properties that enhance the ability of these minerals to trap potential pollutants within the soil. In some cases, this renders the pollutant immobile (a good effect insofar as this prevent the pollutant from spreading) but also prolongs the time of the pollutant in the soil (a bad effect insofar as the pollutant is fixed and cannot be desorbed thereby making remediation processes more difficult).

The adsorption and desorption of pollutants in soils largely depends on: (i) the type of mineral composition of the soil; (ii) the presence or absence of clay minerals; (iii) the pH; (iv) the redox conditions; and (v) the available chemical species. Inorganic ions, such as phosphate derivatives, HPO_4^{2-}, nitrate derivatives, NO_3^-, chloride derivatives, Cl^-, and sulfate derivatives, SO_4^{2-} and organic ligands, such as citrate, oxalate, fulvic, and dissolved organic carbon (DOC) can affect pollutant behavior in soil.

Inorganic ions can influence adsorption by interactions with metals (metalloids). For example, metals (and metalloids-elements that have properties that are intermediate between the properties of metals and the properties of nonmetals.) complexed with such ions exhibit less sorption affinity to soils than free ions, but free states of some ions (such as phosphate ions, e.g., PO_4^{3-}) in soil increase net negative surface charge and therefore increase the sorption of cationic metal(loid)s. Soil organic matter, which is often estimated and expressed as soil organic carbon, plays an important role in the sorption of pollutants. For example, humic substances have a high affinity for metal cations (M^+, M^{2+}, M^{3+}) and the ability of heavy metals (metals with a high density and high atomic number) to interact (bind) with humic substances is attributed to: (i) the surface charge; (ii) the particle size; (iii) the diffusion coefficient of the humic substance; and (iv) the content of oxygen-containing functional groups, including hydroxyl derivatives (–OH), carboxylic acid derivatives (–COOH), thiol derivatives (–SH), as well as aldehyde derivatives (–CH=O) or ketone derivatives (>C=O).

Finally, soil texture is an important property that influence the behavior and reactivity of the soil. In order to determine the soil texture and to estimate the absorptive capacity of the soil, it is essential to know the constituents of the soil in terms of the amount of non-clay minerals (such as sand and silt) and the clay minerals (clay and, often the amount of silt-silt can be fine sand, clay, or other material carried by running water and deposited as a sediment) in terms of the %w/w of sand,% w/w silt, and% w/w clay in the soil (Table 5.10) which can be estimated from the non-clay mineral and clay mineral content. However, this simplified approach to determining texture will may not have the required accuracy because of the ranges of the constituents (Table 5.10) and if the soil contains a high amounts of gypsum ($CaSO_4$)-soils that contain high amounts of gypsum ($CaSO_4$) typically show a pinkish-white hue. A more accurate test method (ASTM D7928) may be necessary using the soil triangle which is a diagram which that illustrates how soils can be classified based on the percent of sand, silt, and clay in each (Figure 5.4).

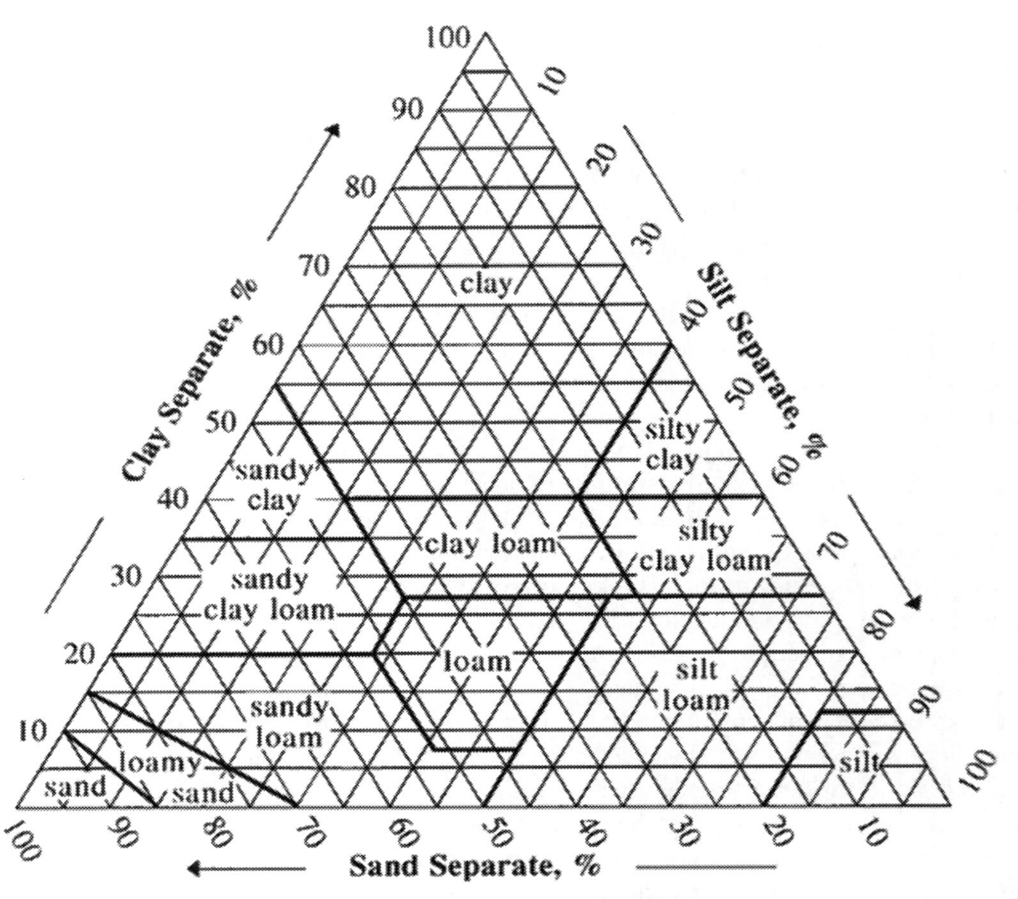

FIGURE 5.4 The soil texture triangle with examples of description of soil texture.

Source: United States Department of Agriculture, Washington, DC; https://www.nrcs.usda.gov/wps/portal/nrcs/detail/soils/survey/?cid=nrcs142p2_054167.

KEYWORDS

- **anthropogenic stress**
- **cation exchange capacity**
- **dissolved organic carbon**
- **geosphere**
- **minerals**

REFERENCES

Aboudi, M. S. C., Hanafiah, M. M., & Chowdhury, A. F. K., (2017). Environmental characteristics of clay and clay-based minerals. *Geology, Ecology, and Landscapes, 1*(3), 155–161.

Anderson, M. A., & Rubin, A. J., (1981). *Adsorption of Inorganics at Solid Liquid Interfaces*. Ann Arbor Science Publishers, Ann Arbor, Michigan.

ASTM D7928, (2020). *Standard Test Method for Particle-Size Distribution (Gradation) of Fine-Grained Soils Using the Sedimentation (Hydrometer) Analysis*. ASTM International, West Conshohocken, Pennsylvania.

Bourke, J. B., Felsot, A. S., Gilding, T. J., Jensen, J. K., & Seiber, J. N., (1992*). Pesticide Waste Management.* Symposium Series No. 510. American Chemical Society, Washington, DC.

Chesworth, W., (2008). *Encyclopedia of Soil Science*. Springer, Dordrecht, The Netherlands.

Christopherson, R. W., (2008). *Geosystems: An Introduction to Physical Geography* (7th edn.). Pearson Prentice-Hall, New York.

Crossman, N. D., Bryan, B. A., De Groot, R. S., Lin, Y. P., & Minang, P. A., (2013). Land science contributions to ecosystem services. Current opinion in environmental sustainability. *Human Settlements and Industrial Systems, 5*(5), 509–514.

Hillier, S., (1995). Erosion, sedimentation, and sedimentary origin of clays. In: Velde, B., (ed.), *Origin*, and *Mineralogy of Clays* (pp. 162–219). Springer-Verlag, New York. Page.

Jazib, J., (2018). *Basics of Environmental Sciences*. Iqra Publishers, New Delhi, India.

Kramer, P. J., & Boyer, J. S., (1995). *Water Relations of Plants and Soils*. Academic Press Inc., San Diego, California.

Kumar, P., & Mina, U., (2018). *Fundamentals of Ecology and Environment.* Pathfinder Publications, New Delhi, India.

Mills, J. S., Hatert, F., Nickel, E. H., & Ferraris, G., (2009). The standardization of mineral group hierarchies: Application to recent nomenclature proposals. *European Journal of Mineralogy, 21*(5), 1073–1080.

Montenari, M., (2016). *Stratigraphy and Timescales*. Amsterdam: Academic Press, Elsevier, Amsterdam, The Netherlands.

Nickel, E. H., (1995). The definition of a mineral. *The Canadian Mineralogist, 33*(3), 689, 690.

Rafferty, J. P., (2011). *Minerals*. Rosen Publishing Group, New York.

Sarkanen, K. V., & Ludwig, C. H., (1971). *Lignins: Occurrence, Formation, Structure, and Reactions*. John Wiley & Sons Inc., New York.

Sigel, H., & Sigel, A., (1995). *Metal Ions in Biological Systems.* Marcel Dekker In., New York.

Speight, J. G., (1996). *Environmental Technology Handbook*. Taylor & Francis, Washington, DC.

Speight, J. G., (2020). *Global Climate Change Demystified*. Scrivener Publishing, Beverly, Massachusetts.

Stevenson, F. J., (1994). *Humus Chemistry*. John Wiley & Sons Inc., New York.

Theng, B. K. G., (1974). *The Chemistry of Clay-Organic Reactions*. John Wiley & Sons Inc., New York.

Velde, B., (1995). composition and mineralogy of clay minerals. In: Velde, B., (ed.), *Origin*, and *Mineralogy of Clays* (pp. 8–42). Springer-Verlag, New York.

Verburg, P. H., Erb, M. K. H., Mertz, O., & Espindola, G., (2013). Land system science: Between global challenges and local realities. current opinion in environmental sustainability. *Human Settlements and Industrial Systems, 5*(5), 433–443.

Vernon, R. H., (2004). *A practical Guide to Rock Microstructure.* Cambridge University Press, Cambridge University, Cambridge, United Kingdom.

Voroney, R. P., & Heck, R. J., (2007). The soil habitat. In: Paul, E. A., (ed.), *Soil Microbiology, Ecology, and Biochemistry* (3rd edn.). Elsevier, Amsterdam, The Netherlands.

Wenk, H. R., & Bulakh, A., (2004). *Minerals: Their Constitution and Origin.* Cambridge University Press, Cambridge University, Cambridge, United Kingdom.

PART III
Emissions and Emissions Management

CHAPTER 6

Sources and Types of Chemicals

6.1 INTRODUCTION

The potential for pollution of the environment by chemicals starts during the production stage of the chemical (or chemicals). The typical process involves combining one or more feedstocks in a series of unit process operations (Speight, 2020). Commodity chemicals tend to be synthesized in a continuous reactor, while specialty chemicals usually are produced in batches. The yield of the chemical will partially determine the kind and quantity of by-products and releases-a release of a chemical, in the context of this book, refers to a chemical that is emitted to the air, discharged to water, or disposed of in some type of land disposal unit.

In spite of numerous safety protocols that are in place and the care taken to avoid environmental incidents that are harmful to the environment, every industry suffers accidents that lead to contamination by chemicals. It is therefore often helpful to be aware of the nature (the chemical and physical properties) of the chemical contaminants and the products arising therefrom (when ecosystem parameters interact with the chemicals) in order to understand not only the nature of the chemical contamination but also chemical changes to the contaminants following from which, cleanup methods can be chosen (Table 6.1) (Chapter 8).

In the past, the existence and source of such information was unknown and, if known, was not always consulted. When the existence and sources of the relevant information are known, decisions must be made in order for environmental scientists and engineers to make an informed, and often quick, decision on the next steps, even if it is decided at a later time not to use the information for a particular application. However, on the basis that it is better to know than to not know, knowing related to the relevant data gives investigators and analysts the ability to assess whether or not a chemical discharge into the environment should be addressed or whether the environment can take care of itself through biodegradation of the chemical. This is especially true for scientists and engineers involved in site cleanup operations, assessment of ecological risk, and assessment of ecological damage. Because of the plethora of modern data bases relating to the properties of chemicals, there can be no reasons (or excuses) for not knowing or understanding the fundamental aspects of the behavior of chemicals that pollute the environment.

The Science and Technology of the Environment. James G. Speight (Author)

TABLE 6.1 General Classification of Chemical Pollutants Based on Chemical Structure

Inorganic Chemical Pollutants
• Acids and bases are used in a variety of industrial applications as well as in chemical laboratories.
• Ammonia is a poisonous gas if released in higher amounts.
• Inorganic fertilizers (such as nitrate derivatives and phosphate derivatives) used largely in agriculture and gardening.
• Metals and metal salts-usually from mining and smelting activities, as well as disposal of mining wastes.
• Oxides of nitrogen and sulfur resulting from vehicle emissions, industrial processes, and other human activities
• Sulfide derivatives (such as pyrite, (FeS_2) are usually mined minerals and once disposed of in the environment, they may generate sulfuric acid in the presence of precipitation water and microorganisms.
• Perchlorate (ClO_4^-) includes the perchloric acid and the various perchlorate salts) which is used in a variety of applications and is persistent in the environment.
Organic Chemical Pollutants
• Chlorinated solvents which are used in industrial degreasing processes, as well as in dry cleaning, and in various household products.
• Crude oil and refined products (such as gasoline, diesel fuel, kerosene, mineral spirit, motor oil, lubricating oil).
• Detergents which include a variety of chemical compounds with surface activity.
• Solvents (such as acetone, methyl ethyl ketone, toluene, benzene, xylene) used in industry as well as in many household products.
• Alcohols (such as ethanol, methanol, isopropanol) are used in a large variety of applications and household products.
• PAHs (polyaromatic hydrocarbons or polynuclear aromatic hydrocarbon derivatives, PNAs) that occur in crude oil and crude oil products, crude oil, but are also a result of combustion processes.
• PCBs (polychlorinated biphenyl ethers) which are now banned but were used in transformers and are already present in large amounts in the environment.
• Pesticides/insecticides/herbicides which are commonly used in agriculture and may contain toxic organic chemicals and organo-metallic compounds metals (such as mercury and arsenic).
• Phenol derivatives which are usually an indication of wastewater and a result of industrial processes.
• Trihalomethanes (such as chloroform, dibromochloromethane, bromochloromethane, bromoform) which are common products of water chlorination.

Once discharged into an ecosystem, some chemicals are removed from the ecosystem by natural events, such as attack by indigenous bacteria (biodegradation) or by increasing the concentration of natural-occurring bacteria to remove the chemical from the ecosystem (bioremediation) (Speight and Arjoon, 2012). However, there are chemicals that are known as persistent pollutants (POPs) which are compounds that are resistant to environmental degradation through the various chemical and biological processes (Jacob, 2013).

POPs, as the name implies, are not easily degraded in the environment due to their stability and low decomposition rates and, thus, have a long life in various ecosystems and often require other forms of removal such as physical or chemical methods of cleanup as well as the addition of non-indigenous microbes for cleanup (Speight, 1996; Speight and Arjoon, 2012). POPs and, also have the ability for long-range transport and environmental contamination by POPs is extensive, even in areas where these chemicals have never been used, and will remain in these environments for a considerable time (even years) and after restrictions implemented due to their resistance to degradation.

POPs, like any chemical pollutant, can enter an ecosystem through the gas phase, the liquid phase, or solid phase and which can resist degradation and are mobile over considerable distances (especially in the gas phase or through transportation in river systems) before being re-deposited in a location that is remote to the location of their introduction into the ecosystem). Furthermore, pollutants can arise from various sources (Table 6.2) and can be present as vapors in the atmosphere or bound to (adsorbed on) the surface of soil or mineral particles and also have variable solubility in water.

TABLE 6.2 Examples of the Sources of Chemical Pollution in Soil

Acid Rain
Caused when pollutants present in the air mixes up with the rain and fall back on the ground.
The polluted water could dissolve away some of the important nutrients found in soil and change the structure of the soil.
Agricultural Activity
Chemical utilization has increased significantly since the increased use of pesticides and fertilizers.
Pesticides and fertilizers are not produced in nature and cannot be broken down by natural forces.
Some chemicals can damage the composition of the soil and make it easier to erode by water and air.
Industrial Activity
A significant contributor to pollution in the 20th Century, especially since the amount of mining and manufacturing has increased.
If the by-products are contaminated and are not sent for disposal in a manner that can be considered environmentally safe, the adverse effects can linger in the soil for months, years, or even decades.

Many POPs are currently (or were in the past) arose from the extensive use of agrochemicals (agricultural chemicals) such as pesticides, herbicides, and biocides), solvents, pharmaceuticals, and various industrial chemicals (Chapter 3). Although some POPs arise naturally, for example from various biosynthetic pathways, most are products of human industry and tend to have higher concentrations and are eliminated more slowly. If not removed and because of their properties, POPs will bioaccumulate and have significant impacts on and the flora and fauna of the environment. The most frequently used measure of the potential for bioaccumulation and persistence of an compound in the environment are the result of the physicochemical properties (such partition coefficients and reaction rate constants) (Mackay et al., 2001).

Furthermore, the capacity of the environment to absorb the effluents and other impacts of process technologies is not unlimited, as some would have us believe. The environment should be considered to be an extremely limited resource, and discharge of chemicals into it should be subject to severe constraints. Indeed, the declining quality of raw materials, dictates that more material must be processed to provide the needed fuels. And the growing magnitude of the products and effluents from industrial processes has moved above the line where the environment has the capability to absorb such process effluents without disruption.

This chapter presents a brief survey of the various types of chemicals that have been (in some cases, continue to be) released into the environment. This will offer a guide to behavior in terms of the chemical transformations that can occur after the release of these chemicals.

6.2 INORGANIC CHEMICALS

By way of definition, environmental pollution is the contamination of the physical and biological components of the Earth system (atmosphere, aquasphere, and geosphere) to such an extent that the normal environmental processes are adversely affected. Thus, inorganic pollution is the introduction of inorganic chemicals into the environment that cause harm or discomfort to the floral and faunal species or that damage the environment which can come in the form of inorganic chemicals or byproducts arising from these chemicals such as radiant energy as well as noise, heat, or light (Speight, 2017a, 2018).

Environmental pollution is often divided into pollution of (i) the air, (ii) the water, and (iii) the land or soil. While the blame for much of the pollution is caused by the inorganic chemical industry, domestic sources of pollution include waste and automobile exhaust. Individuals and chemical and petroleum companies contribute to the pollution of the atmosphere by releasing inorganic and organic gases and particulates into the air (Speight, 2017b). Furthermore, pollution by inorganic chemicals is another case of habitat destruction (Speight, 2017a, 2018) and involves (predominantly) chemical destruction rather than the more obvious physical destruction. The overriding theme of the definition is the ability (or inability) of the environment to absorb and adapt to changes that are caused by human activities.

Pollution of the environment by inorganic chemicals is caused by the release of inorganic chemical waste that causes detrimental effects on the environment. Inorganic chemicals also include metals and metal particles, and these types of contaminants can be found in storm-water runoff from urban development and will accumulate in drainage systems or low lying areas of land, such as marshy areas and wetlands. Many of these contaminants eventually are the result of discharge into waterways with little prior treatment to remove chemicals. Some contaminants, such as mercury, may bioaccumulate in animal tissues and be occur in fish and eventually end up as part of the human diet.

Many inorganic contaminants typically result from the leaching of a contaminated source zone, such as waste disposal and mine-tailing sites, into both surface and groundwater. The contaminant from these sites will move in the groundwater flow direction creating a plume of dissolved phase-contaminated water. The pollutants can take the form of overloading the environment (or an ecosystem) by returning (or disposing of) naturally-occurring chemicals to the environment in amounts that exceed the natural abundance of the chemicals and which are considered contaminants when in excess of natural levels. Inorganic pollution is also the addition of any substance or form of energy (such as radioactivity) to the environment at a rate faster than the environment can accommodate it by dispersion, breakdown, recycling, or storage in some benign form.

Thus, environmental pollution occurs when the environment is unable to accept, process, and neutralize any harmful by-product of human activities (such as the gases that contribute to acid rain such as the various gases from the combustion of fuels which cannot be disposed of without any damage to the environmental system and the transformation of the gases in the atmosphere to acid rain (Chapter 7):

$$2[C]_{fuel} + O_2 \rightarrow 2CO$$
$$[C]_{fuel} + O_2 \rightarrow CO_2$$
$$2[N]_{fuel} + O_2 \rightarrow 2NO$$
$$[N]_{fuel} + O_2 \rightarrow NO_2$$
$$[S]_{fuel} + O_2 \rightarrow SO_2$$
$$2SO_2 + O_2 \rightarrow 2SO_3$$
$$SO_2 + H_2O \rightarrow H_2SO_3 \text{ (sulfurous acid)}$$
$$SO_3 + H_2O \rightarrow H_2SO_4 \text{ (sulfuric acid)}$$
$$NO + H_2O \rightarrow HNO_2 \text{ (nitrous acid)}$$
$$3NO_2 + 2H_2O \rightarrow HNO_3 \text{ (nitric acid)}$$

Thus, inorganic pollution occurs when (i) the natural environment is incapable of decomposing the *un*naturally generated chemicals (anthropogenic pollutants), and (ii) when there is a lack of knowledge on the means by which these chemicals can treated for disposal. This leaves the environment (or any ecosystem) subject to any negative impacts on crucial environmental chemistry. However, it is important to realize that some elements are in fact essential nutrients required by organisms and that they can be deficient in some soils or diets; examples include zinc (Zn), boron (B), copper (Cu), and selenium (Se). A consequence of this is that additions of nutrients will at first be beneficial for the growth and reproduction of organisms. It is only when they become excessive that negative impacts will occur. This typically occurs because the ability of the coil to hold these substances is overwhelmed, along with the ability of the organisms in the ecosystem to cope with the oversupply.

Rather than deal with the individual pollutants, combustion ash is an example of the types of complex mixtures of inorganic chemicals that can cause pollution. Therefore, this section focuses on the various types of pollutants such as: (i) typical inorganic pollutants; (ii) agricultural chemicals, also known as agrochemicals; (iii) ash, which includes fly ash and bottom ash; (iv) and boiler slag.

6.2.1 INORGANIC POLLUTANTS

Significantly elevated levels of inorganic pollutants are often in soils from previously heavily industrialized areas. These sites produced wastes contaminated with heavy metals, arsenic, and antimony. Following the decline of heavy manufacturing, many industrial sites have been abandoned and remain so without remedial works to their contaminated and often phytotoxic substrates.

Toxic metals, such as lead (Pb), are problematic contaminants in soils, surface waters, recent sediments, and aerosols, mainly in urban environments. Toxic metals occur naturally in mining areas, but the mining activities reinforce their release into the environment. Because of their limited mobility, toxic metals are less frequent in groundwater and often form cations (such as divalent lead, Pb^{2+}) that are prone to adsorption, or they precipitate as difficult-to-dissolve oxide derivatives, sulfide derivatives, or other minerals. Some metal cations are more mobile in the environment, such as cadmium (Cd^{2+}), which is also highly toxic. Mercury is extremely toxic and specific bacteria can convert it into the highly mobile methyl mercury ion ($HgCH_3^+$) or volatile dimethyl mercury [$Hg(CH_3)_2$]. In karst systems, sediments in conduits and caves can act as reservoirs for toxic metals and other contaminants; during high-flow events, they can be mobilized and conducted toward large springs.

By way of explanation, a karst system is formed from the dissolution of soluble rocks such as limestone ($CaCO_3$), dolomite ($CaCO_3 . MgCO_3$), and gypsum ($CaSO_4$). It is characterized by underground drainage systems with sinkholes and caves. Subterranean drainage may limit surface water, with few to no rivers or lakes. However, in regions where the dissolved bedrock is covered (perhaps by debris) or confined by one or more superimposed non-soluble rock strata, distinctive karst features may occur only at subsurface levels and can be totally missing above ground.

Arsenic (As) is a toxic metal that causes large-scale groundwater contamination and diseases in several regions of the world. Natural arsenic contamination occurs in arid to semiarid inland basins and in reducing alluvial aquifers, that is, in flat, low-lying areas, where groundwater is sluggish and often exploited by deep pumping wells. Arsenic-rich water is also found in mining areas; in the past, it was also used for insecticides, some of which might still be in use in some regions.

The nitrate ion (NO_3^-) is a common groundwater contaminant, mainly in agricultural areas, where different nitrogen-containing compounds are used as fertilizers and while nitrate typically is not very harmful, high concentrations over long periods of time are unfavorable to human health. High nitrate levels often go along with another type of agricultural contamination, such as fecal bacteria or pesticides, and are therefore a bad sign for the general water quality. As nitrate is a nutrient, increased concentrations also influence the ecosystem. Under oxidizing conditions, nitrate behaves conservatively; under reducing conditions, it is microbiologically transformed into nitrogen gas (denitrification). Effluent waters contain high levels of ammonia (NH_4^+), which is often converted to nitrate under oxidizing conditions.

Dissolved salts are natural water constituents, but increased levels caused by human activities can render the water unusable. The application of road salt locally affects soils and vegetation but rarely causes large-scale groundwater contamination. Salt mining is another contamination source. However, the severest problems are due to improper agricultural and water management practices such as: (i) salinization by irrigation; and (ii) seawater intrusion caused by over-pumping of coastal aquifers, often also for irrigation purposes.

Inorganic pollutants typically result from the leaching of a contaminated source zone, such as waste disposal and mine-tailing sites, into both surface and groundwater. The contaminant from these sites will move in the groundwater flow direction creating a plume of dissolved phase-contaminated water. Remediation of these sites may require characterization of both the source zone and the plume. The contaminant plumes can be up to many kilometers in length and of varying width depending on the groundwater flow velocity and flow direction, and the age of the plume. Also, the plume shape can be variable depending on the hydraulic conductivity distribution in the subsurface and the groundwater flow regime.

6.2.2 AGROCHEMICALS

Inorganic agrochemicals (agricultural chemicals) are the various inorganic chemical products that are used in agriculture. In most cases, the term inorganic *agrochemical* refers to the inorganic chemicals that are used as fertilizers, as opposed to the broader range of organic agrochemicals that are used as insecticides, pesticides, herbicides, fungicides, and nematicides (i.e., chemicals that are used to kill roundworms). Inorganic pesticides were formerly used in large quantities, especially as fungicides. However, they have largely been replaced by various organic (carbon-containing) pesticides (Section 6.3.2).

Most inorganic fertilizers contain varying amounts of nitrogen, phosphorus, and potassium, which are inorganic (nonliving) nutrients that plants need to grow. However, the inorganic pesticide nano-formulations refer to those nanosized compounds of inorganic nature that display pesticide activity. There are diverse inorganic compounds used to formulate inorganic nano-pesticides, including metals (such as silver, Ag, and copper, Cu), metal oxides (such as silica, SiO_2, titanium dioxide, TiO_2, zinc oxide, ZnO, and alumina, Al_2O_3), carbon nanotubes, or combinations thereof.

6.2.3 ASH

Ash is the general term that refers to whatever waste is leftover after a mineral-containing carbonaceous fuel is coal is combusted, usually in a coal-fired power plant and typically contains arsenic, mercury, lead, and many other heavy metals (Table 5.13). As an example, coal ash (as produced in power generating facilities) is commonly divided into two subcategories based on particle size: (i) fly ash; and (ii) bottom ash (Speight, 2013a, b, 2014). Fly ash, bottom ash, and boiler slag are presented here in a separate section because of the composition of these types of waste products, none of which can be classed as a single inorganic chemical.

Fly ash (also known as pulverized fuel ash) in the United Kingdom, is one of the coal combustion products and is composed of the fine particles that are driven out of the boiler with the flue gases. Ash that falls in the bottom of the boiler is called bottom ash. Fly ash, together with bottom ash removed from the bottom of the boiler, is more commonly known as coal ash. Bottom ash and *boiler slag* are the coarse, granular, incombustible by-products (Tables 5.14 and 5.15) that are collected from the bottom of furnaces that burn coal for the generation of steam, the production of electric power, or both. Boiler slag is the melted form of coal ash that can be found both in the filters of exhaust stacks and the boiler at the bottom of the stack. Most these coal by-products are produced at coal-fired electric utility generating stations, although considerable bottom ash and/or boiler slag are also produced from many smaller industrial or institutional coal-fired boilers and from coal-burning independent power production facilities. The type of by-product (i.e., bottom ash or boiler slag) produced depends on the type of furnace used to burn the coal.

6.2.3.1 FLY ASH

The most voluminous and well-known constituent is fly ash, which makes up more than half of the coal leftovers. Fly ash particles are the lowest density particle in ash and pas upward from the combustor into the exhaust stacks of the power plant. Filters within the stacks capture approximately 99% w/w of the ash, attracting it with opposing electrical charges and the captured fly ash is recyclable. The fine particles bind together and solidify, especially when mixed with water, making them an ideal ingredient in concrete and wall-board. The recycling process also renders the toxic materials within fly ash safe for use.

In modern coal-fired power plants, fly ash is generally captured by electrostatic precipitators or other particle filtration equipment before the flue gases reach the chimneys. Depending upon the source and makeup of the coal being burned, the components of fly ash vary considerably, but all fly ash includes substantial amounts of silica (silicon dioxide, SiO_2) (both amorphous and crystalline), aluminum oxide (alumina, Al_2O_3) and calcium oxide (CaO), the main mineral compounds in coal-bearing rock strata. The constituents depend upon the specific coal-bed make-up but may include one or more of the following elements or substances found in trace concentrations (up to hundreds ppm): arsenic, beryllium, boron, cadmium, chromium, hexavalent chromium, cobalt, lead, manganese, mercury, molybdenum, selenium, strontium, thallium, and vanadium (Speight, 2013).

In the past, fly ash was generally released into the atmosphere but air pollution control standards now require that it be captured prior to release by fitting pollution control equipment. In the United States, fly ash is generally stored at coal power plants or placed in landfills. Approximately 43% w/w of the fly ash is recycled, often used as a pozzolan to produce hydraulic cement or hydraulic plaster and a replacement or partial replacement for Portland cement in concrete production. Pozzolans ensure the setting of concrete and plaster and provide concrete with more protection from wet conditions and chemical attack. In the case that fly or bottom ash is not produced from coal, for example, when solid waste is used to produce electricity in an incinerator, this kind of ash may contain higher levels of contaminants than coal ash. In that case, the ash produced is often classified as hazardous waste.

Fly ash material solidifies while suspended in the exhaust gases and is collected by electrostatic precipitators or filter bags. Since the particles solidify rapidly while suspended in the exhaust gases, fly ash particles are generally spherical in shape and range in size from 0.5 μm to 300 μm. The major consequence of the rapid cooling is that few minerals have time to crystallize, and that mainly amorphous, quenched glass remains. Nevertheless, some refractory phases in the pulverized coal do not melt (entirely), and remain crystalline. In consequence, fly ash is a heterogeneous material. SiO_2, Al_2O_3, Fe_2O_3 and occasionally CaO are the main chemical components present in fly ashes. The mineralogy of fly ashes is diverse.

The main phases encountered are a glass phase, together with quartz, mullite, and the iron oxides: hematite, magnetite, and/or maghemite. Other phases often identified are cristobalite, anhydrite, free lime, periclase, calcite, sylvite, halite, portlandite, rutile, and anatase. The calcium-bearing minerals anorthite, gehlenite, akermanite, and various calcium silicates and calcium aluminates identical to those found in Portland cement can be identified in Ca-rich fly ashes. The mercury content can reach 1 ppm, but is generally included in the range 0.01 to 1 ppm for bituminous coal. The concentrations of other trace elements vary as well according to the kind of coal combusted to form it. In fact, in the case of bituminous coal, with the notable exception of boron, trace element concentrations are generally similar to trace element concentrations in unpolluted soils.

Two classes of fly ash are defined: Class F fly ash and Class C fly ash (ASTM C618) in which the chief difference is the amount of calcium, silica, alumina, and iron in the ash. The chemical properties of the fly ash are largely influenced by the chemical content of the coal burned (i.e., anthracite, bituminous coal, and lignite). The particle size distribution of raw fly ash tends to fluctuate constantly, due to changing performance of the coal mills and the boiler performance, which may influence the method of disposal of the ash.

Fly ash contains trace concentrations of heavy metals and other substances that are known to be detrimental to health in sufficient quantities. Potentially toxic trace elements in coal include arsenic, beryllium, cadmium, barium, chromium, copper, lead, mercury, molybdenum, nickel, radium, selenium, thorium, uranium, vanadium, and zinc. Approximately 10% of the mass of coals burned in the United States consists of unburnable mineral material that becomes ash, so the concentration of most trace elements in coal ash is approximately 10 times the concentration in the original coal.

Crystalline silica and lime along with toxic chemicals represent exposure risks to human health and the environment. Exposure to fly ash through skin contact, inhalation of fine particulate dust and ingestion through drinking water may present health risks. Fly ash contains crystalline silica which is known to cause lung disease, in particular silicosis. Also, lime (CaO) reacts with water (H_2O) to form calcium hydroxide [$Ca(OH)_2$], giving fly ash a pH somewhere between 10 and 12, a medium to strong base. This can also cause lung damage if present in sufficient quantities.

6.2.3.2 BOTTOM ASH

The most common type of coal-burning furnace in the electric utility industry is dry bottom pulverized coal boiler. When pulverized coal is burned in a dry bottom boiler, approximately 80% w/w of the unburned material or ash is entrained in the flue gas and is captured and recovered as fly ash. The remaining 20% w/w of the ash is dry bottom ash, a dark gray, granular, porous that is collected in a water-filled hopper at the bottom of the furnace. When a sufficient amount of bottom ash drops into the hopper, it is removed by means of high-pressure water jets and conveyed by sluiceways either to a disposal pond or to a decant basin for dewatering, crushing, and stockpiling for disposal or use.

Bottom ash is the coarser component of coal ash, comprising approximately 10% of the waste. Rather than floating into the exhaust stacks, it settles to the bottom of the power plant boiler. Bottom ash is not quite as useful as fly ash, although power plant owners have tried to develop options for beneficial use options, such as structural fill and road-base material. However, the bottom ash remains toxic when recycled and can leak heavy metals into the groundwater.

6.2.4 BOILER SLAG

Boiler slag is a by-product produced from a wet-bottom boiler, which is a special type of boiler designed to keep bottom ash in a molten state before it is removed. These types of boilers (slag-tap and cyclone boilers) are much more compact than pulverized coal boilers used by most large utility generating stations and can burn a wide range of fuels and generate a higher proportion of bottom ash than fly ash (50 to 80% w/w bottom ash compared to 15 to 20% w/w bottom ash for pulverized coal boilers). With wet-bottom boilers, the molten ash is withdrawn from the boiler and allowed to flow into quenching water. The rapid cooling of the slag causes it to immediately crystallize into a black, dense, fine-grained glassy mass that fractures into angular particles, which can be crushed and screened to the appropriate sizes for several uses.

There are two types of wet-bottom boilers: (i) the slag-tap boiler and (ii) the cyclone boiler. The slag-tap boiler burns pulverized coal and the cyclone boiler burns crushed coal. In each type, the bottom ash is kept in a molten state and tapped off as a liquid. Both boiler types have a solid base with an orifice that can be opened to permit the molten ash that has collected at the base to flow into the ash hopper below. The ash hopper in wet-bottom furnaces contains quenching water. When the molten slag comes in contact with the quenching water, it fractures instantly, crystallizes, and forms pellets. The resulting boiler slag, often referred to as black beauty, is a coarse, hard, black, angular, glassy material.

Since boiler slag is angular, dense, and hard, it is often used as a wear-resistant component in surface coatings of asphalt in road paving. Finer-sized boiler slag can be used as blasting grit and is commonly used for coating roofing shingles. Other uses include raw material for the manufacture of cement, and in colder climates, it is spread onto icy roads for traction control. Because there are so many uses and such a limited supply, most of the boiler slag produced in the United States is used and even imported some from other countries.

6.3 ORGANIC CHEMICALS

As a result of the increasing concern related to pollution (especially pollution by a variety of chemicals), there has been a global ban on those organic chemicals that were particularly harmful and toxic to the environment among which were many pesticides, herbicides, and fungicides which are historically or commercially important (Hites, 2007) and required the participating governments to take measures to eliminate or reduce the release of persistent organic pollutants in the environment (Speight, 1996, 2017b, 2018; Hites, 2007; Manahan, 2010; Speight and Arjoon, 2012).

As a commencement to this process of data examination and ingestion, this chapter introduces the terminology of environmental technology as it pertains to the sources and types of organic pollutants (Table 6.3). Briefly, a *contaminant*, which is not usually classified as a pollutant unless it has some detrimental effect, can cause deviation from the normal composition of an environment. A *receptor* is an object (animal, vegetable, or mineral) or a locale that is affected by the pollutant. A chemical waste is any solid, liquid, or gaseous waste material that, if improperly managed or disposed of, may pose substantial hazards to human health and the environment.

TABLE 6.3 Examples of the Types of Organic Chemical Contaminants

Source	Waste Type
Chemical manufacturing	Spent solvents
	Reactive materials
	Ignitable materials*
Cleaning agents	Spent solvents
	Flammable solvents*
Construction industry	Spent solvents
	Flammable solvents*
Cosmetics manufacturing	Ignitable materials*
	Flammable solvents*
Crude oil recovery and refining	Drilling mud spills
	Spilled solvents
	Flammable solvents*
	Process sludge
Furniture manufacturing and refinishing	Ignitable materials*
	Spent solvents
	Flammable solvents*
Leather products	Waste solvents
	Flammable solvents*
Power generation	Gases and coal dust
	Combustion waste (ash and slag)
Printing industry	Spent solvents
	Flammable solvents*
Vehicle maintenance	Ignitable materials
	Spent solvents
	Flammable solvents*

*Ignitable materials and flammable solvents often fall into the hazardous waste category.

At any stage of the management process, a chemical waste may be designated by law as a hazardous waste (Syed, 2006). Improper disposal of these waste streams, such as organic solvents (Table 6.4), in the past has created hazards to human health and the need for expensive cleanup operations (Tedder and Pohland, 1993). Correct handling of these chemicals (NRC, 1981), as well as dispensing with many of the myths related to chemical processing (Kletz, 1990) can mitigate some of the problems that will occur, especially problems related to the flammability of organic liquids (Table 6.5), that will occur when incorrect handling is practiced.

TABLE 6.4 Effects of Organic Solvents

Solvent	Affected Parts of Human Body
Alcohols	
Methyl alcohol (methanol and toxic metabolites)	Optic nerve
Isopropyl alcohol	Central nervous system
Aliphatic Hydrocarbon Derivatives	
Pentanes, hexanes, heptanes, octanes	Central nervous system, liver
Aromatic Hydrocarbon Derivatives	
Benzene	Blood, immune system
Toluene	Central nervous system
Xylene	Central nervous system
Glycols	
Ethylene glycol (and toxic metabolites)	Central nervous system
Halogenated Aliphatic Hydrocarbon Derivatives	
Methylene chloride	Central nervous system, respiratory system
Chloroform	Liver
Carbon tetrachloride	Liver and kidneys

TABLE 6.5 Flammability of Selected Organic Liquids

Liquid	Flash Point (°C)*	% v/v in Air LFL	UFL**
Diethyl ether	–43	1.9	36
Acetone	–20	2.6	13
Methanol	12	6.0	37
Naphthalene	157	0.9	5.9
Pentane	–40	1.5	7.8
Toluene	4	1.3	7.1

*Closed-cup flash point test, ASTM D93, ASTM D3278; see also ASTM D92; ASTM D1310).

The flashpoint is the lowest temperature at which a liquid will form a vapor in the air near its surface that will or briefly ignite (flash) when exposed to an open flame. The flashpoint is a general indication of the flammability or combustibility of a liquid. Below the flashpoint, insufficient vapor is available to support combustion, and at some temperature above the flashpoint (dependent upon the properties of the liquid), the liquid will produce sufficient vapor to support combustion-this temperature is known as the fire point (see Table 6.7).

**LFL: lower flammability limit; UFL: upper flammability limit at 25°C (77°F).

6.3.1 *AEROSOLS*

An aerosol is a suspension of liquid or solid particles in a gas, with particle diameters in the range of 10^{-9} to 10^{-4} meters. In atmospheric science, however, the term aerosol traditionally refers to suspended particles that contain a large proportion of condensed matter other than water, whereas clouds are considered as separate phenomena (Pöschl, 2005). Aerosols give rise to a class of compounds known as volatile organic compounds (VOCs) which can arise from various sources.

In addition to the emissions of VOCs from vegetation, large quantities of organic compounds are emitted into the atmosphere from anthropogenic (man-made) sources, largely from combustion of crude oil products such as gasoline and diesel fuels and from other sources such as solvent use and the use of consumer products (Speight, 2005, 2014). Although the theory is that the combustion of such chemicals (alkanes in particular) should proceed completely to produce quantitative yields of carbon dioxide and water, this is not always the case:

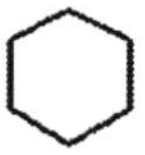

$$C_6H_{12} + 9\,O_2 \longrightarrow 6\,CO_2 + 6\,H_2O + \text{energy}$$

cyclohexane

$$C_6H_{10} + 8.5\,O_2 \longrightarrow 6\,CO_2 + 5\,H_2O + \text{energy}$$

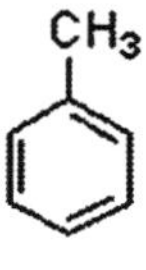

cyclohexene

$$C_7H_8 + 9\,O_2 \longrightarrow 7\,CO_2 + 4\,H_2O + \text{energy}$$

toluene

The presence of heteroatoms (non-carbon and non-hydrogen atoms such as nitrogen, and/or oxygen, and/or sulfur) complicates the process by converting the heteroatoms to the gases oxide (such as nitrogen oxides and sulfur oxides) that are gaseous pollutants. In addition, incomplete combustion (i.e., combustion in a dearth of oxygen will produce polynuclear aromatic producers that appear in the atmosphere (as PM) or in the soil.

In the atmosphere, organic compounds are partitioned between the gaseous and particulate phases, with the chemicals being at least partially in the gas-phase for liquid-phase vapor pressures of at least 10^{-6} Torr at atmospheric temperature (1 Torr is a unit of pressure based on an absolute scale and is 1/760 of a standard atmosphere; thus 1 Torr = 1 mm Hg pressure). In the atmosphere, these gaseous organic compounds are transformed by photolysis and/or reaction with hydroxyl ions (OH^-), nitrate ions (NO_3^-), and ozone (O_3). Emissions of organic compounds and their subsequent *in situ* atmospheric transformations lead to a number of adverse effects, including (i) the formation-in the presence of oxides of nitrogen, NOx-of ozone, a criteria air pollutant, (ii) the formation of secondary organic minute particles-aerosols-resulting in loss of visibility and risks to human health, and (iii) the *in situ* atmospheric formation of toxic air contaminants, including, for example, formaldehyde HCHO), peroxy-acetyl nitrate and nitrated aromatic species(Yang et al., 2019).

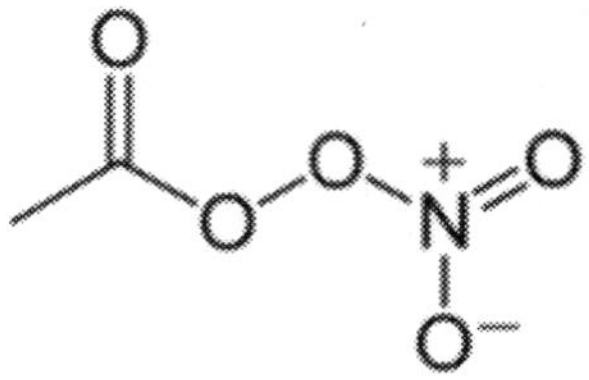

Peroxy-acetyl nitrate

Atmospheric aerosol particles originate from a wide variety of natural and anthropogenic sources, including biomass. Primary particles are directly emitted as liquids or solids from sources such as biomass burning, volcanic eruptions, and wind-driven or traffic-related suspension of road, soil, and mineral dust, sea salt, and biological materials (such as plant fragments, microorganisms, and pollen) (Oliveira et al., 2011; Speight and Singh, 2014).

This also include the commercial conversion of biomass to a variety of chemicals, some of which may not be beneficial when released to the environment (Kim and Holtzapple, 2005, 2006, a; Speight, 2011; Speight and Singh, 2014). Another example of a primary aerosol is the carbonaceous soot formed during incomplete combustion processes-diesel soot is a typical example of this form of primary carbonaceous aerosol. In spite of claims to the contrary, diesel fuel is not a clean fuel it might be clean insofar as sulfur content is concerned but try following a diesel fuel vehicle under full load and/or up an incline when the emission of black fumes becomes very evident and uncomfortable.

Another example of primary aerosols (although not organic in nature) is the fly ash from coal combustion systems or the incineration of wastes. This material is typically generated by high-temperature processes that result in the production of condensed inorganic materials formed as small spherical beads, typically high in silicates or iron oxides. Anthropogenic sources of primary aerosol and particulate material include mechanical abrasion that produces construction and industrial dusts, as well as abrasion of tires and pavement materials on roads.

Secondary aerosol particles, on the other hand, are formed by gas-to-particle conversion in the atmosphere (new particle formation by nucleation and condensation of gaseous precursors). The most important aerosol-generating inorganic gases that are released into the atmosphere are the nitrogen oxides (nitric oxide and nitrogen dioxide) and sulfur dioxide by the combustion of various fuels. Nitric oxide (NO) is oxidized to nitrogen dioxide (NO_2) and subsequently to nitric acid (HNO_3) by the reaction of hydroxyl radical with nitrogen dioxide. The nitric acid can in turn react with ammonia to form ammonium nitrate, a white solid. The chemical equations are often subject to debate but can be represented simply as:

$$NO + O_2 \rightarrow NO_2$$
$$NO + O_3 \rightarrow NO_2 + O_2$$
$$NO_2 + OH \rightarrow HNO_3$$
$$HNO_3 + NH_3 \rightarrow NH_4NO_3$$

Sulfur dioxide-a common product from the combustion of sulfur-containing organic fuels-in the atmosphere reacts with hydroxyl radical in the gas phase and with hydrogen peroxide and ozone in the aqueous phase to form sulfuric acid (H_2SO_4), which is water-soluble and also has a very low vapor pressure, so it rapidly forms aerosol once it is formed. Sulfuric acid can also react with ammonia to form ammonium bisulfate (NH_4HSO_4) and ammonium sulfate [$(NH_4)_2SO_4$]. These species are all fairly water-soluble, and at high relative humidity values they will grow by adding water vapor to their surfaces.

Thus, airborne particles undergo various physical and chemical interactions and transformations (often referred to as atmospheric aging) which involves changes of particle composition, particle size, and particle structure through chemical reaction, gas uptake, and restructuring. Particularly efficient particle aging occurs in clouds, which are formed by condensation of water vapor on preexisting aerosol particles. Most clouds re-evaporate, and modified aerosol particles are again released from the evaporating cloud droplets or ice crystals (cloud processing). If, however, the cloud particles cause precipitation that reaches the surface of the Earth, not only the condensation nuclei but also other aerosol particles that are scavenged on the way to the surface are removed from the atmosphere (wet deposition). Particle deposition without precipitation of airborne water particles (dry deposition)-is less important on a global scale but is highly relevant with respect to air quality. Depending on the properties of the aerosol and meteorological conditions, the characteristic residence times (lifetimes) of aerosol particles in the atmosphere can range from hours to weeks (Pöschl, 2005).

Depending on the origin of organic aerosols, the components can be classified as primary or secondary. Primary organic aerosol components are directly emitted in the condensed phase (as liquid particles or as solid particles) or as semi-volatile vapors which are condensable under atmospheric conditions. The main sources of primary organic aerosol particles and components are natural and anthropogenic biomass burning (forest fires, slashing, and burning, domestic heating), fuel combustion (domestic heating, industrial operations, traffic density), and wind-driven or the traffic-related suspension of soil and road dust, biological materials (such as plant debris, animal debris, pollen, and spores,), sea spray, and spray from other surface waters that contain dissolved organic chemicals.

Secondary organic aerosol components are formed by chemical reaction and gas-to-particle conversion of VOCs (VOCs) in the atmosphere, which may proceed through different chemical and physical pathways, such as: (i) new particle formation; (ii) formation of semi-volatile organic compounds-SVOCs-by gas-phase reactions and participation of the SVOCs in the nucleation and growth of new aerosol particles; (iii) gas-particle partitioning, which results in the formation of SVOCs by gas-phase reactions and uptake through adsorption or by absorption by preexisting aerosol or cloud particles; and (iv) heterogeneous or multiphase reactions: formation of low-volatile organic compounds

(LVOCs) or non-volatile organic compounds (NVOCs) by chemical reaction of volatile organic compounds or SVOCs at the surface or in the bulk of aerosol or cloud particles.

Thus, in summary, aerosols caused by either the entry of organic chemicals into the environment or the reactivity of organic chemicals in the environment are of major importance for atmospheric chemistry and physics, the biosphere, and climate. The airborne solid and liquid particles in the nanometer (1×10^{-9} meter) to the micrometer (1×10^{-6} meter) size range influence the energy balance of the Earth, the hydrological cycle, atmospheric circulation, and the abundance of greenhouse gases and reactive trace gases. Moreover, aerosols play an important role in the reproduction of biological organisms and the primary parameters that determine the environmental effects of aerosol particles are (i) the concentration of the particles, (ii) particle size, (iii) particle structure, and (iv) the chemical composition of the particles. These parameters, however, are spatially and temporally highly variable.

6.3.2 AGROCHEMICALS

Agrochemicals (agricultural chemicals, agrichemicals) are the various chemical products that are used in agriculture. In most cases, the term *agrochemical* refers to the broad range of pesticide chemicals, including insecticide chemicals, herbicide chemicals, fungicide chemicals, and nematicides chemicals (chemicals used to kill roundworms). The term may also include synthetic fertilizers, hormones, and other chemical growth agents, as well as concentrated stores of raw animal manure.

Typically, agrochemicals are toxic and when stored in bulk storage systems may pose significant environmental risks, even explosions (Guglielmi, 2020). However, in many countries, the use of agrochemicals has become highly regulated and government-issued permits for purchase and use of approved agrichemicals may be required. Significant penalties can result from misuse, including improper storage resulting in chemical leaks, chemical leaching, and chemical spills. Wherever these chemicals are used, proper storage facilities and labeling, emergency clean-up equipment, emergency clean-up procedures, safety equipment, as well as safety procedures for handling, application, and disposal are often subject to mandatory standards and regulations.

While agrochemicals increase plant and animal crop production, they can also damage the environment. Excessive use of fertilizers has led to the contamination of groundwater with nitrate, a chemical compound that in large concentrations is poisonous to humans and animals. In addition, the runoff (or leaching from the soil) of fertilizers into streams, lakes, and other surface waters (the aquasphere) can increase the growth of algae, which can have an adverse effect on the life-cycle of fish and other aquatic animals (Table 6.6).

Pesticides that are sprayed on entire fields using equipment mounted on tractors, airplanes, or helicopters often drift away (due to wind or air convection patterns) from the targeted field, settling on nearby plants and animals. Some older pesticides, such as the powerful insecticide DDT (dichlorodiphenyltrichloroethane), remain active in the environment for many years (Table 6.6), contaminating virtually all wildlife, well water, food, and even humans with whom it comes in contact.

TABLE 6.6 Harmful Chemicals Identified by the United Nations Environment Program Governing Council*

Aldrin: An insecticide used in soils to kill insects such as termites, grasshoppers, and western corn rootworm.

Chlordane: An insecticide used to control termites and on a range of agricultural crops; a chemical that remains in the soil with a reported half-life of one year.

Chlordecone: A synthetic chlorinated organic compound that is primarily used as an agricultural pesticide.

Dichlorodiphenyltrichloroethane (DDT); used as insecticide during WWII to protect against malaria and typhus; after the war, used as an agricultural insecticide; can persist in the soil for 10 to 15 years after application.

Dieldrin: A pesticide used to control termites, textile pests, insect-borne diseases and insects living in agricultural soils; half-life is approximately five years.

Dioxin Derivatives: by-products of high-temperature processes, such as incomplete combustion and pesticide production also emitted from the burning of hospital waste, municipal waste, and hazardous waste as well as automobile emissions, combustion of peat, coal, and wood.

Endosulfans: Insecticides used to control pests on crops such coffee, cotton, rice, and sorghum and soybeans, tsetse flies, ectoparasites of cattle; also used as a wood preservative.

Endrin: An insecticide sprayed on the leaves of crops, and used to control rodents; half-life is up to 12 years.

Heptachlor: A pesticide primarily used to kill soil insects and termites, along with cotton insects, grasshoppers, other crop pests, and malaria-carrying mosquitoes.

Hexabromocyclododecane (BHC): A brominated flame retardant used as a thermal insulator in the building industry; persistent, toxic, and ecotoxic with bioaccumulative properties and long-range transport properties.

Hexabromodiphenyl Ether (hexaBDE) and Heptabromodiphenyl Ether: The main components of commercial octabromodiphenyl ether (octa-BDE); highly persistent in the environment.

Hexachlorobenzene: A fungicide used as a seed treatment, especially on wheat to control the fungal disease bunt; also, a by-product produced during the manufacture of chlorinated solvents and other chlorinated compounds.

α-Hexachlorocyclohexane (α-HCH) and β-hexachlorocyclohexane (β-HCH); insecticides as well as by-products in the production of lindane; highly persistent in the water of colder regions.

Lindane, also known as *gamma*-hexachlorocyclohexane, (γ-HCH), gammaxene, Gammallin, and sometimes incorrectly called benzene hexachloride (BHC): a chemical variant of hexachlorocyclohexane that has been used as an agricultural insecticide.

Mirex: An insecticide used against ants and termites or as a flame retardant in plastics, rubber, and electrical goods; half-life is up to 10 years.

Pentachlorobenzene (PeCB): A pesticide and also used in polychlorobiphenyl products, dyestuff carriers, as a fungicide, a flame retardant, and a chemical intermediate.

Perfluorooctane Sulfonic Acid (PFOS) Salts of the Acid: Used in the production of fluoropolymers; extremely persistent in the environment through bioaccumulation and biomagnification.

Polychlorinated Biphenyls (PCBs): Used as heat exchange fluids in electrical transformers and capacitors; also used as additives in paint, carbonless copy paper, and plastics; a half-life up to 10 years.

Polychlorinated Dibenzofurans: By-products of high-temperature processes, such as incomplete combustion after waste incineration, pesticide production, and polychlorinated biphenyl production.

Tetrabromodiphenyl Ether (tetraBDE) and Pentabromodiphenyl Ether (pentaBDE): Industrial chemicals and the main components of commercial pentabromodiphenyl ether (pentaBDE).

Toxaphene: An insecticide used on cotton, cereal, grain, fruits, nuts, and vegetables, as well as for tick and mite control in livestock; a half-life up to 12 years in soil.

*Listed alphabetically and not by effects; all chemicals listed are harmful to flora and fauna, including humans.

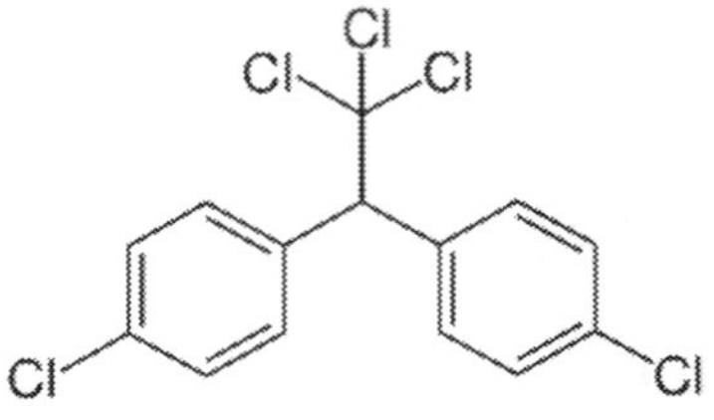

DDT (dichlorodiphenyltrichloroethane

Although many of these pesticides have been banned (Chapter 1), some newer pesticides still cause severe environmental damage.

There is now an awareness of the health hazards of pesticides and related chemicals due to the pioneering work that commenced in the latter half of the 20th century and has continued into the 21st century (Carson, 1962; Carson and Mumford, 1988, 1995, 2002). These materials are carefully regulated, and the safety requirements for every pesticide product are spelled out in detail. Most fertilizers have been in an opposite category, considered useful, safe, and inert.

These and other environmental effects have prompted the search for non-chemical methods of enhancing soil fertility and dealing with crop pests.

6.3.3 CHEMICAL WASTE

Chemical waste is waste is a general term and covers many types of materials but is generally recognized as a waste that is composed of harmful chemicals. Thus, by this definition, organic chemical waste is composed of harmful chemicals. The wastewater treatment industry describes sludge as a waste stream resulting from wastewater treatment. This description does not provide a physical or chemical basis for a definition but is adequate for the needs of wastewater treatment plant (WWTP) operations (i.e., anyone can distinguish between the treated effluent, which flows out of the plant, and the material that is pumped, scraped or skimmed from the treatment units).

However, this definition of sludge although adequate for the wastewater treatment industry, is inadequate for distinguishing between waste forms because it is qualitative and it ties the designation of sludge to its origins in wastewater treatment and, this, two materials with identical characteristics may be defined as sludge or as liquid (i.e., treated effluent) based on facility-specific treatment processes and regulatory requirements governing the off-site discharge of the effluent. In order to develop a quantitative definition of sludge which will also help with correlating waste forms with disposal options, a simple and measurable characteristic of all three waste forms needs to be identified. One option is to define sludge on the basis of the solids content (Figure 6.1).

If the waste is an aqueous solution, the type and composition of the liquid waste depends on the source. In urban areas, the main sources are households, commercial establishments, and industries. Accurate information is needed about the characteristics of liquid wastes in order to establish proper waste management processes to deal with them (Syed, 2006).

The composition of liquid waste, also known as wastewater, is highly varied and depends principally on its source. In towns and cities, the three main sources are residential, commercial, and industrial areas. This is particularly true if the liquid waste is classed as a hazardous waste.

Form of waste	*Total solids*
	% w/w
Solid	100
Sludge	2-50
Liquid	equal to or less than 1

FIGURE 6.1 Simplified distinction between solid waste, sludge, and liquid waste.

However, all chemical wastes are not hazardous wastes and an organic chemical waste may or may not be classed as hazardous waste. An organic chemical hazardous waste is a gaseous, liquid, or material that displays either a hazardous characteristic or is specifically listed by name as a hazardous waste. There are four characteristics chemical wastes may have to be considered as hazardous are (i) ignitability, (ii) corrosivity, (iii) reactivity, and (iv) toxicity (Chapter 10). This type of hazardous waste must be categorized as to its identity, constituents, and hazards so that it may be safely handled and the disposal method defined.

In urban areas, the liquid wastes from residential areas are often referred to as domestic wastewaters. These wastewaters come from the day-to-day living of humans and include those from food preparation, washing, bathing, and toilet usage. Thus, different terms may be used to describe wastewater from these various domestic sources. For example, blackwater and greywater are produced from domestic dwellings with access to a piped water supply and also from business premises and the various institutions, such as schools and health centers, found in residential areas. The term *sewage* is used to describe a combination of all these types of liquid waste, frequently also with surface run-off. In many towns and cities in the world, sewage is collected in underground sewers that carry the effluents to a sewage treatment works where the sewage is treated (cleaned) by various physical and biological processes before being discharged into a river or lake. The quantity and type of liquid waste generated in a residential area depends on several factors, such as population size, standard of living, rate of water consumption, habits of the people and the climate. It also depends on the number and type of institutions such as schools and health centers in the area.

The wastewaters from commercial areas-comprising business establishments, shops, open market places, restaurants, and cafes-will mostly resemble those from households. This is because only human-related activities are undertaken in such areas, as opposed to other activities such as industrial production. Effluent from restaurants and cafes may contain high levels of oil from cooking processes, but this can be overcome by using a grease trap in the outlet pipes.

A grease trap consists of a small tank or chamber which slows the speed of effluent flow. In the grease trap, fats, oils, and grease float to the top of the wastewater and form a layer of scum that is contained within the tank. This can then be removed and disposed of as solid waste. Relatively clean water exits from the grease trap for disposal.

The quantity of wastewater generated per person in a commercial area will be less than it would be at home because the only time spent there is during the working day, and so activities such as bathing are not usually undertaken at these establishments.

In industrial areas liquid wastes are generated by processing or manufacturing industries and service industries, such as vehicle repair shops. The type of industry determines the composition of the waste. The wastewaters from facilities that produce food products will not be harmful to humans, but those from other industries may contain a variety of chemical compounds, some of which may be hazardous (and therefore potentially harmful). Industrial wastewaters which contain hazardous substances must be treated, and the substances removed before the wastewater is discharged to the environment. The presence of hazardous materials is one way in which industrial wastewaters are often different from domestic wastewaters. Another difference is that the flow rate can vary dramatically in some industries, for example, where production rates vary with the season, such as in the processing of certain food crops.

Although not a form of liquid waste in the same way as wastes from residential, commercial, and industrial areas, stormwater is also a form of wastewater. Stormwater can be contaminated with many different types of pollutants such as fecal matter, soil, rubber from vehicle tire wear, litter, and oil from vehicles. Where there is a sewerage network (a system of sewers), stormwater may be channeled into the sewers, or it may flow into open ditches.

In terms of a general classification of liquid waste, the waste can be described according to their physical, chemical, and biological characteristics, such as: (i) solids content and type; (ii) temperature; (iii) odor; and (iv) organic matter.

Wastewater may contain particles of solid material carried along in the flow. These may be settleable solids or suspended solids. Solids that are capable of settling to the bottom of the waste when the speed of flow is reduced, for example, will do so when the wastewater is stored in a tank. Suspended solids are small particles that remain in suspension in the water; they do not dissolve in the wastewater but are carried along in it. The solids content can be measured by filtering out and weighing the solids in a given volume of water and can be expressed in terms of milligrams of solid matter per liter of water, in units of milligrams per liter.

In many cases, the wastewater may be warmer than the ambient temperature because warm or hot water may be included in the waste stream from domestic activities such as showering or from industrial processing. Also wastewater typically (but not always) has an odor, usually due to generation of gases as a result of biodegradation of the contents of the wastewater. By definition, the organic matter is any substance that is derived from living organisms, such as human and animal wastes, food waste, paper, and agricultural wastes.

In fact, wastewater from many different sources contains organic matter, which is a frequent cause of pollution in surface waters. If organic matter is released into a river or lake, bacteria, and other micro-organisms that are naturally present in freshwater will degrade the waste and in the process they use dissolved oxygen from the water. If there is a lot of organic matter, then most or all of the dissolved oxygen may be used up, thus depriving other life forms in the water of this essential element. The oxygen taken up in degrading the organic matter (oxygen demand) which can be determined by a series of test methods (which give the biochemical oxygen demand: BOD) which involve measuring the amount of oxygen used, usually over a period of five days, as the organic matter in the wastewater breaks down. The result is the amount of oxygen used in degrading the organic matter in the wastewater, which is expressed in milligrams per liter.

Wastewater also contains inorganic chemicals which includes a wide range of different chemicals as well as inert solids like sand and silt. Many inorganic chemicals are dissolved in the water, and although some are harmless, others are pollutants that can damage aquatic life such as fish and other organisms that live in water. One example is ammonia (NH_3) which is present in human and animal excreta. Like organic matter, ammonia is broken down in the environment by natural processes. If ammonia is released into a river, it is converted by the action of bacteria to nitrate (NO_3), which requires oxygen and the conversion is limited if there are excessive quantities of ammonia. Other examples of inorganic chemicals in wastewaters are chloride (from salt), phosphates (from chemical fertilizers and from human and animal wastes), and metal compounds (from mining operations or metal-plating plants).

Liquid wastes may also contain many different types of bacteria and other micro-organisms originating from human wastes and other sources. Many of these bacteria are beneficial and are responsible for the biodegradation of organic components of the wastes; others may be pathogenic. The presence of bacteria in wastewater is to be expected but can be a problem if the waste is not kept separate from people or if it contaminates clean water or food.

In terms of quantity by weight, more wastes than all others combined are those from categories designated by hazardous waste numbers preceded by F and K. The F categories are those wastes from nonspecific sources. The K-type hazardous wastes are those from specific sources produced by industries such as, in the context of organic chemicals, the manufacture of organic chemicals, pesticides, explosives, as well as from processes such as wood preservation crude oil refining or wood preservation.

As stated above, an organic chemical waste is considered hazardous if it exhibits one or more of the following characteristics: ignitability, corrosivity, reactivity, and toxicity. Under the authority of the Resource Conservation and Recovery Act (RCRA) and the United States Environmental Protection Agency (EPA), a hazardous substance has one or more of the above characteristics.

Briefly, ignitability is that characteristic of chemicals that are volatile liquids and the vapors are prone to ignition in the presence of an ignition source. Non-liquids that may catch fire from friction or contact with water and which burn vigorously or are persistently ignitable compressed gases and oxidizers also fall under the mantle of ignitable chemicals. Examples include solvents, friction-sensitive substances, and pyrophoric solids that may include catalysts and metals isolated from various refining processes. Organic solvents are indigenous to the crude oil industry and released to the atmosphere as vapor and can pose a significant inhalation hazard. Improper storage, use, and disposal can result in the contamination of land systems as well as groundwater and drinking water (Barcelona et al., 1990; Speight, 2005).

Often, the term ignitable chemical (ignitable organic chemical, such as naphtha or gasoline) is used in the same sense as the term flammable organic chemical insofar as it is a chemical that will burn readily but a combustible organic chemical (any higher boiling hydrocarbon product of refining but which can include naphtha or gasoline) often requires relatively more persuasion to burn, i.e., the chemical is less flammable. Most crude oil products that are likely to burn accidentally are low-boiling liquids that liquids that form vapors that are usually denser than air and thus tend to settle in low spots. The tendency of a liquid to ignite is measured by a test in which the liquid is heated and periodically exposed to a flame until the mixture of vapor and air ignites at the surface of the liquid. The temperature at which this occurs is called the flashpoint (Speight, 2015).

There are several standard tests for determining the flammability of materials. For example, the upper and lower concentration limits for the flammability of chemicals and waste can be determined by standard test methods (ASTM D4982; ASTM E681) as can the combustibility and the flashpoint (ASTM D1310; ASTM E176; ASTM E502). With these definitions in mind, it is possible to divide ignitable materials into four subclasses (Table 6.7). However, there are two important concepts to be considered defined and these are: (i) the flammability limit; and (ii) the flammability range.

TABLE 6.7 Categories of Flammable Chemicals

Type	Description
Combustible liquid	Has a flashpoint in excess of 37.8°C (100°F), but below 93.3°C (200F).
Combustible gases	Substances that exist entirely in the gaseous phase at 0°C (32°F) and 1 atmosphere pressure (14.7 psi) pressure.
Flammable compressed gas	Any liquefied hydrocarbon gas or crude oil product) that meets specified criteria for lower flammability limit, flammability range, and flame projection.
–	Liquefied petroleum gas (commonly referred to as LPG) is an example.
Flammable liquid	A liquid having a flash point below 37.8°C (100°F) (ASTM D92; ASTM D1310; see also ASTM D93, ASTM D3278).
Flammable solid	A solid that can ignite from friction or from heat remaining from its manufacture, or which may cause a serious hazard if ignited.
–	Explosive materials are not included in this classification.

Values of the vapor/air ratio below which ignition cannot occur because of insufficient fuel define the lower flammability limit. Similarly, values of the vapor/air ratio above which ignition cannot occur because of insufficient air define the upper flammability limit. The difference between upper and lower flammability limits at a specified temperature is the flammability range. In addition, explosions that are not due to the flammability of an organic chemical can also occur. Dust explosions (ASTM E789) can occur during catalytic reactor shutdown and cleaning are due production of finely divvied solids through attrition. Many catalyst dusts can burn explosively in air. Thus, control of dust generated by catalyst attrition is essential (Mody and Jakhete, 1988). Organic chemicals that catch fire spontaneously in air without an ignition source are called pyrophoric organic *chemicals*, all of which may occur on a refinery site. Moisture in air is often a factor in *spontaneous ignition.*

Reactive chemicals are those that tend to undergo rapid or violent reactions under certain conditions. Such substances include those that react violently or form potentially explosive mixtures with water, such as some of the common oxidizing agents. Explosives (Sudweeks et al., 1983; Austin, 1984) constitute another class of reactive chemicals. For regulatory purposes, those substances are also classified as reactive that react with water, acid, or base to produce toxic fumes, particularly hydrogen sulfide or hydrogen cyanide.

Heat and temperature are usually important factors in reactivity since many reactions require energy of activation to get them started. The rates of most reactions tend to increase sharply with increasing temperature, and most chemical reactions give off heat. Therefore, once a reaction is started in a reactive mixture lacking an effective means of heat dissipation, the rate will increase exponentially with time (doubling with every 10° rise in temperature), leading to an uncontrollable event. Other factors that may affect the reaction rate include the physical form of reactants, the rate and degree of mixing of reactants, the degree of dilution with a non-reactive medium (e.g., an inert solvent), the presence of a catalyst, and pressure.

Corrosivity is that characteristic of chemicals that exhibit extremes of acidity or basicity or a tendency to corrode steel. Such chemicals, as used in various refining (treating) processes are acidic and are/or capable of corroding metal such as tanks, containers, drums, and barrels. On the other hand, *reactivity* is a violent chemical change (an explosive substance is an obvious example) that can result to pollution and/or harm to indigenous flora and fauna. Such wastes are unstable under ambient conditions insofar as they can create explosions, toxic fumes, gases, or vapors when mixed with water.

As with flammability, there are many tests that can be used to determine corrosivity (ASTM D1838; ASTM D2251). Most corrosive substances belong to at least one of the four following non-organic chemical classes: (i) strong acids; (ii) strong bases; (iii) oxidants; or (iv) dehydrating agents, which are all are used in the refining industry (Speight, 2005). For example, sulfuric acid is a prime example of a corrosive substance (ASTM C694). As well as being a strong acid, concentrated sulfuric acid is also a dehydrating agent and oxidant. The heat generated when water and concentrated sulfuric acid are mixed and illustrates the high affinity of sulfuric acid for water. If this is done incorrectly by adding water to the acid, localized boiling and spattering can occur and result in personal injury. The major destructive effect of sulfuric acid on skin tissue is removal of water with accompanying release of heat. Contact of sulfuric acid with tissue results in tissue destruction at the point of contact. Inhalation of sulfuric acid fumes or mists damages tissues in the upper respiratory tract and eyes. Long term exposure to sulfuric acid fumes or mists has caused erosion of teeth, as well as destruction of other parts of the body!

Finally, toxicity (defined in terms of a standard extraction procedure (EP) followed by chemical analysis for specific substances) is a characteristic of all chemicals whether or not the chemicals are crude oil or non-crude oil in origin. Toxic wastes are harmful or fatal when ingested or absorbed and, when such wastes are disposed of on land, the chemicals may drain (leach) from the waste and pollute groundwater. Leaching of such chemicals from contaminated soil may be particularly evident when the area is exposed to acid rain. The acidic nature of the water may impart mobility to the waste by changing the chemical character of the waste or the character of the minerals to which the waste species are adsorbed.

Toxicity is of the utmost concern in dealing with chemicals and their disposal. This includes both long-term chronic effects from continual or periodic exposures to low levels of toxic chemicals and acute effects from a single large exposure (Zakrzewski, 1991). Not all toxins are immediately apparent. For example, living organisms require certain metals for physiological processes. These metals, when present at concentrations above the level of homeostatic regulation can be toxic (ASTM E1302). In addition, there are metals that are chemically similar to, but higher in molecular weight than, the essential metals (heavy metals). Metals can exert toxic effects by direct irritant activity, blocking functional groups in enzymes, altering the conformation of biomolecules, or displacing essential metals in a metallo-enzyme.

6.3.4 FLAME RETARDANTS

Flame retardant chemicals are used in commercial and consumer products (such as furniture and building insulation) to meet flammability standards. Not all flame retardants present concerns, but the following types often do present concerns: (i) halogenated flame retardants, also known as organo-halogen flame retardants that contain chlorine or bromine bonded to carbon; and (ii) organo-phosphorous flame retardants that contain phosphorous bonded to carbon.

Flame retardants inhibit or delay the spread of fire by suppressing the chemical reactions in the flame or by the formation of a protective layer on the surface of a material. They may be mixed with the base material (additive flame retardants) or chemically bonded to it (reactive flame retardants). Mineral flame retardants are typically additive while organohalogen and organophosphorus compounds can be either reactive or additive.

Many flame retardants, while having measurable or considerable toxicity, degrade into compounds that are also toxic, and in some cases, the degradation products may be the primary toxic agent. For example, halogenated compounds with aromatic rings can degrade into dioxin derivatives, particularly when heated, such as during production, a fire, recycling, or exposure to the sun. In addition, polybrominated diphenyl ethers with higher numbers of bromine atoms, such as decabromodiphenyl ether (decaBDE), are less toxic than pentabromodiphenyl ether derivatives with lower numbers of bromine atoms (Table 6.6). However, as the higher-order pentabromodiphenyl ether derivatives degrade biotically or abiotically, bromine atoms are removed, resulting in more toxic pentabromodiphenyl ether derivatives.

In addition, when some of the halogenated flame retardants such as pentabromodiphenyl ether derivatives are metabolized, they form hydroxylated metabolites that can be more toxic than the parent compound. These hydroxylated metabolites, for example, may compete more strongly to bind with transthyretin or other components of the thyroid system, can be more potent estrogen mimics than the parent compound, and can more strongly affect neurotransmitter receptor activity.

When products with flame retardants reach the end of their usable life, they are typically recycled, incinerated, or landfilled. Recycling can contaminate workers and communities near recycling plants, as well as new materials, with halogenated flame retardants and their breakdown products. Electronic waste, vehicles, and other products are often melted to recycle their metal components, and such heating can generate toxic dioxins and furans. Brominated flame retardants may also change the physical properties of plastics, resulting in inferior performance in recycled products. Poor-quality incineration similarly generates and releases high quantities of toxic degradation products. Controlled incineration of materials with halogenated flame retardants, while costly, substantially reduces release of toxic byproducts.

Many products containing halogenated flame retardants are sent to landfills. Additive, as opposed to reactive, flame retardants are not chemically bonded to the base material and reach out more easily. Brominated flame retardants, including pentabromodiphenyl ether derivatives, have been observed leaching out of landfills in some countries. Landfill designs must allow for leachate capture, which would need to be treated, but these designs can degrade with time.

6.4 INDUSTRIAL CHEMICALS

Organic chemistry chemicals are some of the important starting materials for a great number of major chemical industries (Chapter 3). The production of organic chemicals as raw materials or reagents for other applications is a major sector of manufacturing polymers, pharmaceuticals, pesticides, paints, artificial fibers, food additives, etc. Organic synthesis on a large scale, compared to the laboratory scale, involves the use of energy, basic chemical ingredients from the petrochemical sector, catalysts, and after the end of the reaction, separation, purification, storage, packaging, distribution, etc. During these processes, there are many problems of health and safety for workers in addition to the environmental problems caused by their use and disposition as waste.

The industrial organic chemical sector produces organic chemicals used as either chemical intermediates or end-products (Chapter 3) (Sheldon, 2010). This categorization corresponds to Standard Industrial Classification (SIC) code 286 established by the Bureau of Census to track the flow of goods and services within the economy. The 286 category includes gum and wood chemicals (SIC 2861), cyclic organic crudes and intermediates, organic dyes and pigments (SIC 2865), and industrial organic chemicals not elsewhere classified (SIC 2869). By this definition, the industry does not include plastics, drugs, soaps, and detergents, agricultural chemicals or paints, and allied products which are typical end-products manufactured from industrial organic chemicals.

The industrial organic chemical industry uses feedstocks derived from natural gas and crude oil (approximately 90%) and from recovered coal tar condensates generated by coke production (approximately 10%) (Chapter 3) (Speight, 2013a, b, 2014, 2019). The chemical industry produces raw materials and intermediates, as well as a wide variety of finished products for industry, business, and individual consumers. The important classes of products within SIC code 2861 are hardwood and softwood distillation products, wood, and gum naval stores, charcoal, natural dyestuffs, and natural tanning materials.

The chemicals industry is very diverse, comprising (i) basic or commodity chemicals, (ii) specialty chemicals derived from basic chemicals, such as adhesives and sealants, catalysts, coatings, electronic chemicals, and plastic additives, (iii) products derived from life sciences, such as pharmaceuticals, pesticides, and products of modern biotechnology, and (iv) consumer care products, such as soap, detergents, bleaches, hair, and skincare products, and fragrances. The modern global chemicals industry produces tens of thousands of substances (some in volumes of millions of tons, but most of them in quantities of less than 1,000 tons per year). The substances can be mixed by the chemicals industry and sold and used in this form, or they can be mixed by downstream customers of the chemicals industry (e.g., retail stores which sell paint). It is important to note that most of the output from chemical companies is used by other chemical companies or other industries (e.g., metal, glass, electronics), and chemicals produced by the chemicals industry are present in countless products used by consumers (e.g., automobiles, toys, paper, clothing).

The chemical industry involves the use of chemical processes such as chemical reactions and refining methods to produce a wide variety of solid, liquid, and gaseous materials. Most of these products serve to manufacture other items, although a smaller number go directly to consumers. Solvents, pesticides, lye, washing soda, and Portland cement, provide a few examples of products used by consumers. The industry includes manufacturers of and organic-industrial chemicals, petrochemicals, agrochemicals, polymers, and rubber (elastomers), oleo-chemicals (oils, fats, and waxes), explosives, fragrances, and flavors (Table 6.6) (Chapter 3).

Chemical processes such as chemical reactions operate in chemical plants to form new substances in various types of reaction vessels. In many cases, the reactions take place in special corrosion-resistant equipment at elevated temperatures and pressures with the use of catalysts. The products of these reactions are separated using a variety of techniques including distillation, especially fractional distillation, precipitation, crystallization, adsorption, filtration, sublimation, and drying (Speight, 2020).

The processes and product or products are usually tested during and after manufacture by dedicated instruments and on-site quality control laboratories to ensure safe operation and to assure that the product will meet required specifications. More organizations within the industry are implementing chemical compliance software to maintain quality products and manufacturing standards. The products are packaged and delivered by many methods, including pipelines, tank cars, and tank trucks (for both solids and liquids), cylinders, drums, bottles, and boxes. Chemical companies often have a research-and-development laboratory for developing and testing products and processes. These facilities may include pilot plants, and such research facilities may be located at a site separate from the production plant(s).

Industrial organic chemical manufacturers use and generate both large numbers and quantities of chemicals (Chapter 3). The types of pollutants a single facility will release depend on the feedstocks, processes, equipment in use and maintenance practices. These can vary over a short period of time (sometimes on an from hour to hour basis) and can also vary with the part of the process that is underway. For example, for batch reactions in a closed vessel, the chemicals are more likely to be emitted at the beginning and end of a reaction step (associated with vessel loading and product transfer operations), than during the reaction.

Industrial organic synthesis followed a largely *stoichiometric* line of evolution that can be traced back to the synthesis of mauveine by Perkin, the subsequent development of the dyestuffs industry based on coal tar, and the fine chemicals and pharmaceuticals industries, which can be regarded as spin-offs from the dyestuffs industry. Consequently, fine chemicals and pharmaceuticals manufacture, which is largely the domain of synthetic organic chemists, is rampant with classical stoichiometric processes (Chapter 3).

The desperate need for more catalytic methodologies in industrial organic synthesis is nowhere more apparent than in oxidation chemistry. For example, as any organic chemistry textbook will note that the reagent of choice for the oxidation of secondary alcohols to the corresponding ketones, a pivotal reaction in organic synthesis, is the Jones reagent. The latter consists of chromium trioxide and sulfuric acid and is reminiscent of the phloroglucinol process referred to earlier. The introduction of the storage-stable pyridinium chlorochromate (PCC) and pyridinium dichromate (PDC) in the 1970s, represented a practical improvement, but the stoichiometric amounts of carcinogenic chromium(VI) remain a serious problem. Obviously, there is a definite need in the fine chemical and pharmaceutical industry for catalytic systems that are green and scalable and have broad utility.

6.4.1 VOLATILE ORGANIC COMPOUNDS

Organic compounds that evaporate easily are collectively referred to as volatile organic compounds (VOCs). Microbial volatile organic compounds (MVOCs) are a variety of compounds formed in the metabolism of fungi and bacteria (Korpi et al., 2009). Typically, a VOC is an organic compound that will evaporate at the temperature of use and which, by a photochemical reaction, will cause oxygen in the air to be converted into smog-promoting ozone under favorable climatic conditions.

Thus, VOCs are gases that are emitted into the air from products or processes (Table 6.8). Some are harmful by themselves, including some that cause cancer. In addition, they can react with other gases and form other air pollutants after they are in the air. Many VOCs have been classified as toxic and carcinogenic. They are found in a wide variety of commercial, industrial, and residential products including fuel oils, gasoline, solvents, cleaners, and degreasers, paints, inks, dyes, refrigerants, and pesticides. When VOCs are spilled or improperly disposed of, they can soak into the soil and eventually end up in groundwater. VOCs are found in almost all-natural and synthetic materials and are commonly used in fuels, fuel additives, solvents, perfumes, flavor additives, and deodorants. Potential health hazards and environmental degradation resulting from the widespread use of VOCs has prompted increasing concern among scientists, industry, and the general public (Table 6.9).

Typically, VOCs are organic chemicals that have a high vapor pressure at ordinary room temperature. The high vapor pressure results from a low boiling point, which causes large numbers of molecules to evaporate from the liquid form or sublime from the solid form of the compound and enter the surrounding air. VOCs are numerous, varied, and ubiquitous. They include both human-made and naturally occurring chemical compounds. Many VOCs are dangerous to human health or cause harm to the environment. Anthropogenic VOCs are regulated by law, especially indoors, where concentrations are the highest. Harmful VOCs typically are not acutely toxic, but have compounding long-term health effects.

TABLE 6.8 Sources of Volatile Organic Compounds

Indoor Sources*
• Air fresheners
• Car exhaust in an attached garage
• Cleaning products, varnishes, wax
• Furniture or building products such as flooring, carpet, pressed wood products
• Gasoline, fuel oil
• Hobby products such as glue
• Office equipment including printers and copiers
• Paint, paint remover
• Pesticides
• Personal care products such as cosmetics
• Tobacco smoke
• Wood burning stoves
Outdoor Sources*
• Crude oil production and refining
• Diesel emissions
• Gasoline emissions
• Industrial emissions.
• Natural gas production and refining
• Woodburning

*Listed alphabetically.

TABLE 6.9 Harmful Effects of Selected Constituents of Volatile Organic Compounds (VOCs)

Constituent	Effects
Benzene	Exposure to benzene (C_6H_6) can cause skin and respiratory irritation, and long-term exposure can lead to cancer and blood, developmental, and reproductive disorders.
Toluene	Long-term exposure to toluene ($C_6H_5CH_3$) can cause skin and respiratory irritation, headaches, dizziness, birth defects and damage to the nervous system.
Ethylbenzene	Long-term exposure to ethylbenzene ($C_6H_5CH_2CH_3$) can cause irritation of the throat and eyes, and dizziness and long-term exposure can cause blood disorders.
Xylene	High levels of exposure to the xylene isomers [($C_6H_4(CH_3)_2$] can have numerous short-term impacts, including nausea, gastric irritation, and neurological effects.
	Long-term exposure to the xylene isomers can negatively impact the nervous system.
n-Hexane	Exposure to n-hexane [n-$CH_3(CH_2)_4CH_3$] can cause dizziness, nausea, and headaches.
	Long-term exposure to n-hexane can lead to numbness, muscular atrophy, blurred vision. and fatigue.

Non-methane volatile organic compounds (NMVOCs) are a collection of organic compounds that differ widely in their chemical composition but display similar behavior in the atmosphere. NMVOCs are emitted into the atmosphere from a large number of sources including combustion activities, solvent use and production processes. NMVOCs contribute to the formation of ground-level (tropospheric) ozone. In addition, certain NMVOC species or species groups such as benzene and 1,3 butadiene are hazardous to human health. Quantifying the emissions of total NMVOCs provides an indicator of the emission trends of the most hazardous NMVOCs.

6.4.2 WOOD SMOKE

Wood smoke forms when wood and is made up of a complex mixture of gases and fine particles (particulate matter, PM). In addition to particle pollution, wood smoke contains several toxic harmful air pollutants, including benzene, formaldehyde, acrolein, and polycyclic aromatic hydrocarbon derivatives (PAHs, also known as polynuclear aromatic hydrocarbon derivatives, PNAs).

Wood smoke is by far the most compelling argument against wood heating. In cold climates, a frigid, stagnant air mass traps the smoke close to the ground. In the process of wood combustion, wood vaporizes when heated into gases and tar particles. If the temperature is high enough the tar particles vaporize into chemicals and carbon particles. If the temperature is higher still and there is oxygen present, the gases and particles burn in bright flames.

Chemically, wood is approximately 50% w/w carbon and the rest is mostly made up of oxygen and hydrogen. When a piece of wood is heated, it starts to smoke and turn black at the same time. This is because the other products vaporize under intense heat faster than the carbon burns, so smoking leaves much of the carbon behind until only charcoal remains. The smoke that vaporizes out of the wood is a cloud of nasty, gooey little droplets of a tar-like liquid. Chemically, these droplets are actually big, gooey, complicated hydrocarbon molecules that take a number of different forms, mostly bad.

When wood is burned correctly in a bright, hot, turbulent fire, what is observed is the tar droplets rising off the wood into a zone of extreme heat where they re-vaporize, undergoing thermal decomposition into (for the most part) gaseous, constituents. This leaves carbon dioxide, some carbon monoxide and a number of other gases, water vapor and some not quite completely oxidized hydrocarbon products (typically identified as particulate emissions.

The sentiment that wood smoke, being a natural substance, must be benign to humans is still sometimes heard. It is now well established, however, that wood-burning stoves and fireplaces as well as wildfires and agricultural fires emit significant quantities of known health-damaging pollutants, including several carcinogenic compounds (Naeher et al., 2007). Two of the principal gaseous pollutants in wood smoke, carbon monoxide (CO) and the oxides of nitrogen (NOx), add to the atmospheric levels of these regulated gases emitted by other combustion sources. Thus, there are two issues related to wood smoke and they are (i) whether or not woodsmoke should be regulated and/or managed separately, even though some of the individual constituents are already regulated in many jurisdictions and (ii) whether or not woodsmoke particles pose different levels of risk than other ambient particles of similar size.

The particulate matter (PM) wood smoke is extremely small and therefore are not filtered out by the nose or the upper respiratory system. Instead, these small particles end up deep in the lungs where they remain for months, causing structural damage and chemical changes. The carcinogenic chemicals in wood smoke adhere to these tiny particles, which enter deep into the lungs.

6.5 EFFECTS ON THE ENVIRONMENT

Pollution by chemicals can be caused by chemicals from a variety of sources and can involve a variety of health effects from simple digestive problems to chemical intoxication which is caused by exposure to chemical pollutants and can have immediate effects or delayed effects, which may appear after weeks or even months after the exposure occurred. Various chemical pollutants may accumulate in the aquatic sediments over longer periods of time. Thus, if the requested testing procedure are not performed, chemical pollution in the ocean water could pose serious health risks to the ecosystem and ultimately could cause mild or deadly chemical intoxication in humans after the consumption of contaminated seafood.

Organisms in an ecosystem will (more than likely) respond differently to the frequency and duration of a given environmental change. For example, if some individual organisms in a population have adaptations that allow them to survive and to reproduce under new environmental conditions, the population will continue, but the genetic composition will have changed (Darwinism). On the other hand, some organisms have the ability to adapt to the environment (i.e., to adjust their physiology or morphology in response to the immediate environment) so that the new environmental conditions are less (certainly no more) stressful than the previous conditions. Such changes may not be genetic (Lamarckism).

In terms of anthropogenic stress (the effect of human activity on other organisms), there is the need for the identification and evaluation of the potential impacts of proposed projects, plans, programs, policies, or legislative actions upon the physical-chemical, biological, cultural, and socioeconomic components of the environment. This activity is also known as environmental impact assessment (EIA) and refers to the interpretation of the significance of anticipated changes related to a proposed project. The activity encourages consideration of the environment and arriving at actions that are environmentally compatible.

Identifying and evaluating the potential impact of human activities on the environment requires the identification of mitigation measures. *Mitigation* is the sequential consideration of the following measures: (i) avoiding the impact by not taking a certain action or partial action; (ii) minimizing the impact by limiting the degree or magnitude of the action and its implementation; (iii) rectifying the impact by repairing, rehabilitating, or restoring the affected environment; (iv) reducing or eliminating the impact over time by preservation and maintenance operations during the life of the action; and (v) compensating for the impact by replacing or providing substitute resources or environments.

Nowhere is the effect of anthropogenic stress felt more than in the development of natural resources of the earth. Natural resources are varied in nature and often require definition. For example, in relation to mineral resources, for which there is also descriptive nomenclature (ASTM C294), the terms related to the available quantities of the resource must be defined. In this instance, the term resource refers to the total amount of the mineral that has been estimated to be ultimately available. Reserves are well-identified resources that can be profitably extracted and utilized by means of existing technology.

In some cases, environmental pollution is a clear-cut phenomenon, whereas in others it remains a question of degree. The ejection of various materials into the environment is often cited as pollution, but there is the ejection of the so-called beneficial chemicals that can assist the air, water, and land to perform their functions. However, it must be emphasized that the ejection of chemicals into the environment, even though they are indigenous to the environment, in quantities above the naturally occurring limits can be extremely harmful. In fact, the timing and the place of a chemical release are influential in determining whether a chemical is beneficial, benign, or harmful! Thus, what may be regarded as a pollutant in one instance can be a beneficial chemical in another instance. The phosphates in fertilizers are examples of useful (beneficial) chemicals while phosphates generated as by-products in the metallurgical and mining industries may, depending upon the specific industry, be considered pollutants (Chenier, 1992). In this case, the means by which such pollution can be prevented must be recognized (Breen and Dellarco, 1992). Thus, increased use of the resources of the Earth as well as the use of a variety of chemicals that are non-indigenous to the Earth have put a burden on the ability of the environment to tolerate such materials.

Finally, some recognition must be made of the term carcinogen since many of the environmental effects referenced in this text can lead to cancer. *Carcinogens* are cancer-causing substances, and there is a growing awareness of the presence of carcinogenic materials in the environment. A classification scheme is provided for such materials (Table 6.10) (Zakrzewski, 1991; Milman and Weisburger, 1994). The number of substances with which a person comes in contact are in the tens of thousands, and there is not a full understanding of the long-term effects of these substances in their possible propensity to cause genetic errors that ultimately lead to carcinogenesis. Teratogens are those substances that tend to cause developmental malformations.

TABLE 6.10 Carcinogenicity Classification Scheme as Determined by the United States Environmental Protection Agency

Group	Description
A	Human carcinogen
B1	Probable human carcinogen; limited human data are available
B2	Probable human carcinogen
C	Possible human carcinogen
D	Not classifiable as a human carcinogen
E	No carcinogenic activity in humans

Pollution is the introduction of indigenous (beyond the natural abundance) and non-indigenous (artificial) gaseous, liquid, and solid contaminants into an ecosystem. The atmosphere and water and land systems have the ability to cleanse themselves of many pollutants within hours or days, especially when the effects of the pollutant are minimized by the natural constituents of the ecosystem. For example, the atmosphere might be considered to be self-cleaning as a result of rain. However, removal of some pollutants from the atmosphere (e.g., sulfates, and nitrates) by rainfall results in the formation of acid rain that can cause serious environmental damage to ecosystems within the water and land systems (Pickering and Owen, 1994).

Briefly, lakes in some areas of the world are now registering a low pH (acidic) reading because of excess acidity in rain. This was first noticed in Scandinavia and is now prevalent in eastern Canada and the northeastern United States. Normal rainfall has a pH of 5.6 and the slight acidity (neutral water has a pH equal to 7.0) because of carbon dioxide (CO_2) in the air that, with water, forms carbonic acid (H_2CO_3):

$$CO_2 + H_2O \rightarrow H_2CO_3$$

The increased use of hydrocarbon fuels in the last five decades is slowly increasing the concentration of carbon dioxide in the atmosphere, which produces more carbonic acid leading to an imbalance in the natural carbon dioxide content of the atmosphere that, in turn, leads to more acidity in the rain. In addition, there is a so-called greenhouse effect (Figure 6.2) and the average temperature of the earth may be increasing.

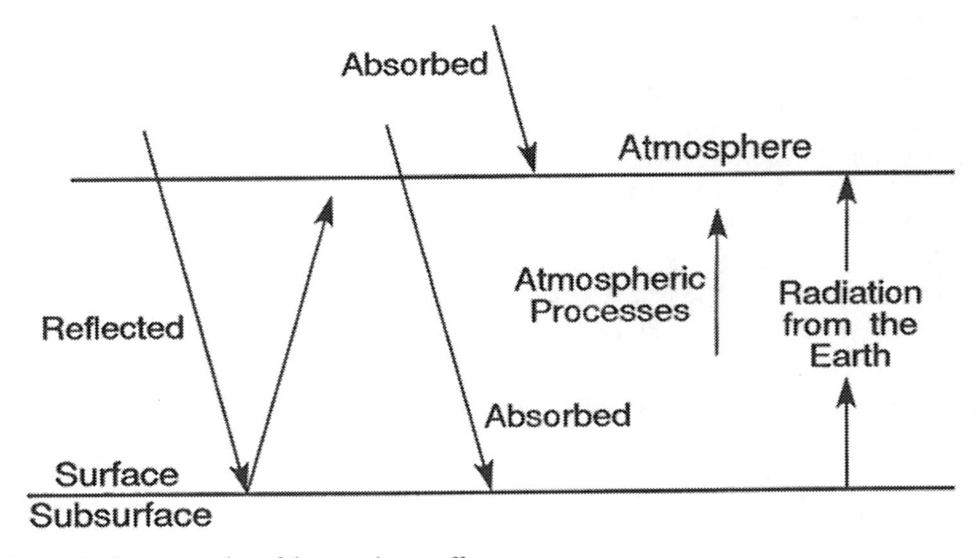

FIGURE 6.2 Representation of the greenhouse effect.
Source: Reproduced with permission from: Speight (1996). © Taylor and Francis.

In addition, excessive use of fuels with high sulfur and nitrogen content cause sulfuric and nitric acids in the atmosphere from the sulfur dioxide and nitrogen oxide products of combustion that can be represented simply as:

$$SO_2 + H_2O \rightarrow H_2SO_3$$
sulfurous acid

$$2SO_2 + O_2 \rightarrow 2SO_3$$

$$SO_3 + H_2O \rightarrow H_2SO_4$$
sulfuric acid

$$NO + H_2O \rightarrow HNO_2$$
nitrous acid

$$2NO + O_2 \rightarrow NO_2$$

$$NO_2 + H_2O \rightarrow HNO_3$$
nitric acid

A pollutant is a substance (for simplicity, most are referred to as chemicals) present in a particular location when it is not indigenous to the location or is in a greater-than-natural concentration. The substance is often the product of human activity. The pollutant, by virtue of its name, has a detrimental effect on the environment, in part or in toto. Pollutants can also be subdivided into two classes: (i) primary pollutants; and (ii) secondary pollutants:

$$\text{Source} \rightarrow \text{Primary Pollutant} \rightarrow \text{Secondary Pollutant}$$

Primary pollutants are those pollutants emitted directly from the source. In terms of atmospheric pollutants by crude oil constituents, examples are hydrogen sulfide, carbon oxides, sulfur dioxide, and nitrogen oxides from refining operations (see above). On the other hand, secondary pollutants are produced by interaction of primary pollutants with another chemical or by dissociation of a primary pollutant, or other effects within a particular ecosystem. Again, using the atmosphere as an example, the formation of the constituents of acid rain is an example of the formation of secondary pollutants.

The question of classifying nitrogen dioxide and sulfur trioxide as primary pollutants often arises, as does the origin of the nitrogen. In the former case, these higher oxides can be formed in the upper levels of the combustors. The nitrogen, from which the nitrogen oxides are formed, does not originate solely from the fuel but may also originate from the air used for the combustion.

KEYWORDS

- **Environmental Protection Agency**
- **low-volatile organic compounds**
- **non-volatile organic compounds**
- **resource conservation and recovery act**
- **volatile organic compounds**
- **wastewater treatment plant**

REFERENCES

ASTM C694, (2016). *Standard Test Method for Weight Loss (Mass Loss) of Sheet Steel During Immersion in Sulfuric Acid Solution*. Annual Book of Standards, ASTM International, West Conshohocken, Pennsylvania.

ASTM D1310, (2016). *Standard Test Method for Flash Point and Fire Point of Liquids by Tag Open-Cup Apparatus*. Annual Book of Standards, ASTM International, West Conshohocken, Pennsylvania.

ASTM D1838, (2016). *Standard Test Method for Copper Strip Corrosion by Liquefied Petroleum (LP) Gases*. Annual Book of Standards, ASTM International, West Conshohocken, Pennsylvania.

ASTM D2251, (2016). *Standard Test Method for Metal Corrosion by Halogenated Organic Solvents and Their Admixtures*. Annual Book of Standards, ASTM International, West Conshohocken, Pennsylvania.

ASTM D4982, (2016). *Standard Test Methods for Flammability Potential Screening Analysis of Waste*. Annual Book of Standards, ASTM International, West Conshohocken, Pennsylvania.

ASTM D92, (2016). *Standard Test Method for Flash and Fire Points by Cleveland Open Cup Tester*. Annual Book of Standards, ASTM International, West Conshohocken, Pennsylvania.

ASTM E1302, (2016). *Standard Guide for Acute Animal Toxicity Testing of Water-Miscible Metalworking Fluids*. Annual Book of Standards, ASTM International, West Conshohocken, Pennsylvania.

ASTM E176, (2016). *Standard Terminology of Fire Standards*. Annual Book of Standards, ASTM International, West Conshohocken, Pennsylvania.

ASTM E502, (2016). *Standard Test Method for Selection and Use of ASTM Standards for the Determination of Flash Point of Chemicals by Closed Cup Methods*. Annual Book of Standards, ASTM International, West Conshohocken, Pennsylvania.

ASTM E681, (2016). *Standard Test Method for Concentration Limits of Flammability of Chemicals (Vapors and Gases)*. Annual Book of Standards, ASTM International, West Conshohocken, Pennsylvania.

ASTM E789, (2016). *Standard Test Method for Explosibility of Dust Clouds*. Annual Book of Standards, ASTM International, West Conshohocken, Pennsylvania.

Austin, G. T., (1984). In: *Shreve's Chemical Process Industries* (5th edn.). McGraw-Hill, New York. Chapter 22.

Barcelona, M., Wehrmann, A., Keeley, J. F., & Pettyjohn, J., (1990). *Contamination of Ground Water*. Noyes Data Corp., Park Ridge, New Jersey.

Carson, P., & Mumford, C., (1988). *The Safe Handling of Chemicals in Industry, 1, 2*. John Wiley & Sons Inc., New York.

Carson, P., & Mumford, C., (1995). *The Safe Handling of Chemicals in Industry, 3*. John Wiley & Sons Inc., New York.

Carson, P., & Mumford, R., (2002). *Hazardous Chemicals Handbook* (2nd edn.). Butterworth-Heinemann, Oxford, United Kingdom.

Carson, R., (1962). *Silent Spring*. Houghton Mifflin Company, Houghton Mifflin Harcourt International, Geneva, Illinois.

Chenier, P. J., (1992). *Survey of Industrial Chemistry* (2nd edn.). VCH Publishers Inc., New York.

Guglielmi, G., (2020). *Why Beirut's Ammonium Nitrate Blast Was So Devastating*. Nature-News. August 10.

Hites, R. A., (2007). *Elements of Environmental Chemistry* (pp. 155–179). John Wiley & Sons Inc., Hoboken, New Jersey. Chapter 6.

https://www.nature.com/articles/d41586-020-02361-x (accessed on 20 November 2021).

Jacob, J., (2013). A review of the accumulation and distribution of persistent organic pollutants in the environment. *International Journal of Bioscience, Biochemistry, and Bioinformatics, 3*(6), 657–661.

Kim, S., & Holtzapple, M. T., (2005). Lime pretreatment and enzymatic hydrolysis of corn stover. *Bioresource Technology, 96*(18), 1994–2006.

Kim, S., & Holtzapple, M. T., (2006a). Effect of structural features on enzyme digestibility of corn stover. *Bioresource Technology, 97*(4), 583–591.

Kim, S., & Holtzapple, M. T., (2006b). Delignification kinetics of corn stover in lime pre-treatment. *Bioresource Technology, 97*(5), 778–785.

Kletz, T. A., (1990). *Improving Chemical Engineering Practices* (2nd edn.). Hemisphere Publishers, Taylor& Francis, Washington, DC.

Korpi, A., Järnberg, J., & Pasanen, A. L., (2009). Microbial volatile organic compounds. *Critical Reviews in Toxicology, 39*(2), 139–193.
Mackay, D., McCarty, L. S., & MacLeod, M., (2001). On the validity of classifying chemicals for persistence, bioaccumulation, toxicity, and potential for long-range transport. *Environmental Toxicology and Chemistry, 20*(7), 1491–1498.
Manahan, S. E., (2010). *Environmental Chemistry* (9th edn.). CRC Press, Taylor & Francis Group, Boca Raton, Florida.
Milman, H. A., & Weisburger, E. K., (1994). *Handbook of Carcinogen Testing* (2nd edn.). Noyes Data Corp., Park Ridge, New Jersey.
Mody, V., & Jakhete, R., (1988). *Dust Control Handbook.* Noyes Data Corp., Park Ridge, New Jersey.
Naeher, L. P., Brauer, M., Lipsett, M., Zelikoff, J. T., Simpson, C. D., Koenig, J. Q., & Smith, K. R., (2007). Woodsmoke health effects: A review. *Inhalation Toxicology, 19*, 67–106.
NRC, (1981). *Prudent Practices for Handling Hazardous Chemicals in Laboratories*. National Academy Press, National Research Council, Washington, DC.
Oliveira, B. F., Ignotti, E., & Hacon, S. S., (2011). A systematic review of the physical and chemical characteristics of pollutants from biomass burning and combustion of fossil fuels and health effects in Brazil. *Cad Saude Publica, 27*(9), 1678–1698.
Pickering, K. T., & Owen, L. A., (1994). *Global Environmental I*ssues. Routledge Publishers, New York.
Pöschl, U., (2005). Atmospheric aerosols: composition, transformation, climate, and health effects. *Angew. Chem. Int. Ed., 44,* 7520–7540.
Sheldon, R., (2010). Introduction to green chemistry, organic synthesis and pharmaceuticals. In: Dunn, P. J., Wells, A. S., & Williams, M. T., (ed.), *Green Chemistry in the Pharmaceutical Industry*. Wiley-VCH Verlag GmbH & Co. KGaA, Weinheim, Germany.
Speight, J. G., & Arjoon, K. K., (2012). *Bioremediation of Petroleum and Petroleum Products*. Scrivener Publishing, Salem, Massachusetts.
Speight, J. G., & Singh, K., (2014). *Environmental Management of Energy from Biofuels and Biofeedstocks*. Scrivener Publishing, Beverly, Massachusetts.
Speight, J. G., (1996). *Environmental Technology Handbook*. Taylor & Francis, Washington, DC.
Speight, J. G., (2005). *Environmental Analysis and Technology for the Refining Industry*. John Wiley & Sons Inc., Hoboken, New Jersey.
Speight, J. G., (2011). *The Biofuels Handbook.* Royal Society of Chemistry, London, United Kingdom.
Speight, J. G., (2013a). *The Chemistry and Technology of Coal* (3rd edn.). CRC Press, Taylor & Francis Group, Boca Raton, Florida.
Speight, J. G., (2013b). *Coal-Fired Power Generation Handbook.* Scrivener Publishing, Beverly, Massachusetts.
Speight, J. G., (2014). *The Chemistry and Technology of Petroleum* (5th edn.). CRC Press, Taylor & Francis Group, Boca Raton, Florida.
Speight, J. G., (2015). *Handbook of Petroleum Product Analysis* (2nd edn.). John Wiley & Sons Inc., Hoboken, New Jersey,
Speight, J. G., (2017a). *Environmental Inorganic Chemistry for Engineers*. Butterworth-Heinemann, Elsevier, Oxford, United Kingdom.
Speight, J. G., (2017b). *Environmental Organic Chemistry for Engineers*. Butterworth-Heinemann, Elsevier, Oxford, United Kingdom.
Speight, J. G., (2018). *Reaction Mechanisms in Environmental Engineering: Analysis and Prediction.* Butterworth-Heinemann, Elsevier, Oxford, United Kingdom.
Speight, J. G., (2019). *Natural Gas: A Basic Handbook* (2nd edn.). Gulf Publishing Company, Elsevier, Cambridge, Massachusetts.
Speight, J. G., (2020). *Handbook of Industrial Hydrocarbon Processes* (2nd edn.). Gulf Publishing Company, Elsevier, Cambridge, Massachusetts.
Sudweeks, W. B., Larsen, R. D., & Balli, F. K., (1983). In: Kent, J. A., (ed.), *Riegel's Handbook of Industrial Chemistry* (p. 700). Van Nostrand Reinhold, New York.
Syed, S., (2006). Solid and liquid waste management. *Emirates Journal for Engineering Research, 11*(2), 19–36.

Tedder, D. W., & Pohland, F. G., (1993). *Emerging Technologies in Hazardous Waste Management III.* Symposium Series No. 518. American Chemical Society, Washington, D.C.

Yang, X., Luo, F., Li, J., Chen, D., Ye, E., Lin, W., & Jin, J., (2019). Alkyl and aromatic nitrates in atmospheric particles determined by gas chromatography-tandem mass spectrometry. *Journal of the American Society for Mass Spectrometry, 30*, 2762–2770.

Zakrzewski, S. F., (1991). *Principles of Environmental Toxicology.* ACS Professional Reference Book, American Chemical Society, Washington, DC.

CHAPTER 7

Introduction to the Environment

7.1 INTRODUCTION

The release or disposal of a contaminant (a chemical or a mixture of chemicals) into the environment occurs in several ways. Sources (domestic and/or industrial facilities) may release chemical waste into the air or water or dispose of it on land, per the local, state (provincial), or national regulatory requirements.

Contamination of the environment is a global issue and there is no single company that should shoulder all of the blame. Past laws and regulations (or the lack thereof) allowed unmanaged disposal of contaminants and discharge of contaminants into the environment. These companies were not breaking the law. It is a matter of there being insufficient laws (the fault of various levels of government) enacted to protect the environment. Moreover, the inappropriate management of such chemical waste has resulted in negative impacts on the environment. Thus, contaminants (with varying levels of toxicity) are found practically in all ecosystems on earth along with the adverse effects of these contaminants: (i) on biodiversity; (ii) on agricultural production; and/or (iii) on water resources. At the end of the life cycles of the contaminant, the contaminant is recycled or sent for disposal.

Briefly, environmental pollution can be classified using the following criteria: (i) natural pollution resulting from volcanoes, hurricanes, earthquakes, sand storms, and (ii) artificial pollution resulting from human activities: industry, agriculture, and domestic activities (Table 7.1). A second method of general classifications can be made according to the type of the pollutants, into: (i) physical pollution, such as radiation, (ii) chemical pollution, such as by combustion products (carbon monoxide, carbon dioxide and nitrogen oxides, sulfur compounds, nitrates, phosphates, and heavy metals (Wuana and Okieimen, 2011). The third and final method of pollution is a sub-division according to the polluted medium, into (i) air pollution resulting from gases, powders from factories, vehicle emissions, and odors from agricultural activities, (ii) water pollution, discharges of industrial residues such as metals and salts, and (iii) soil pollution, resulting from non-ecological tourism, waste grounds, car cemeteries, foams, and insecticides.

The potential for emissions from the manufacture and use of chemicals is high but, because of economic necessity, the potential for emissions is reduced by recovery of the chemicals. In some cases, the manufacturing process is operated as a closed system that allows little or no emissions to escape to the environment (air, water, or land). Emission

The Science and Technology of the Environment. James G. Speight (Author)

sources from the various processes include heaters and boilers; valves, flanges, pumps, and compressors; storage and transfer of products and intermediates; wastewater handling; and emergency vents. Regular maintenance of the process equipment reduces the potential for these emissions. However, the emissions that do reach the atmosphere from industrial courses are generally gaseous emissions that are controlled by a sequence of gas cleaning operations, including an adsorption process or an or absorption process. In addition, emission of particulate matter (PM), which could also lead to an environmental issue, since the particulate materials emitted is usually extremely small (typically a collection of particulates in the micron range) also requires (and is subject to) efficient treatment for removal (Mokhatab et al., 2006; Speight, 2007, 2014, 2017; Kidnay et al., 2011; Bahadori, 2014).

TABLE 7.1 Examples of Inorganic Pollutants

Pollutant	Formula	Sources	Polluted Medium
Ammonia	NH_3	Industry	Air, soil
–	–	Agriculture	Air, soil
Ammonium salts	NH_4^+	Farms, factories	Soil, water
Carbon monoxide	CO	Transports	Air
Hydrogen chloride	HCl	Industry	Air
–	–	Transports	Air
Hydrogen fluoride	HF	Industry	Air
Hydrogen sulfide	H_2S	Industry	Soil, water
–	–	Anaerobic fermentation	Soil, water
Lead salts*	Pb	Heavy industry	Air
–	–	Transports	Air
Mercury salts*	Hg	Industry	Soil, water
Nitrate salts	NO_3^-	Farms, factories	Soil, water
Nitrite salts	NO_2^-	Farms, factories	Soil, water
Nitrogen dioxide	NO_2	Volcanoes	Air
–	–	Industry	Air
–	–	Transports	Air
Sulfur dioxide	SO_2	Volcanoes	Air
–	–	Industry	Air
–	–	Transports	Air
Zinc salts*	Zn	Industry	Air, water

*May also appear as the metal.

Evaluating the release of various contaminants can assist in the identification of any potential concerns and gain a better understanding of potential consequences that the release of the contaminants may pose (Table 7.2). However, it is important to consider that the quantity of releases is not necessarily an indicator of health impacts posed by the contaminants. Many factors can affect trends in releases at facilities, including production rates, management practices, the composition of raw materials used, and the installation of control technologies.

TABLE 7.2 Examples of Common Pollutants*

Pollutant	Polluted Medium
Ammonia, NH_3	Air, soil
Carbon dioxide, CO_2	Air
Carbon monoxide, CO	Air
Hydrogen chloride, HCl	Air
Hydrogen fluoride, HF	Air
Hydrogen sulfide, H_2S	Soil, water
Lead, Pb	Water, soil
Mercury, Hg	Water, soil
Nitrogen dioxide, NO_2	Air
Sulfur dioxide, SO_2	Air

*Listed alphabetically.

Both the concentration of the contaminants and how they enter a waterway vary greatly. As with some contaminants, natural global cycling has always been a primary contributor to the presence of contaminants in air, water, and soil (or sediments). This process can involve transfer of a contaminant between the atmosphere, the hydrosphere, and the lithosphere, where it may eventually be transported and deposited onto surface water and soil. Major anthropogenic sources of contaminants in the environment have been: (i) mining operations; (ii) industrial processes; (iii) combustion of fossil fuels, especially charcoal; (iv) production of cement; as well as (v) incineration of municipal, industrial, and medical wastes. Alternatively, industries such as forest processing, meat processing, and dairy processing, wastewater treatment may discharge wastewater that can potentially contain inorganic chemical contaminants: examples are bleach (hypochlorite derivative), curing agents (which include nitrate derivatives, $-NO_3$, and nitrite derivatives, $-NO_2$), and certain metals like mercury, copper, chromium, zinc, iron, arsenic, and lead). Prior treatment of these contaminants before discharge is now strictly regulated and controlled via the resource consenting process, and will vary depending on the type, quantity, and the potential environmental reactivity of the contaminant to be discharged.

In the case of the discharge of any contaminant into the environment is proposed, it is necessary to set standards for acceptable concentrations in air, water, soil, and flora as well as fauna-if there are any such acceptable concentrations. Monitoring of these contaminants in an ecosystem and any subsequent biological effects must be undertaken to ensure that the standards as set in any regulation are realistic and provide protection of the environment from any adverse effects. Furthermore, considerable attention must involve the prediction of the behavior and effects of a contaminant from the properties of that substance (Chapter 6).

Also, this concept that the molecular characteristics of the molecule (such as the three-dimensional structure and the presence of functional groups) govern the physical and chemical properties of the compound which in turn influence the effects of the contaminant on the ecosystem as well as the transformation and distribution of the contaminant from the original ecosystem where the contaminant was first released (Chapter 8). In many cases, the transformation and distribution in the environment as well as any effects on the floral and faunal species can be predicted from the physical properties and/or the chemical properties contaminant. However, the prediction of biological effects may also involve a complex set of contaminant-floral interactions and/or contaminant-faunal interactions.

In the past, especially during the 19th Century and the first half of the 20th Century, when environmental legislation was not in place or, if in place, was not enacted, the disposal of chemical waste to all types of environmental ecosystems including air through both fugitive emissions (emissions of gas or vapor from process equipment due to leaks and other inadvertent release of contaminants) and direct emissions (emissions from sources that are owned or controlled by the reporting entity), water (direct discharge and runoff) and land.

The pollutants that are most likely to present ecological risks are those that are: (i) reactive chemicals, which have the potential to react with any part of the ecosystem and undergo transformation; (ii) contaminants-especially bio-accumulative chemicals-that accumulate in floral and faunal tissues even when concentrations in the ecosystem remain relatively low; and (iii) highly toxic chemicals, which can cause adverse effects to the ecosystem itself and to the floral and faunal members of the ecosystem at comparatively low doses; and (iv) synthetic chemicals that contain halogen atoms (particularly fluorine, chlorine, or bromine) are often resistant to degradation in the environment or within organisms. Briefly and by way of explanation, bioaccumulation is a process by which persistent environmental pollution leads to the uptake and accumulation of one or more contaminants, by organisms in an ecosystem.

Finally, the environmental investigator should be aware of the influence of minerals on the dispersion of pollutants in the environment. Minerals (of which more than 3,000 individual minerals are known) are substances formed naturally in the Earth. Have a definite chemical composition and structure (Chapter 5). Some are rare and precious such as gold (which occurs as the element) and diamond (a crystalline form of carbon), while others are more ordinary and ubiquitous, such as quartz (silica, SiO_2). Certain minerals directly cause environmental hazards ranging from air and water pollution to contamination within residential communities. Other minerals affect the dispersion of pollutants because of the adsorptive capacity of the minerals or the tendency of the minerals to interact chemically with the pollutant(s) thereby promoting the chemical and physical transformation of the pollutant(s) in the environment (Chapter 8).

Of particular interest in this respect are the clay minerals which are abundant in nature and because of the (i) large surface area, (ii) the high porosity, (iii) the surface charge, and (iv) functional groups on the mineral surface, clay minerals function as natural adsorbents and (Yuan et al., 2013). Furthermore, the surfaces of clay minerals can vary from hydrophilic to hydrophobic making them adsorbents and carriers not only for inorganic (ionic) chemicals but also for non-ionic organic chemicals (Chapter 5).

7.2 RELEASE OF CHEMICALS INTO THE ENVIRONMENT

For the purposes of this text, it is assumed that any contaminants released into the environment are harmful to the environment. In fact, it is not only safe, but necessary, to assume that any contaminants (except chemicals that are indigenous to the ecosystem into which they exist but in quantities that do not exceed the indigenous amounts) have the potential to be harmful to the environment and also to human health.

As already stated (Chapter 1), in the case of the various pollutants released into the environment by human activity (Figure 7.1) (Speight, 1996), persistent pollutants (POPs) among the most dangerous to environmental flora and fauna are (i) pesticides, (ii) various industrial chemicals, or (iii) the unwanted by-products of industrial processes that have been found to share several disturbing characteristics. These characteristics include: (i) persistence in the environment, which means that the contaminants resist degradation in air, water, and sediments and these types of pollutants (POPs) are highly toxic and long-lasting, and cause an array of adverse effects on the flora and fauna of an ecosystem., (ii) bio-accumulation, which is the tendency for contaminants to accumulate in living tissues at concentrations higher than the concentration in the surrounding environment, and (iii) long-range transport, which means that the contaminants can travel a considerable distance from the source of release through the air, water, and migratory animals, often leading to the contamination ecosystems that are distances removed from any known source of the contaminating chemical.

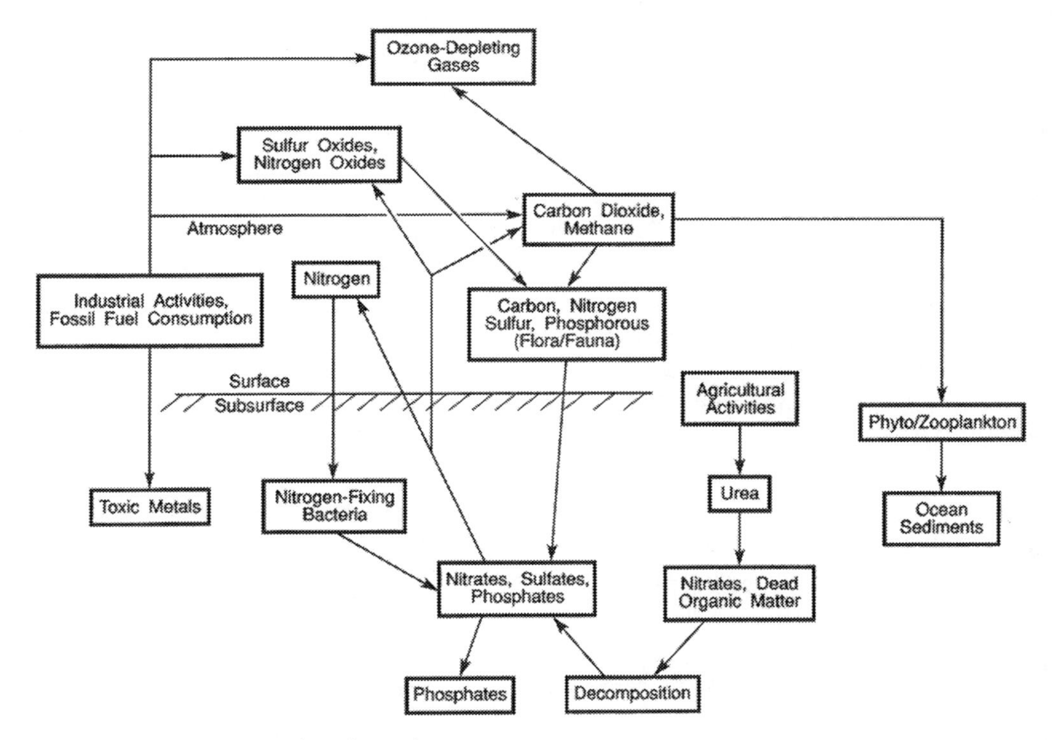

FIGURE 7.1 Pollutant entry into the environment.
Source: Reproduced with permission from: Speight (1996). © Taylor and Francis.

Thus, many contaminants that have toxic, carcinogenic, mutagenic, or teratogenic (causing developmental malformations) effects on environmental flora and fauna are designated either as (i) acutely hazardous waste or as (ii) toxic waste by the United States Environmental Protection Agency (EPA) (https://www.epa.gov/hw). Materials containing any of the toxic constituents so listed are to be considered hazardous waste, unless, after considering the following factors it can reasonably be concluded that (i) the waste is not capable of posing a substantial present or potential hazard to public health; or (ii) the waste is not capable of posing a substantial present or potential hazard to the environment when improperly treated, stored, transported or disposed of, or otherwise managed.

Contaminants released from various sources are ultimately dispersed among, and can at times accumulate in, various environmental systems (such as the air, the water, and the land). Some contaminants that are released are likely to contribute primarily to local ecosystems unless transportation effects are involved. However, other contaminants that are more persistent can be distributed over much greater distances. Most of the contaminants released from the various sources into the air (Chapter 3) do not remain in air but are deposited into or on to surface water (Chapter 4) and soil (Chapter 5).

Furthermore, the accidental release of non-hazardous chemicals and hazardous chemicals into the environment has occurred. It is a situation that, to paraphrase chaos theory: no matter how well the preparation, the unexpected is always inevitable. It is, at this point, that the environmental scientist and the environmental engineer must identity (through careful analysis) the nature of the contaminants and their potential effects on the ecosystem(s). Thus, the predominance of one contaminant or any particular class of contaminants may offer the environmental scientist or engineer an opportunity for predictability of behavior of the contaminant contaminant(s) after consideration of the chemical and physical properties of the contaminant and the effect of the contaminant of the floral and faunal species in an ecosystem.

Thus, when a spill of chemicals occurs the primary processes determining the fate of chemicals are: (i) dispersion, (ii) dissolution, (iii) emulsification, (iv) evaporation, (v) leaching, (vi) sedimentation, (vii) spreading, and (viii) by wind. These processes are influenced by the physical and properties of the chemicals (especially if the chemicals are constituents of a mixture), spill characteristics, environmental conditions, and chemical and physical properties of the spilled material after it is undergone any form of chemical transformation.

Also, it is essential to recognize that the release of pollutants into the environment can also occur as a result of natural hazards such as earthquakes, hurricanes, tsunamis, and floods. A natural hazard can trigger the release of a contaminant from a source which can exacerbate the impact of a natural disaster on the environment and on human health.

For the purposes of this text, the terms listed in the following sub-sections encompass the phenomena (presented alphabetically) which give rise to the proliferation of chemicals through the natural environment and various ecosystems.

7.2.1 ABSORPTION AND ADSORPTION

Absorption is a physical or chemical process in which atoms, ions, or molecules enter the bulk phase of a liquid or solid material. In this case (as opposed to the adsorption process) the guest atoms, ions, or molecules are taken up by the volume, not by the surface (as in the case for adsorption). The absorption process means that a substance captures and transforms energy insofar as the absorbent distributes the guest atoms, ions, or molecules throughout whole and adsorbent only distributes it through the surface. Thus, absorption involves the whole volume of the material, although adsorption often precedes absorption (Table 7.3).

TABLE 7.3 Comparison of Adsorption and Absorption.

	Adsorption	Absorption
Characteristic	A surface phenomenon	A bulk phenomenon
		Accumulation at surface
		Accumulation in bulk of the solid or liquid
Concentration	Different at surface to bulk	Same throughout
Reaction rate	Increases to equilibrium	Occurs at a uniform rate
Reaction type	Exothermic process	Endothermic process
Temperature	Unaffected	Unaffected

Chemical absorption (sometimes referred to as chemisorption or reactive absorption) is a chemical reaction between the absorbent and the absorbing substances and often combines with physical absorption. This type of absorption depends upon the stoichiometry of the reaction and the concentration of its reactants. They may be carried out in different units, with a wide spectrum of phase flow types and interactions.

Hydrophilic solids, which includes many solids of biological origin, can readily absorb water. Polar interactions between water and the molecules of the solid favor partition of the water into the solid, which can allow significant absorption of water vapor even in relatively low humidity.

On the other hand, adsorption is a surface phenomenon that involves the adhesion of atoms, ions, or molecules from a gas, liquid or dissolved solid (the adsorbate) to a surface which results in a film of the adsorbate on the surface of the adsorbent. Adsorption might be considered as a sedimentation process that is encourage or enhanced by the properties of the chemical and the adsorbing properties of the surface to which the chemical is adsorbed.

Adsorption (and the reverse effect, desorption) of chemical pollutants in soils largely depend on: (i) the type of mineral; (ii) the presence or absence of clay minerals; (iii) the pH; (iv) the redox conditions; and (v) the available chemical species. Inorganic ions, such as phosphate derivatives, HPO_4^{2-}, nitrate derivatives, NO^{3-}, chloride derivatives, Cl^-, and sulfate derivatives, SO_4^{2-} and organic ligands, such as citrate, oxalate, fulvic, and dissolved organic carbon (DOC) can affect pollutant behavior in soils.

The term sorption includes both absorption and adsorption processes, while *desorption* is the reverse of either absorption or adsorption.

7.2.2 DISPERSION

The dispersion of chemicals into the environment can occur in three ways: (i) dispersion during combustion processes, such as fuel conversion or waste incineration: chemicals released include polycyclic aromatic hydrocarbon (PAH) derivatives, dioxin derivatives, an furan derivatives; (ii) dispersion after the unintentional release of chemicals following accidents or by slow leakage such as the release of polychlorinated biphenyl derivatives from electrical installations; and (iii) the intentional dispersion of chemicals as, for example, during the use of pesticides and other agricultural chemicals. Thus, the disposal of a chemical into a watercourse and subsequent transportation of the chemical is referred to as dispersion. The dispersed chemicals may remain in the water column and eventually be adsorbed by minerals, soil or sediment.

In the context of the geosphere, dispersion is a process that occurs, for example, in soil that is particularly vulnerable to erosion by water. In a soil layer where clay minerals are saturated with sodium ions (sodic soil), the soil can break down very easily into fine particles and wash away which can lead to a variety of soil and water quality problems, including: (i) large losses of soil losses by gully erosion and tunnel erosion; (ii) structural degradation of the soil, clogging, and sealing where dispersed particles settle; and (iii) turbidity in water due to suspended soil particles which also cause transportation of nutrients from the land. G erosion involves the creation of a gully created by running water, eroding sharply into soil and typically occurs on a hillside. On the other hand, tunnel erosion occurs where channels and tunnels develop beneath the surface.

The dispersion of chemicals released into the environment has been the focus of much attention because of the realization that the dispersion behavior of chemicals can be markedly different when different chemicals are considered. Accidents that involve chemicals give rise to a new class of problems in dispersion prediction for the following reasons: (i) the material is, in almost all cases, stored as a solid but may also emit a gas after a spill and exposure to the air; (ii) the modes of release can vary widely and geometry of the source can take many forms and the initial momentum of the spill may be significant; (iii) in some cases, a chemical transformation also takes place as a result of reaction with water vapor in the ambient atmosphere.

Emissions of many chemicals of concern occur into the air initially, from where they are dispersed into other media. Many chemicals emitted into the air, for instance from combustion processes, tend to become associated with PM. Removal from the air occurs through a range of complex processes involving photodegradation, and particle sedimentation and/or precipitation, (known respectively as dry deposition and wet deposition). Chemicals may undergo several cycles of physical and/or chemical transformation (Chapter 7), which can also make chemicals more accessible to photochemical or biodegradation.

In addition, the physical properties of the chemical may result in one or more interactions with the surrounding ecosystem, especially if the chemical is reactive and has the potential to react quickly with the air, water, or soil. This reactivity will influence the dispersibility of the chemical and, moreover, if the release occurs over a short time scale, compared to the steady-state releases characteristic of many chemical releases problems, this can give rise to the complication of predicting dispersion for time-varying releases. There is also the uncertainty of individual predictions resulting from variability about the behavior of a mixture. Also, the dispersing chemical, which is typically denser than air, may form a low level cloud that is sensitive to the effects of either natural or man-made obstructions in the surrounding topography.

Wind transport (Aeolian transport-relocation by wind) can also occur and is particularly relevant when dust from solids. Dust becomes airborne when winds traversing arid land pick up small particles such as dust, and other debris and send the particles skyward after which the movement of pollutants in the atmosphere is caused by transport, dispersion, and deposition.

7.2.3 DISSOLUTION

Dissolution is the process whereby chemicals (such as gases, liquids, or solids) dissolve into a liquid (such as water or another solvent) by which these chemicals in the original state (gas, liquid, or solid) become a solute (dissolved component). In the cases of liquids and gases, the molecules must be able to form non-covalent intermolecular interactions with those of the solvent for a solution to form. Dissolution is of fundamental importance in all chemical processes, especially the solubility of chemicals in water which can be enhanced by the occurrent of acid rain.

Generally, many chemicals (especially the inorganic chemicals) are polar and are usually soluble in water but insoluble in organic solvents (such as. diethyl ether, dichloromethane, chloroform, and hexane). The solubility (dissolution) characteristics of chemical species in water is complex and very much dependent upon the structure and properties of the chemicals(s). The chemical may be a gas, a liquid, or a solid (the solute) that dissolves in the water-the solubility depends on the physical and chemical properties of the chemicals as well as on temperature, pressure, and pH (acidity or alkalinity of the water).

Furthermore, the extent of solubility of the chemical can range from infinitely soluble (without limit) to poorly soluble (such as some lead salts)-the term insoluble is (in the world of chemistry) a common threshold to describe a chemical as insoluble is a solubility less than 0.1 g per 100 mL of water. However, the solubility of a chemical in water should not be confused with the ability of water to dissolve the chemical because apparent solubility of the chemical might also occur because of a chemical reaction (reactive solubility).

Solubility of a chemical in water applies not only to environmental chemistry but to areas of chemistry, such as (alphabetically): biochemistry, geochemistry, chemistry, and physical chemistry. In all cases, the solubility of the chemical depends on the physical conditions (temperature, pressure, and concentration of the chemical) and the enthalpy and entropy relating to the water and the chemicals concerned. Water is, by far, the most common solvent in chemistry and is a solvent for a wide range of chemicals, especially those chemicals that readily form ions. This is a crucial factor in which acidity and alkalinity play a role as in much of the area known as environmental chemistry.

In addition, the term dissolved chemicals (or TDS) is used as a broad classification for chemicals of varied origin and composition within aquatic systems (the aquasphere. The source of the dissolved chemical in freshwater systems and in marine systems depends on the body of water. When water contacts highly ionizable chemicals, these components can drain into rivers and lakes as a dissolved chemical. Whatever the source of the dissolved chemical, it is also extremely important in the transport of metals in aquatic systems-certain metals can form extremely strong metallic complexes with a dissolved solute which enhances the solubility of the metal in aqueous systems while also reducing the bioavailability of the metal.

In terms of chemicals that do not occur naturally in the environment, knowledge of the structure and properties can be used to examine relationships between the solubility properties of the chemical and its structure, and vice versa. In fact, structure dictates function which means that by knowing the structure of a chemical it may be (but not always) possible to predict the properties of the chemical such as its solubility, acidity or basicity, stability, and reactivity. In the context of the environmental distribution of a chemical, predicting the solubility of a molecule is a useful component of knowledge.

Dissolved chemicals are associated with the various characteristics of drinking water (such as taste, smell, color, and odor) and the quality of drinking water largely determines the acceptability of water. For health-related contaminants-a contaminant that is unsafe for one species of flora or fauna should be considered unsafe for all, until proven otherwise-the general characteristics (taste, smell, and odor) are subject to social, economic, and cultural considerations.

On the other hand, the presence of non-dissolved) suspended solids in water gives rise to turbidity, which may not only be offensive to the general senses but can, in a more serious aspect, protect bacteria and viruses from the action of disinfectants.

7.2.4 EMULSIFICATION

An emulsion is a mixture of two or more liquids that are immiscible and the physical structure of an emulsion is based on the size of the droplets-most emulsions contain droplets with a mean diameter on the order of 1 μm (1 micron, 1×10^{-6} meter) and greater. However, mini-emulsions and nano-emulsions can be formed with droplet sizes in the 100 to 500 nm range (nanometer, 1×10^{-9} meter) range and highly stable microemulsions can be prepared having droplets as small as a few nanometers. The stability of an emulsion refers to the ability of the emulsion to resist change in form and properties and there are four types of instability in emulsions which are: (i) flocculation, (ii) creaming, and (iii) coalescence.

Flocculation occurs when there is an attractive force between the droplets which enables the droplets to form a flocculant mass (flocs) in the fluid. Creaming occurs when the droplets rise to the top of the emulsion under the influence of buoyancy while coalescence occurs when the droplets combine to form a larger droplet. Use of a surface-active agent (surfactant) can increase the stability of an emulsion insofar as the size of the droplets does not change significantly with time, giving rise to a stable emulsion.

The demulsification process can be slowed with the help of emulsifiers (surface-active chemicals that have hydrophilic properties) that are used to be stabilize emulsions and promote dispersing the dispersed phase to form microscopic (invisible to the human eye) droplets which accelerates the decomposition of the pollutant in the water.

7.2.5 EVAPORATION

Evaporation (the opposite of the opposite of condensation and sublimation) is the process by which water changes from a liquid to a gas or vapor and is the primary pathway that water moves from the liquid state back into the water cycle as atmospheric water vapor. The process is the phenomenon that occurs when a liquid is transformed to gas without chemical alteration and the conversion of the solid to a liquid and thence the liquid to a gas. Evaporation should not be confused with sublimation in which a solid is transferred directly to a gas:

Evaporation:

Liquid → gas

Sublimation:

Solid → gas

Both of these processes are relevant to the environmental chemical cycle since both processes involve the disappearance of, for example, a contaminant in the water or on the land, *but* the contaminant does appear in the atmosphere.

7.2.6 LEACHING

Leaching is a natural process by which water-soluble substances (such as water-soluble chemicals or hydrophilic chemicals) are washed out from soil or waste disposal areas (such as a landfill). These leached out chemicals (leachates) can cause pollution of surface waters (ponds, lakes, rivers, and the sea) and sub-surface water groundwater aquifers).

In the environment, leaching is the process by which contaminants are released from, say, the solid phase into the water phase under the influence of dissolution, desorption, and complexation that are affected by pH, redox chemistry, and biological activity. The process itself is ever-present, and any material, such as a rock, l exposed to contact with water will lose components from its surface or its interior depending on the porosity of the material. In terms of the effect of pH (acidity or alkalinity), the leachability of a chemical; (i.e., the ability of the chemical to be leached from the source can be enhanced by the occurrence of acid rain which can enhance the solubility of the chemical in the acidified water.

Leaching is affected by: (i) the texture of the solid that holds the chemical to be leached; (ii) structure of the chemical to be leached; and (iii) water content of the solid which holds the chemical. If the solid holding the chemical is soil, the soil texture such as the relative portions of sand, silt, and clay minerals affect the movement of water through the soil. **Coarse textured** soil contains more sand particles which have large pores, and the soil is highly permeable, which allow the water to move rapidly through the pore system. However, contaminants such as phosphate pesticides carried by water through coarse-textured soil are more likely to reach the groundwater. On the other hand, **clay textured** soils have low permeability and tend to retain more water and adsorb more chemicals from the water which: (i) can slow the downward movement of the contaminants chemicals; (ii) assist in increasing the chance of degradation and adsorption to soil particles; and (iii) reduces the chance of the contaminants entering the groundwater.

In addition, the amount of water that is already present in the soil has an influence on whether rain or irrigation results in the recharging of groundwater and possible leaching of chemicals into an aquifer. Soluble chemicals are more likely to reach groundwater when the water content of the soil approaches or is at saturation.

7.2.7 SEDIMENTATION

Sedimentation is the tendency for chemicals in suspension in water to settle out of the water. In geology, sedimentation is a term that is often used as the opposite of erosion, i.e., the terminal end of sediment transport. Settling is the falling of suspended particles through the liquid, whereas sedimentation is the termination of the settling process. The sedimentation process sedimentation occurs when particles in suspension to settle out of the fluid in which they are entrained and come to rest against a barrier, which is typically the basement of a waterway. Settling is due to the motion of the particles through the fluid in response to the forces acting on them, which is an ecosystem, can be due to gravity. The term *sedimentation* is often used as the opposite of erosion. Settling occurs when suspended particles fall through the liquid, whereas sedimentation is the termination of the settling process.

Sedimentation can be generally classified into three different categories that are all applicable to chemical contaminants, thus: (i). Type 1 sedimentation is characterized by particles that settle discretely at a constant settling velocity and typically these particles settle as individual particles and do not flocculate or stick to other during settling; (ii) Type 2 sedimentation is characterized by particles that flocculate during sedimentation and because of this the particle size is constantly changing and therefore their settling velocity is changing; and (iii) Type 3 sedimentation (also known as zone sedimentation) involves particles that are at a high concentration (for example, greater than 1,000 mg/L) such that the particles tend to settle as a mass and a distinct clear zone and sludge zone are present. Zone settling occurs in active sludge sedimentation and sedimentation of sludge thickeners.

In terms of the sedimentation of chemical contaminants, some of the contaminants may be adsorbed on the suspended material (especially if the suspended material is a clay mineral or another highly adsorptive mineral) and deposited to the bottom of the water system. This phenomenon typically occurs in shallow water system where PM is abundant and the water is subjected to intense mixing, usually through turbulence. Simultaneously, the process of bio-sedimentation can also occur when plankton and other organisms absorb the chemical. The suspended forms of the contaminants undergo intense chemical and biological (microbial) decomposition in the water. However, this situation radically changes when the suspended chemical reaches the lake bed, river bed or sea bed and the decomposition rate of the chemical(s) buried on the bottom abruptly drops. The oxidation processes slow down, especially under anaerobic conditions in the bottom environment and the chemical(s) accumulated inside the sediments can be preserved for many months and even years.

7.2.8 SPREADING

Spreading is a form of dispersion in which the movement of chemical contaminants through the subsurface is difficult to predict since different types of contaminants react differently with soils, sediments, and other geologic materials and commonly travel along different flow paths and at different velocity. Chemical contaminants infiltrate (through solubility in water) into the subsurface and migrate downward by gravity through the vadose zone (the zone that extends from the top of the ground surface to the water table). However, when low permeability formations (such as those formations containing clay minerals) soil units are encountered, the solution of the contaminants can also spread laterally along the permeability contrast.

Any interactions between the soil and the contaminant are important for assessing the fate and transport of the contaminant in the subsurface, especially in a groundwater system. Contaminants that are highly soluble, such as salts and any potentially ionic derivatives can move readily from surface soil to saturated materials below the water table and often occurs during and after rainfall events. Those contaminants that may not be highly soluble may have considerably longer residence times in the surface strata (the soil zone). Some contaminants are readily adsorbed by soil particles and slowly dissolve during precipitation events, resulting in migration as an aqueous solution or aqueous suspension. Once below the water table, the contaminants are also subject to dispersion (mechanical mixing with uncontaminated water) and diffusion (dilution in water because of a concentration gradient). Many contaminants begin to spread immediately after entry into the environment.

7.2.9 SUBLIMATION

Sublimation is an endothermic phase transition that involves the transition of a substance directly from the solid phase to the gas phase without passing through an intermediate liquid phase (Table 7.4). The term sublimation refers to a physical change of state and is not used to describe the transformation of a solid to a gas in a chemical reaction.

TABLE 7.4 Examples of Phase Transformations

From:	To:	Process
Gas	Gas	N/A*
Gas	Liquid	Condensation
Gas	Solid	Deposition
Liquid	Gas	Evaporation or boiling
Liquid	Liquid	N/A*
Liquid	Solid	Freezing
Solid	Gas	Sublimation
Solid	Liquid	Melting
Solid	Solid	Transformation**

*May involve a chemical change depending upon the temperature and pressure.

**For example, a change in crystal structure.

The process is occurring at temperatures and pressures that are below the triple point of a chemical in a phase diagram. The reverse process of sublimation is the *deposition* process in which some chemicals pass directly from the gas phase to the solid phase, also without passing through an intermediate liquid phase. As an examples of anon-sublimation process, the dissociation on heating of solid ammonium chloride (NH_4Cl) into ammonia (NH_3) and hydrogen chloride (HCl) is *not* sublimation but a chemical reaction:

$$NH_4Cl \rightarrow NH_3 + HCl$$

Sublimation requires additional energy and is an endothermic change and the enthalpy of sublimation (also referred to as the heat of sublimation) can be calculated by adding the enthalpy of fusion and the enthalpy of vaporization.

7.3 DISTRIBUTION IN THE ENVIRONMENT

The movement of released chemicals into the Earth systems is caused by transport, dispersion, and deposition of the chemicals (or, in some cases transformed chemicals). Furthermore, the distribution of chemical contaminants in the environment is influenced by the physical properties of the chemicals and the minerals (such as the clay minerals) with which the chemicals come into contact. Different types of chemicals react differently with soils, sediments, and other geologic materials and commonly travel along different flow paths and at different velocities. For example, using soil as the example, contaminants that are highly soluble, such as ionic salts (e.g., sodium chloride, NaCl) move readily from surface soils to saturated materials below the water table, which often occurs during and after rainfall events. Other contaminants that are not highly soluble may have considerably longer residence times in the soil zone. Once below the water table, chemicals are also subject to dispersion (mechanical mixing with uncontaminated water) and diffusion (dilution by concentration gradients).

Typically, chemicals released from various sources are ultimately dispersed among, and can at times accumulate in, various environmental compartments (such as soil as well as various floral and faunal species). Some contaminants may contribute primarily to environmental compartments on a local scale but other contaminants that are more persistent in the environment, can be distributed over much greater distances-even up to a regional scale, a national scale, or an international scale.

Thus, understanding the potential environmental impact of dispersed chemicals requires an understanding of the relative contribution of the various sources of the pollutants as well as the types of the (Chapter 5) and the potential for a pollutant to undergo chemical (or physical) transformation in the environment (Chapter 7). Therefore, an investigation of a potential contaminant must account for the transport of the contaminant through ecosystems.

To effectively monitor changes in the environmental behavior of chemicals that are of most concern, it is extremely important to understand how these chemicals typically behave in natural systems and in specific ecosystems. Equally important is an understanding of how these chemicals might respond to specific best management practices.

An effective monitoring program explicitly considers how these chemicals may change as they move from a source into the groundwater, surface water, or into the soil. This includes an understanding of how a specific chemical may be introduced or mobilized within an ecosystem, how the chemical moves or through an ecosystem the transformations that may occur during this process.

However, the entry of chemicals into the environment and the distribution of these chemicals within the environment is often complex, and there have been many occasions when a significant amount of a chemical (or a mixture of chemicals) has entered an ecosystem and the effects of contamination are well defined. Generally, the assumption is that the chemical (or a mixture thereof) does not rapidly diffuse away, but remains in the immediate vicinity at a noticeably high concentration or perhaps moves, but in such a way that concentration levels of the chemical remain high as it moves. Such cases would normally occur when large quantities of a substance were being stored, transported or otherwise handled in concentrated form. Thus, due to leakages, spills, improper disposal and accidents during transport, compounds have become subsurface contaminants that threaten various ecosystems.

Methods are available that can be used to use to calculate factors available that can be used to understand the entry of chemical in the environment and the method of dispersal. For example, the amount of a pollutant available for exposure depends on its persistence and the potential for its bioaccumulation as well as the exposure route.

By way of explanation, an exposure route refers to the way that a released enters chemical enters the flora and fauna as a result of the release. The route of the potential uptake of the chemical is an important attribute of an exposure event. In addition, any chemical (including both organic chemicals and inorganic chemicals) that is introduced into the environment can be considered to be capable of accumulation (or bioaccumulation) if the chemical has a degradation half-life in excess of thirty days or if the chemical has a bioconcentration factor (BCF) greater than 1,000 or if the log K_{ow} (the octanol-water partition coefficient of the chemical; Chapter 4) is greater than 4.2:

$$BCF = (\text{concentration in biota})/(\text{concentration in ecosystem})$$

The octanol/water partition coefficient (Kow) is the ratio of the concentration of a chemical in the octanol phase relative to the concentration of the chemical in the aqueous phase of a two-phase octanol/water system:

$$K_{ow} = (\text{concentration in octanol phase})/(\text{concentration in aqueous phase})$$

The BCF indicates the degree to which a chemical may accumulate in biota (flora and fauna). However, measurement of bioconcentration is typically made on faunal species and is distinct from transport in the food chain, bioaccumulation, or biomagnification. The BCF is a constant of proportionality between the chemical concentration in flora or fauna in an ecosystem. It is possible, for many chemicals, to estimate the BCF from the octanol-water partition coefficients (K_{ow}):

$$\text{Log(bioconcentration factor)} = m\log K_{ow} + b$$

In terms of actual numbers, for many lipophilic chemicals, the BCF can be calculated using the regression equation:

$$\log BCF = -2.3 + 0.76 \times (\log K_{ow})$$

Furthermore, empirical relationships between the octanol-water partition coefficients and the BCF can be developed on a chemical by chemical basis.

Indeed, an analysis of the amount of chemical waste formed in various processes has revealed that the amount of waste generated in some processes may be in excess of the amount of the desired product and such over production of waste was not exceptional in the chemical industry. As a means of measuring the amount of waste vis-à-vis the amount of products, the E-factor (environmental factor) (kilograms of waste per kilogram of product) was introduced as an indication of the environmental footprint of the manufacturing process in various segments of the chemical industry (Chapter 7). This factor is derived from the chemicals and fine chemicals industries as a measure of the efficiency of the manufacturing process: Thus, simply:

$$E = \text{kilograms of waste/kilogram of product}$$

The factor can be conveniently calculated from the number of tons of raw materials purchased and the number of tons of product sold, the calculation being for a particular product or a production site or even a whole company. A higher E-factor means more waste and, consequently, a larger environmental footprint-thus, since mass cannot be created resulting in a negative E-factor, the ideal E-factor for any process is zero.

However, in the context of environmental protection and to be all-inclusive, the E-factor is the total mass of raw materials plus ancillary process requirements minus the total mass of product, all divided by the total mass of product. Thus, the E-factor should represent the actual amount of waste produced in the process, defined as everything but the desired product and takes the chemical yield into account and includes reagents, solvent losses, process aids, and (in principle) even the fuel necessary for the process. Moreover, use of the E-factor has been widely adopted by many of the chemical industries and the pharmaceutical industries in particular. Thus, a major aspect of process development recognized by process chemists and by process engineers is the need for determining an E-factor-whether or not it is called by that name (i.e., the E-factor) but chemicals is compared to chemical out has become a major yardstick in many of the chemicals industries.

In terms of properties and transportation of the chemical, PM occupies a somewhat different lace to the more conventional chemical contaminants insofar as the transport characteristics of particles depend on the particle size. Fine and coarse particles in ambient air differ in their chemical composition, solubility, acidity, sources, and formation processes, atmospheric lifetime, infiltration indoors, and transport distances. Most airborne particles are small (less than 0.1 μm in diameter, 0.2 micron; 1 micron = 1 meter × 10^{-6}), but most of the particle volume (and mass) is found in particles with diameters greater than 0.1 μm. The size distribution of airborne particles is often multimodal.

Particles formed as a result of the chemical reaction(s) of gases (which are secondary particles because the direct emission from a source is a gas such as sulfur dioxide, SO_2, or nitric oxide, NO) that is subsequently converted to a low vapor pressure substance (such as sulfuric acid or, nitric acid) that subsequently nucleates or condenses. Examples include sulfate derivatives, some low VOCs, and ammonium salts.

These types of transformations can take place: (i) locally, during prolonged stagnations of ambient air; or (ii) during transport over long distances, and are affected by moisture, sunlight, temperature, and the presence or absence of fogs and clouds. In general, particles formed from these types of secondary processes will be more uniform in space and time than those that result from primary emissions. Particles directly emitted by sources, referred to as primary particles, are also found in the fine particle fractions (the most common being particles from combustion sources that are less than 1.0 μm in aerodynamic diameter). In contrast to fine particles, most of the coarse particle fraction of ambient aerosol originated as particles emitted directly to the atmosphere, and some particles generated during combustion processes, such as fly ash and soot, might also be found in the coarse fraction.

Every particle in the atmosphere tends to settle to the ground through the effects of gravity, but the tendency to settle is opposed or abetted by other effects including electrostatic and aerodynamic forces. The outcome is that particles deposit to the ground at velocities that depend primarily on their particle diameter and density. For coarse particles, controlled primarily by gravity, the deposition velocity is proportional to the square of the particle diameter. For fine particles, deposition is controlled more by electrostatic and other effects than by gravity, so that they deposit more rapidly than would be expected from gravity and their size alone. The result is that fine particles with an aerodynamic diameter between 0.1 and 1.0 μm have the minimum deposition velocity of particles. Such fine particles will remain suspended for much longer times (on the order of days to weeks for fine particles as opposed to minutes to hours for coarse particles) and will travel much farther (as far as thousand miles) than the coarse particle fraction.

7.4 IMPACT ON THE ENVIRONMENT

Contaminants can enter the atmosphere, the aquasphere, and the geosphere air, water, and soil when they are produced, used, or sent for disposal. The impact of these chemicals on the environment is determined by: (i) the amount of the contaminant that is released; (ii) the type and concentration of the contaminant, and (iii) where the contaminant is found.

The final concentration of a contaminant (a single chemical or a mixture of chemicals) in various environmental systems depends on: (i) the emission rate; (ii) the distribution; and (iii) the fate of the contaminant. Thus, the first step in environmental risk assessment is always to **quantify the emissions of a contaminant into the atmosphere, the water, and the land.**

All chemicals that enter the environment should be categorized and ranked using hazard assessment criteria. This would not only ensure that truly pressing environmental issues are identified and prioritized, but would also maximize the use of limited resources. In the case of soluble chemicals, surrogate data such as persistence and bioaccumulation have been used, in combination with toxicity, for the purpose of hazard categorization. However, for insoluble or sparingly soluble chemicals such as metals and metal compounds, persistence, and bioaccumulation are neither appropriate nor useful. Unfortunately, this is not always recognized by regulators or even by scientists.

The use of persistent, bioaccumulative, and toxic (PBT) criteria for contaminants is used to address the hazards posed by various chemicals. In fact, the criteria and the test methods that are used to evaluate persistence (i.e., the lack of degradability of a chemical) and bioaccumulation (the dispersion of a chemical through knowledge of the water-octanol partition coefficient) were developed to be used in combination with toxicity data in order to reduce the importance given to the use of toxicity data alone. These test methods were based on an understanding of the chemistry of chemicals of concern at the time and of the biological interactions that the chemicals would have with the surrounding biota. Specifically, it was realized that if some chemicals exerted high intrinsic toxicity under standardized laboratory test conditions but did not persist or bioaccumulate, the environmental hazard of such chemicals would be lower.

The phenomenon of contaminant term persistence is measured by determining the lack of degradability of a substance from a form that is biologically available and active to a form that is less available-metals and metal compounds tend to be in forms that are not bioavailable. Thus, rather than persistence, the key criterion for classifying metals and metal compounds should be their capacity to transform into bioavailable form(s). Furthermore, although bioavailability is a necessary precursor to toxicity, it does not inevitably lead to toxicity. Although metals and metal compounds stay in the environment for long periods of time, the risk they may pose generally decreases over time. For example, metals introduced into the aquatic environment are subject to removal/immobilization processes (e.g., precipitation, complexation, and absorption).

Similarly, the use of bioaccumulation has significant limitations for predicting hazard for metals and metal compounds. Generally, either BCFs or bioaccumulation factors (BAFs) are used for this purpose. A BCFs is the ratio of the concentration of a substance in an organism, following direct uptake from the surrounding environment (water), to the concentration of the same substance in the surrounding environment. A BAF considers uptake from food as well. In contrast to organic compounds, uptake of metals is not based on lipid partitioning and many organisms have internal mechanisms (homeostasis) that allow the organism to regulate (usually referred to as bioregulation) the uptake of essential metals and to control the presence of other metals. Thus, if the concentration of an essential metal in the surrounding environment is low and the organism requires more, it will actively accumulate that metal. This will result in an elevated bioconcentration factors (or BAF) value which, while of concern in the case of organic substances, is not an appropriate measure in the case of metals.

The primary determining factor of hazard for metals and metal compounds is therefore toxicity, which requires consideration of dose (indeed, the fundamental tenet of toxicology is the dose makes the poison). Historically, it has been the practice to measure the toxicity of soluble metal salts, or indeed the toxicity of the free metal ion. However, in different media, metal ions compete with different types or forms of organic matter (e.g., fish gills, suspended solids, soil particulate material) to reduce the total amount of metals present in bioavailable form. Toxicity of the bioavailable fraction (i.e., as determined through transformation processes) is the most appropriate and technically defensible method for categorizing and ranking the hazard of metals and metal compounds.

The relative proportion of hazardous constituents present in any collection of chemicals (crude oil-derived products included) is variable and rarely consistent because of site differences. Therefore, the extent of the contamination will vary from one site to another and, in addition, the farther a contaminant progresses from low molecular weight to high molecular weight the greater the occurrence of polynuclear aromatic hydrocarbons, complex ring systems (not necessity aromatic ring systems) as well as an increase in the composition of the semi-volatile chemicals or the non-volatile chemicals. These latter chemical constituents (many of which are not so immediately toxic as the volatiles) can result in long term/chronic impacts to the flora and fauna of the environment. Thus, any complex mixture of chemicals should be analyzed for the semi-volatile compounds which may pose the greatest long term risk to the environment.

Finally, in order to evaluate the impact of a contaminant that has been released to the environment, the contaminant must be characterized in terms of the transport and transformation in that system (atmosphere, water, or land) and the potential for the transport of the chemical from one system to another or from one system to the other two. The assessment should focus on areas with which a released chemical is most likely to have contact. For a meaningful characterization, the environment must be viewed as a series of interacting compartments, and it must be determined whether a chemical will remain and accumulate in the local area of the origin of the chemical. The potential for the chemical to be physically, chemically, or biologically transformed in the system of its origin (such as by hydrolysis, oxidation, or other transformation (Chapter 8) or be transported to another system such as by volatilization or by precipitation. The chemical could also be transferred by deposition and runoff to surface water that provides drinking water.

Characterizing transportation pathways begins at the source of the release of the contaminant. In some situations, the source may be obvious and can be defined and characterized from the concentration of the contaminant in air or in the soil, but in many cases, such as contamination of water supplies, sources, and emissions may be multiple and poorly characterized. However, classification of a potential transportation route should, as much as possible, be based on the released volume, duration of the release, and the rate of emission.

In order to fully understand the impact of a released chemical on the environment, the potential for chemical transformation of the spilled chemical which may occur as a result of biotic or abiotic processes, can significantly reduce the concentration of a substance or alter its structure in such a way as to enhance or diminish its toxicity or change its toxic effect (Chapter 8).

KEYWORDS

- **bioaccumulation factors**
- **bioaccumulative and toxic**
- **bioconcentration factor**
- **dissolved organic carbon**
- **environment**

REFERENCES

Bahadori, A., (2014). *Natural Gas Processing: Technology and Engineering Design.* Gulf Professional Publishing, Elsevier, Amsterdam, Netherlands.

http://dge.stanford.edu/SCOPE/SCOPE_22/SCOPE_22_front%20material.pdf (accessed on 20 November 2021).

https://www.hindawi.com/journals/isrn/2011/402647/ (accessed on 20 November 2021).

Kidnay, A. J., Parrish, W. R., & McCartney, D. G., (2011). *Fundamentals of Natural Gas Processing* (2nd edn.). CRC Press, Taylor & Francis Group, Boca Raton, Florida.

Levy, D. B., Barbarick, K. A., Siemer, E. G., & Sommers, L. E., (1992). Distribution and partitioning of trace metals in contaminated soils near Leadville, Colorado. *Journal of Environmental Quality, 21*(2), 185–195.

Miller, D. R., (1984). In: Sheehan, P. J., Miller, D. R., Butler, G. C., & Bourdeau, P., (eds.), *Effects of Pollutants at the Ecosystem Level.* SCOPE 22. Scientific committee on problems of the environment (SCOPE) of the international council of scientific unions (ICSU). John Wiley & Sons Inc., Hoboken, New Jersey.

Mokhatab, S., Poe, W. A., & Speight, J. G., (2006). *Handbook of Natural Gas Transmission and Processing.* Elsevier, Amsterdam, Netherlands.

Shiowatana, J., McLaren, R. G., Chanmekha, N., & Samphao, A., (2001). Fractionation of arsenic in soil by a continuous flow sequential extraction method. *Journal of Environmental Quality, 30*(6), 1940–1949.

Speight, J. G., (1996). *Environmental Technology Handbook*. Taylor & Francis, Washington, DC.

Speight, J. G., (2007). *Natural Gas: A Basic Handbook.* GPC Books, Gulf Publishing Company, Houston, Texas.

Speight, J. G., (2014). *The Chemistry and Technology of Petroleum* (5th edn.). CRC Press, Taylor & Francis Group, Boca Raton, Florida.

Speight, J. G., (2017). *Handbook of Petroleum Refining*. CRC Press, Taylor & Francis Group, Boca Raton, Florida.

Wuana, R., & Okieimen, F. E., (2011). Heavy metals in contaminated soils: A review of sources, chemistry, risks, and best available strategies for remediation. *ISRN Ecology, 2011*. Article ID 402647; doi:10.5402/2011/402647.

Yuan, G. D., Theng, B. K. G., Churchman, G. J., & Gates, W. P., (2013). Clays and clay minerals for pollution control. In: Bergaya, F., & Lagaly, G., (eds.), *Developments in Clay Science; Handbook of Clay Science* (Vol. 5. pp. 587–644). Elsevier BV, Amsterdam, Netherlands.

CHAPTER 8

Transformation in the Environment

8.1 INTRODUCTION

The major groups of pollutants and the principal reactions that can occur when the pollutants are released into the environment (degradation reactions or transformation reactions) and the pollutant can be converted to a product that is more capable of remaining in the environment and may even prove to be a persistent chemical. Thus, emphasis must be placed on the partial or complete degradation of pollutants as well as any intermediate products that are toxic to floral and faunal organisms so that the original pollutants and the degradation products will not have any adverse effects on the environment. Thus, the chemical transformation of a contaminant is an issue that needs to be given serious consideration because of the changes (often non-benign) that can occur to the pollutant (Manzetti et al., 2014). It would be unusual if the chemical transformation did not show some effect on the properties of the discharged chemical.

The chemical transformations of chemicals released into the environment are, in the context of this book, considered to be the transformation of the released chemical into a product that is still of concern in terms of toxicity. Furthermore, knowledge of the relative amounts of each species present is critical because of the potential for differences in behavior and toxicity (including the possibility of enhanced toxicity), which are of concern because of the potential fate of such chemicals.

As used here, the term *fate* refers to the ultimate disposition of the chemical in the ecosystem, either by chemical or biological transformation to a new form which (hopefully) is non-toxic (degradation) or, in the case of an ultimately persistent pollutants (POPs), by conversion to a less offensive chemicals or even by sequestration in a sediment or other location which is expected to remain undisturbed. Thus latter option-the sequestration in a sediment or other location is not a viable option as for safety reasons, the chemical must be dealt with at some stage of its environmental life cycle and the chemical is likely to manifest its presence at some future date.

Furthermore, pollutants released into the environment are subject to two processes that determine the fate of the chemical in the environment (i) the potential for transportation of the chemical and (ii) the chemical changes that can occur once the chemical has been released to the environment and which depend upon the chemical properties that can affect

The Science and Technology of the Environment. James G. Speight (Author)

transportation processes and physical properties that can play a role in the transformation processes (Table 8.1).

TABLE 8.1 Chemical and Physical Properties That can Influence Transportation Processes

Physical Processes
• Runoff
• Erosion
• Wind
• Leaching
• Movement in Streams or in groundwater
Chemical Processes
• Transport
• Transformation/degradation
• Sorption
• Volatilization
• Biological Processes
Transformation Processes
• Biological transformations due to microorganisms:
o aerobic transformation;
o anaerobic transformation.

In the present context, when an chemical (or a mixture of chemicals) is released into the environment, the issues that need to be considered: (i) the toxicity of the chemical; (ii) the concentration of the released chemical; (iii) the concentration of the toxic chemical in the released material; (iv) the potential of the chemical to migrate to other sites; (v) the potential of the chemical to produce a toxic degradation product, whether or not the toxicity is lower or higher than the toxicity of the released chemical; (vi) the potential of the toxic degradation product or products to migrate to other sites; (vii) the persistence of the chemical in an ecosystem; (viii) the persistence of any toxic degradation product in an ecosystem; (ix) the potential for the toxic degradation product to degrade even further into harmful pollutants or non-harmful pollutants and the rate of degradation; and (x) the degree to which the pollutant or any degradation product of the pollutant can accumulate in an ecosystem.

In order to complete such a list and monitor the behavior and effects of chemicals in an ecosystem, an understanding of chemical transformation processes in which a disposed or discharged chemical might particulate is valuable to any study of the effects on the environment. Chemical transformation processes change the chemical composition and structure of the discharged chemical which can change the properties (and possibly the toxicity) of the chemical and influence behavior and life-cycle of the chemical in the environment.

As an example of chemical transformation of chemicals in the environment, weathering processes are ever-present and include such phenomena as: (i) evaporation; (ii) leaching, which is transfer to the aqueous phase through dissolution; (iii) entrainment, which physical transport along with the aqueous phase; (iv) chemical oxidation; and (v) microbial degradation. The rate of transformation of the chemical is highly dependent on environmental conditions. For example, a product such a low-boiling naphtha solvent (boiling range 30 to 90°C, 86 to 194°F-a mixture of hydrocarbon derivatives produced from crude oil-will evaporate readily when spilled on to the surface of the Earth (specifically a water surface or land surface) and will give the appearance of a reception in the amount that remains. But the low-boiling constituents have not merely *disappeared* or gone away but have transferred from the land or from the water into the atmosphere. On the other hand, naphtha that has been inadvertently released into a formation that lies below a formation of clay minerals will tend to evaporate slowly (the clay can act as a formation trap) and may not be readily detectable.

However, the various chemical transformation processes, which influence the presence and the analysis of chemicals at a particular site, although often represented by simple (and convenient) chemical equations, can be complex (Neilson and Allard, 2012) and the true nature of the chemical transformation process is difficult to elucidate. The extent of transformation is dependent on many factors including the: (i) the properties of the chemical; (ii) the geology of the site; (iii) the climatic conditions, such as temperature, oxygen levels, and moisture; (iv) the type of microorganisms present; and (v) any other environmental conditions that can influence the life of the pollutant. In fact, the primary factor controlling the extent of chemical transformation is the molecular composition of the chemical contaminant.

In the environment, a chemical transformation is the same principle as in the laboratory or in the chemical process industries-the transformation of a substrate to a product-but whether or not the product is benign and less likely to harm the environment (relative to the substrate) or is more detrimental by exerting having a greater impact on the environment depends upon the origin, properties, and reactivity pathways of the starting substrate. Thus, a chemical transformation requires a chemical reaction to lead to the transformation of one chemical to another (Neilson and Allard, 2012).

Transformation reactions may proceed in the forward direction and processed to completion as well as in the reverse direction until they reach equilibrium.

$$A + B \rightarrow C + D$$

$$C + D \rightarrow A + B$$

Reactions that proceed in the forward direction to approach equilibrium are often described as spontaneous, requiring no input of free energy to go forward. Non-spontaneous reactions require input of free energy to go forward (examples application of heat for the reaction to proceed).

In synthetic chemistry, different chemical reactions are used in combinations during chemical synthesis in order to obtain a desired product. Also, in chemistry, a consecutive series of chemical reactions (where the product of one reaction is the reactant of the next reaction) are often catalyzed by a variety of catalysts which increase the rates of biochemical reactions, so that syntheses and decompositions impossible under ordinary conditions can occur at the temperatures, pressures, and reactant concentrations present within a reactor and, by inference, within the environment.

$$A + B \rightarrow C$$

$$C \rightarrow D + E$$

This simplified equation illustrates the potential complexity of chemical reaction and such complexity must be anticipated when a chemical is transformed in an environmental ecosystem.

Thus, the focus in this chapter is upon developing an understanding of the chemical processes that occur when inorganic pollutants or organic pollutants are released into the environment. With this information, the activities that have an effect on the environment the chemistry can be estimated and even predicted.

8.2 REACTIONS OF INORGANIC CHEMICALS

Inorganic pollutants can be released directly into the environment or formed through chemical reactions of original pollutants. Most inorganic pollutants can undergo transformation (chemical) reactions that convert highly toxic substances to less toxic products or even to environmentally inert (benign) products. It is equally possible that the original non-toxic pollutant or pollutants of low-toxicity can be transformed to products of higher toxicity than the original pollutant. Thus, exposure to hazardous inorganic chemicals can occur on time scales that range from minutes to days or even weeks, and a variety of chemistry and physical processes can play an important role in defining the effect of the exposure. It is also possible that there may be the transformation of the original pollutant to a product that is more capable of remaining in the environment and may even prove to be a POP.

Inorganic chemicals are subject to two processes that determine the fate of the chemical in the environment: (i) the potential for transportation of the chemical; and (ii) the chemical changes that can occur once the chemical has been released to the environment and which depend upon a variety of physical and chemical properties some of which may lead to corrosion. Thus, release into the environment of a persistent inorganic pollutant (PIP) leads to an exposure level which ultimately depends on the length of time the chemical remains in circulation, and how many times it is recirculated in some sense, before ultimate termination of the environmental life-cycle of the chemical-the same rationale applied to product formed from the pollutant by any form of chemical transformation. In addition, the potential for transportation and chemical change (either before or after transportation) raises the potential for the chemical to behave in an unpredictable manner.

When an inorganic chemical (or a mixture of inorganic chemicals) is released into the environment, the issues that need to be considered: (i) the toxicity of the inorganic chemical; (ii) the concentration of the released inorganic chemical; (iii) if a mixture is released, the concentration of the toxic inorganic chemical in the mixture; (iv) the potential of the original inorganic chemical to migrate to other sites; (v) the potential of the inorganic chemical to produce a toxic degradation product; (vi) whether or not the toxicity is lower or higher than the toxicity of the released chemical; (vii) the ability of a degradation product or products to migrate to other sites; (viii) the persistence of the inorganic chemical in an ecosystem; (ix) the persistence of any toxic degradation product in an ecosystem; (x) the potential for the toxic degradation product to degrade even further into harmful or non-harmful constituents and the rate of degradation; and (xi) the degree to which the chemical or any degradation product of the chemical can accumulate in an ecosystem.

8.2.1 CHEMICAL TRANSFORMATION

The various chemical transformation processes, which influence the presence and the analysis of inorganic chemicals at a particular site, although often represented by simple (and convenient) chemical equations, can be complex and the true nature of the chemical transformation process is difficult to elucidate. The extent of transformation is dependent on many factors including: (i) the properties of the chemical; (ii) the geology of the site; (iii) the climatic conditions, such as temperature, oxygen levels, and moisture; (iv) the type of micro-organisms present at the site; and (v) any other environmental conditions that can influence the life-cycle of the chemical. In fact, the primary factor controlling the extent of chemical transformation is the molecular composition of the inorganic chemicals contaminant.

Any chemical change can easily be represented with the help of chemical equations which is a simple representation of a chemical reaction. On the basis of cleavage and formation of chemical bonds, most inorganic chemical reactions can be classified into four broad categories: (i) acid-base reactions; (ii) combination reactions; (iii) decomposition reactions; (iv) single displacement reactions; (v) double displacement reactions (Table 8.2).

TABLE 8.2 General Categories of Inorganic Reactions

Reaction Type	Examples
Acid-base reaction*	$HA + BOH \rightarrow H_2O + BA$
	$HBr(aq) + NaOH(aq) \rightarrow NaBr(aq) + H_2O(l)$
	$HCl(aq) + NaOH(aq) \rightarrow NaCl(aq) + H_2O(l)$
Combination reaction	$A + B \rightarrow AB$
	$2Na(s) + Cl_2(g) \rightarrow 2NaCl(s)$
	$8Fe + S8 \rightarrow 8FeS$
	$S + O_2 \rightarrow SO_2$
Decomposition reaction	$AB \rightarrow A + B$
	$2HgO \rightarrow 2Hg + O_2$
Single displacement reaction	$A + BC \rightarrow AC + B$
	$Mg + 2H_2O \rightarrow Mg(OH)_2 + H_2$
	$Cu(s) + 2AgNO_3(aq) \rightarrow 2Ag(s) + Cu(NO_3)_2(aq)$
	$Zn (s) + CuSO_4(aq) \rightarrow Cu (s) + ZnSO_4(aq)$
Double displacement reaction	$AB + CD \rightarrow AD + CB$
	$Pb(NO_3)_2 + 2KI \rightarrow PbI_2 + 2KNO_3$
	$CaCl_2(aq) + 2AgNO_3(aq) \rightarrow Ca(NO_3)_2(aq) + 2AgCl(s)$

*Sometimes categorized as a double displacement reaction.

One other type of reaction that occurs with inorganic chemicals and with organic chemicals is a coagulation reaction-for convenience, it is included here but is equally applicable to inorganic chemicals and to organic chemicals.

In the chemical sense, aggregation (especially particle agglomeration) refers to the formation of assemblages in a suspension and represents a mechanism leading to destabilization of colloidal systems. During this process, particles dispersed in the liquid phase spontaneously form irregular particle clusters, flocs, or aggregates. This phenomenon (also referred to as coagulation or flocculation) and such a suspension is typically unstable-both physically and chemically (Table 8.3). Particle aggregation can be induced by adding salts or another chemical referred to as coagulant or flocculant. Particle aggregation is normally an irreversible process and once particle aggregates have formed; they will not dissociate easily. In the course of aggregation, the aggregates will grow in size, and as a consequence they may settle to the bottom of the container, which is referred to as sedimentation.

TABLE 8.3 Chemical Coagulation Reaction of Selected Inorganic Chemicals

Substrate and Reactant	Reaction
$Na_2Al_2O_4 + Ca(HCO_3)_2 + 2H_2O$	$2Al(OH)_3 + CaCO_3 + Na_2CO_3$
$Al_2(SO_4)_3 \cdot K_2SO_4 + 3Ca(HCO_3)_2$	$2Al(OH)_3 + K_2SO_4 + 3CaSO_4 + 6CO_2$
$Al_2(SO_4)_3 + 6NaOH$	$2Al(OH)_3 + 3Na_2SO_4$
$Al_2(SO_4)_3 \cdot (NH_4)_2SO_4 + 3Ca(HCO_3)_2$	$2Al(OH)_3 + (NH_4)_2SO_4 + 3CaSO_4 + 6CO_2$
$Al_2(SO_4)_3 + 3Na_2CO_3 + 3H_2O$	$2Al(OH)_3 + 3Na_2SO_4 + 3CO_2$
$Al_2(SO_4)_3 + 3Ca(HCO_3)_2$	$2Al(OH)_3 + 3CaSO_4 + 6CO_2$
$Fe_2(SO_4)_3 + 3Ca(HCO_3)_2$	$2Fe(OH)_3 + 3CaSO_4 + 6CO_2$
$FeSO_4 + Ca(OH)_2$	$Fe(OH)_2 + CaSO_4$
$4Fe(OH)_2 + O_2 + 2H_2O$	$4Fe(OH)_3$

Thus, a chemical transformation requires a chemical reaction to lead to the transformation of one chemical to another (Habashi, 1994). Typically, chemical reactions encompass changes that only involve the positions of electrons in the forming and breaking of chemical bonds between atoms, with no change to the nuclei (no change to the elements present), and can often be described by a relatively simple chemical equation; thus:

$$A + B \rightarrow C$$

However, although the various chemical transformation processes that occur and which influence the presence and the analysis of inorganic chemicals at a particular site are often represented by simple (and convenient) chemical equations, the chemical transformation reactions can be much more complex than the equation indicates. In addition, a chemical reaction may proceed in the forward direction and processed to completion as well as in the reverse direction until the reactants and products reach an equilibrium state:

$$A + B \rightarrow C + D$$

$$C + D \rightarrow A + B$$

Thus,

$$A + B \leftrightarrow C + D$$

Reactions that proceed in the forward direction (i.e., from left to right in the above equation) to approach equilibrium often do not require an input of energy (spontaneous reactions). On the other hand, other reactions may require the input of energy to go forward (non-spontaneous reactions require, an example is the application of heat for the reaction to proceed). In inorganic chemical synthesis, different chemical reactions are used in combinations during the reaction in order to obtain a desired product. Also, a consecutive series of chemical reactions (where the product of a reaction is the reactant of a subsequent reaction) are often encouraged to proceed by a variety of catalysts which increase the rates of biochemical reactions without themselves (the catalysts) being consumed in the reaction, so that syntheses and decompositions impossible under ordinary conditions can occur at the temperature, pressure, and concentrations of the reactants present within a reactor and, by inference, in the current content within the environment.

$$A + B \rightarrow C$$

$$C \rightarrow D + E$$

This simplified equation does not truly illustrate the potential complexity of an inorganic chemical reaction, but such complexity must be anticipated when an inorganic chemical is transformed in an environmental ecosystem. Actual examples (listed alphabetically) of the reaction of inorganic chemicals in the environment are presented in the following subsections.

8.2.1.1 HYDROLYSIS REACTIONS

Hydrolysis is a double decomposition reaction in which an inorganic chemical is represented by the formula *AB* in which *A* and *B* are atoms or groups and water (the second reactant) is represented by the formula HOH, the hydrolysis reaction may be represented by the reversible chemical equation:

$$AB + \mathrm{HOH} \rightleftharpoons A\mathrm{H} + B\mathrm{OH}$$

The reactants other than water, and the products of hydrolysis, may be neutral molecules (as in most hydrolysis reactions involving organic compounds) or ionic molecules (charged species), as in hydrolysis reactions of salts, acids, and bases.

Hydrolysis reactions are usually catalyzed by hydrogen ions (H^+) or hydroxyl ions (OH^-) (Table 8.4). This produces the strong dependence on the acidity or alkalinity (pH) of the solution often observed but, in some cases, hydrolysis can occur in a neutral solutions (pH = 7). Adsorption on to a mineral sediment (such as a clay mineral generally reduces the rates of hydrolysis for acid-catalyzed or base-catalyzed reactions (Petrucci et al., 2010).

TABLE 8.4 Examples of Hydrolysis Reactions of Inorganic Chemicals

Substrate	Reaction
Active metals (Ca)	$Ca + 2H_2O \rightarrow H_2 + Ca(OH)_2$
Alkoxides ($NaOC_2H_5$)	$NaOC_2H_5 + H_2O \rightarrow NaOH + C_2H_5OH$
Amides ($NaNH_2$)	$NaNH_2 + H_2O \rightarrow NaOH + NH_3$
Carbides (CaC_2)	$CaC_2 + 2H_2O \rightarrow Ca(OH)_2 + C_2H_2$
Halides ($SiCI_4$)	$SiCl_4 + 2H_2O \rightarrow SiO2 + 4HCl$
Hydrides ($NaAlH_4$)	$NaAlH_4 + 4H_2O \rightarrow 4H2 + NaOH + Al(OH)_3$

8.2.1.2 PHOTOLYSIS REACTIONS

Photolysis (also called photo-dissociation and photo-decomposition) is a chemical reaction in which an inorganic chemical (or an organic chemical) is decomposed by photons. The primary step of a photolysis reaction is:

$$X + h\nu \rightarrow X^*$$

X^*: an electronically excited state of molecule X and subsequently undergo either physical or chemical processes:

Physical processes:

Fluorescence:

$$X^* \rightarrow X + h\nu$$

Collisional deactivation:

$$X^* + M \rightarrow X + M$$

Chemical processes:

Dissociation:

$$X^* \rightarrow Y + Z$$

Isomerization:

$$X^* \rightarrow X$$

Direct reaction:

$$X^* + Y \rightarrow Z_1 + Z_2$$

Intramolecular rearrangement:

$X^* \rightarrow Y$ Ionization $X^* \rightarrow X+ + e$

The general form of photolysis reactions:

$$X + h\nu \rightarrow Y + Z \text{ (h}\nu\text{ represents the energy input)}$$

The rate of reaction is:

$$-d/dt\,[X] = d/dt[Y] = d/dt[Z] = k[X]$$

K is the photolysis rate constant for this reaction in units of s^{-1}.

Photolysis reactions occur in the atmosphere and are reactions by which nitrogen oxides (primary pollutants) react to form secondary pollutants such as peroxyacetyl nitrate derivatives.

The two most important photolysis reactions in the troposphere are the dissociation of ozone (O_3) after which the excited oxygen atom (O^*) can react with water to give the hydroxyl radical:

$$O_3 + h\nu \rightarrow O_2 + O^*$$
$$O^* + H_2O \rightarrow 2OH\cdot$$

The hydroxyl radical can initiate the oxidation of hydrocarbon derivatives in the atmosphere.

Another reaction involves the photolysis of nitrogen dioxide (NO_2) to produce nitric oxide (NO) and an oxygen radical that is a key reaction in the formation of tropospheric ozone:

$$NO_2 + h\nu \rightarrow NO + O$$

The formation of the ozone layer is also caused by photolysis. Ozone in the stratosphere is created by ultraviolet (UV) light striking oxygen molecules (O_2) and splitting the oxygen molecules into individual oxygen atoms (atomic oxygen). The atomic oxygen then combines with unbroken an oxygen molecule to create ozone (O_3):

$$O_2 \rightarrow 2O$$
$$O_2 + O \rightarrow O_3$$

In addition, the absorption of radiation in the atmosphere can cause photo-dissociation of nitrogen (as one of several possible reactions) that can lead to the formation of nitric oxide (NO) and nitrogen dioxide (NO_2) that can act as a catalyst to destroy ozone:

$$N_2 \rightarrow 2N$$
$$O_2 \rightarrow 2O$$
$$CO_2 \rightarrow C + 2O$$
$$H_2O \rightarrow 2H + O$$
$$2NH_3 \rightarrow 3H_2 + N_2$$
$$N + 2O \rightarrow NO_2$$

As shown in the above equations, most photochemical transformations occur through a series of simple steps (Wayne, 2000; Wayne and Wayne, 2005).

8.2.1.3 RADIOACTIVE DECAY

Radioactive decay occurs from radionuclides which are inorganic elements that emit radiation because the atoms of these elements are unstable and disintegrate (decay) as they release energy in the form of radioactive particles or waves. A decay, or loss of energy from the nucleus, results when an atom with an initial type of nucleus (the parent radionuclide or parent radioisotope) transforms into a daughter nuclide. In some decay reactions, the parent element and the daughter nuclide are different chemical elements, and thus the decay process results in the creation of an atom of a different element (nuclear transmutation).

Radionuclides degrade naturally, but the degradation process itself is hazardous to living things. The risk posed by radionuclides is a function of the type and amount of radiation, as well as exposure pathways. The half-life (the time required to reduce the concentration of a chemical to 50% of the initial concentration) of a radioactive element.

8.2.1.4 REARRANGEMENT REACTIONS

A rearrangement reaction falls into a class of (inorganic and organic) reactions where a molecule is rearranged to produce a structural isomer of the original molecule.

Inorganic reactions typically yield products that are in accordance with the generally accepted mechanism of rearrangement reaction. However, in some instances, inorganic reactions do not give exclusively and solely the anticipated products but may lead to other products that arise from unexpected and mechanistically different reaction paths. This type of unexpected product is often referred to as a rearranged product and, while such a product may not be the expected product it may be the major product of the reaction. Thus, the reaction has involved a rearrangement of the expected product to an unexpected product-a rearrangement reaction has occurred. In many cases, molecular rearrangement yields the isomerization product, coupled with some stereochemical changes.

8.2.1.5 REDOX REACTIONS

Redox reactions (reduction-oxidation reactions or sometimes referred to as oxidation-reduction reactions) are reactions in which one of the reactants is reduced and another reactant is oxidized (Table 8.5). Thus, assuming two reactants in the process, in a redox reaction, one reactant is oxidized while the other reactant is reduced by the net transfer of electron from one reactant to the other. As may be expected, the change in the oxidation states of the oxidized species must be balanced by any changes in the reduced species. For example, the production of iron from the iron oxide ore:

$$Fe_2O_3 + 3CO \rightarrow 2Fe + 3CO_2$$

To complicate matters even further, the oxidizing and reducing agents can be the same element or compound, as is the case when disproportionation of the reactive species occurs. For example:

$$2A \rightarrow (A + n) + (A - n)$$

In this equation, *n* is the number of electrons transferred. Disproportionation reactions do not need to commence with a neutral molecule and can involve more than two species with differing oxidation states.

TABLE 8.5 Examples of Reduction and Oxidation Reactions of Inorganic Compounds

Reduction
Chromate
$2CrO_4^- + 3SO_2 + 4H+ \rightarrow Cr_2(SO_4)_3 + 2H_2O$
Permanganate
$MnO_4^- + 3Fe^{2+} + 7H_2O \rightarrow MnO_2 + 3Fe(OH)_3 + 5H^+$
Oxidation
Cyanide
$2CN^- + 5OCl^- + H_2 \rightarrow N_2 + 2HCO_3^- + 5Cl^-$
Iron (Fe^{2}+)
$4Fe^{2+} + O_2 + 10H_2O \rightarrow 4Fe(OH)_3 + 8H^+$
Sulfur dioxide
$2SO_2 + 2O_2 + H_2O \rightarrow 2H_2SO_4$

8.2.2 PHYSICAL TRANSFORMATION

A physical conversion that is often ignored (or not recognized as such) is the change in composition that occurs when precipitation and dissolution reactions exert a major effect on the concentrations of inorganic ions in solution. Also included is the effect of adsorption and absorption on the composition of the pollutant.

The precipitation of a solid phase in environmental systems rarely results in a pure mineral phase. Minor elements, such as cadmium can coprecipitate with major elements such as calcium to form a solid solution. In addition, the equilibrium activity of the component of a solid solution do not correspond to the solubility calculated from the solubility products of pure minerals. Furthermore, it is often impossible to distinguish chemically between a minor component coprecipitated in a solid, and the adsorption of that minor component onto the solid surface.

Also, complexation is an important process that will determine in some cases if mineral solubility limits are reached, the amount of adsorption that occurs, and the redox state that exists in the water. Inorganic chemicals can also form stable, soluble complexes with organic ligands. Ligands in leachates could include synthetic chelating agents (such as ethylenediamine), partially oxidized biodegradation products (such as organic acids), or natural humic materials. In fact, the movement of metals in the subsurface is strongly influenced by the concentration and chemical properties of ligands. However, the types and concentrations of ligands in most waste leachates is largely unknown.

Physical changes are changes affecting the form of an inorganic chemical but not always the chemical composition. A common example of a physical transformation is the settling of suspended sediment particles. Although settling does not actually transform the sediment into something else, it does remove sediment from the control volume by depositing it on the river bed. This process can be expressed mathematically by heterogeneous transformation equations at the river bed and can be considered to be a transformation process.

More pertinent to the current text is the effect of sorption process and dilution on the properties of the contaminant. Inorganic chemicals interact with the environment in different ways, and once a chemical is released into the environment, there are two physical effects that can influence the distribution of the chemicals: (i) adsorption, (ii) absorption, and (iii) dilution.

8.2.2.1 ADSORPTION

Adsorption is the physical accumulation of material (usually a gas or liquid) on the surface of a solid adsorbent and is a surface phenomenon (Calvet, 1989). Typically, adsorption processes remove solutes from liquids based on their mass transfer from liquids to porous solids.

In nature, a variety of potential natural adsorbents exist in the soil-adsorption occurs in many natural, physical, biological, and chemical systems (especially in the environment) where inorganic molecules can adsorb on to minerals (such as clay) or on to charred wood that remains after a forest fire. In fact, clay minerals are particularly good adsorbents and have a high adsorption capacity for inorganic chemicals that have been released into the environment.

A natural clay mineral is not composed of one clay mineral only. Impurities such as calcite ($CaCO_3$), quartz (SiO_2), feldspar ($KAlSi_3O_8/NaAlSi_3O_8/CaAl_2Si_2O_8$), iron oxides ($FeO$ and Fe_2O_3), and humic acids (degradation products of organic materials) are the most common components in addition to the pure clay mineral. Calcite, iron oxides, and humic acids can be removed by chemical treatment, while quartz and feldspar can be removed by sedimentation, depending upon the particle size is larger than the particle size of any clay minerals present.

Physically, clay minerals are typically ultrafine-grained (normally considered to be less than 2 micrometers (<2 microns, $<2 \times 10^{-6}$ meter) in size on standard particle size classifications). In the present context, clay minerals, which can be classified into various chemical groups, such as the silicate clay mineral groups are an important part of many soils, thus rendering the soil capable of having a high adsorption capacity for inorganic chemicals (Chapter 5, Chapter 7). Generally, no two clay minerals are the same and the adsorption capacity will vary accordingly.

Adsorption of an inorganic chemical onto a solid adsorbent on another mineral, is measured by a partition coefficient, which is the ratio of the concentration the inorganic chemical on the solid to the concentration of the chemical in the fluid (usually water) surrounding the solid:

$$K_d = C_{solid}/C_{water}$$

The concentration on the solid has units of mol/kg, and the concentration in the water is mol/L and, thus, the adsorption coefficient (K_d)has units of L/kg.

The adsorption coefficient will often depend on how much of the total mass of the particle is inorganic material. Thus, the absorption coefficient can be corrected by the fraction of inorganic material (f_{om}) in the particles:

$$K_{om} = Kd/f_{om}$$

The ease of desorption of the inorganic is also an important consideration, because in the treatment of wastewaters, materials are used primarily as ion exchangers, while for in situ immobilization, the metals need to be irreversibly bound to the added adsorbent (Zhou and Haynes, 2010).

8.2.2.2 ABSORPTION

Absorption involves the uptake of one chemical substance into the inner structure of another substance-most typically a gas into a liquid solvent. Furthermore, absorption is a physical or chemical phenomenon or a process in which atoms, molecules, or ions enter some bulk phase-gas, liquid or solid material.

Absorption is a different process from *adsorption*, since molecules undergoing absorption are taken up by the volume of the absorbing material and not by the surface of the material (as in the case for adsorption). In the process, the absorbent distributes the material (the absorbate) throughout the whole of the absorbent whereas in the adsorption process, the adsorbent only distributes it (the adsorbate) on the surface.

In chemical absorption (sometimes referred to in the shortened word form as chemisorption), the absorbed material is generally converted to a product different from the starting material. Thus, chemical absorption or reactive absorption involves a chemical reaction between the absorbent (the absorbing substance) and the absorbate (the absorbed substance) and may be combined with the physical absorption phenomenon. This type of absorption depends upon the stoichiometry of the reaction and the concentration of the potential reactants. Thus, chemical absorption or reactive absorption is a chemical reaction between the absorbed and the absorbing substances. Sometimes it combines with physical absorption. This type of absorption depends upon the stoichiometry of the reaction and the concentration of its reactants.

8.2.2.3 DILUTION

Dilution is the process in which an inorganic chemical in an ecosystem becomes less concentrated and there is a decrease in the concentration of a solute in solution.

To dilute a solution means to add more solvent without the addition of more solute. The resulting solution is thoroughly mixed to ensure that all parts of the solution are identical and can be represented by a simple equation:

$$C_1 \times V_1 = C \times V_2$$

In this equation, C_l is the initial concentration of the solute, V_1 is the initial volume of the solution, C_2 is the final concentration of the solute, and V_1 is the final volume of the solution. However, this type of simple equation does not consider any solute-solvent interactions, solute-solute interactions, and solvent-solvent interactions.

Dilution of a pollutant may reduce the risk to the floral and faunal species because the potential individual receptors are likely to be exposed to lower, less toxic concentrations of the hazard. However, dilution does not reduce the mass of the chemical but can spread the pollutant over a larger area of l exposure, albeit in a diluted form that may still be lethal to floral and faunal species. For example, some inorganic pollutants are hazardous (such as the always lethal potassium cyanide, KCN) even at levels that may be too dilute to be detected with standard equipment and analytical methods.

8.3 REACTIONS OF ORGANIC CHEMICALS

Reactions are chemical reactions involving compounds and the basic chemical reaction types are: (i) addition reactions; (ii) elimination reactions; (iii) substitution reactions; (iv) organic redox reactions; and (v) rearrangement reactions (Table 8.6) (March, 1992; Morrison et al., 1992). In synthesis, reactions are used in the construction of new molecules, but in the discipline known as environmental chemistry, these reactions often occur and cause chemical transformation of an pollutant in the environment. Factors governing transformations in the environment are essentially the same as that of any chemical reaction.

TABLE 8.6 Types of Organic Reactions

Reaction Type	Sub-Type	Examples
Addition reactions	Electrophilic addition	Halogenation, hydrohalogenation, hydration
	Free radical addition	
	Nucleophilic addition	
Elimination reactions	–	Condensation in which a molecule of water is eliminated from the reactants
		Dehydration-removal of water
Substitution reactions	Electrophilic substitution	
	Electrophilic aromatic substitution	
	Free radical substitution	
	Nucleophilic aliphatic substitution	
	Nucleophilic aromatic substitution	
	Nucleophilic acyl substitution	
Organic redox reactions		Oxidation-reduction reactions specific to organic compounds
Rearrangement reactions	–	A reaction involving the exchange of bonds between two reacting chemical species, such as A-B + C-D → A-D + C-B
–	1,2 rearrangement	A reaction in which a substituent can move from one atom to another atom, such as movement to an adjacent atom
–	pericyclic reactions	A rearrangement reaction where the intermediate is cyclic

Furthermore, while organic reactions can be organized into several basic reaction types (Table 8.6)-some reactions fit into more than one category-the reactions of organic chemicals can also be categorized on the basis of the type of functional group involved in the reaction as a reactant and the functional group that is formed as a result of this reaction. Thus, functional groups are a key organizing feature of chemistry. By focusing on the functional groups present in a molecule (most molecules have more than one functional group), several of the reactions that the molecule will undergo can be predicted and understood. Thus, functional group transformation/interconversion is the process of converting one functional group into another functional group by any one (or more) of several reactions such as: (i) substitution; (ii) addition; (iii) elimination; (iv) reduction; or (v) oxidation by the use of reagents under different reaction conditions.

In fact, there is no limit to the number of possible reactions, but certain general patterns of reactivity can be used to describe many common or useful reactions. Typically, each reaction typically has a stepwise reaction mechanism that can be used to explain the means by which the reaction occurs, although a detailed description of steps may not always be evident from a list of reactants alone.

8.3.1 ADDITION AND ELIMINATION REACTIONS

An addition reaction is a reaction in which two or more molecules combine to form a product (the adduct) (Morrison et al., 1992). Addition reactions are limited to chemical compounds that have multiple bonds, such as molecules with carbon-carbon double bonds (alkene derivatives, >C=C<), or with triple bonds (alkyne derivatives, –C≡C–). Molecules containing carbon-heteroatom double bonds such as the carbonyl group (>C=O) or the imine group (>C=N–) can also participate in addition reactions.

An addition reaction is the reverse of an elimination reaction, such as the hydration of an alkene derivative to an alcohol derivative (an addition reaction) which can be reversed by the dehydration process (removal of water by an elimination reaction):

Addition reaction:

$$CH_2{=}CH_2 + H_2O \rightarrow CH_3CH_2OH$$

Elimination reaction:

$$CH_3CH_2OH \rightarrow CH_2{=}CH_2 + H_2O$$

The main driver behind addition reactions to alkene derivatives is that alkene derivatives contain the unsaturated >C=C< functional group which characteristically undergoes addition reactions which is the conversion of the weaker π bond into two new, stronger σ bonds. In addition to addition reactions being typical of the unsaturated hydrocarbon derivatives (alkenes and alkynes) and aldehydes and ketones, which have a carbon-to-oxygen double bond.

8.3.2 HYDROLYSIS REACTIONS

A hydrolysis reaction typically involves the direct reaction of a chemical with water in which a chemical bond is cleaved and two new bonds are formed, each one having either the hydrogen component (H) or the hydroxyl component (OH) of the water molecule.

In the process, the hydroxyl group replaces another chemical group on the target molecule and the reaction is usually catalyzed by hydrogen ions or hydroxyl ions. This produces the strong dependence on the acidity (pH = >7.0) or alkalinity (pH = <7.0) of the solution often observed but, in some cases, hydrolysis can occur in a neutral (pH = 7.0) environment. Adsorption onto a mineral sediment (such as a clay sediment that has strong adsorptive powers) generally reduces the rates of hydrolysis for acid- or base-catalyzed reactions. Neutral reactions appear to be unaffected by adsorption, although there is always the possibility that the mineral sediment can cause catalyzed chemical transformation reactions.

The rate of a hydrolysis reaction is typically expressed in terms of the acid-catalyzed, not-catalyzed, and base-catalyzed rate constants. Furthermore, the hydrolysis of compounds is influenced by the composition of the solvent (Lyman et al., 1982) and the rate constants may be much higher in water than in other solvents. In fact, the introduction of a complex mixture of chemicals into the water body of an ecosystem can be expected to produce a significant shift in acidity or alkalinity of the water and, therefore, it would not be surprising to anticipate that hydrolysis would be affected in complex mixtures.

8.3.3 PHOTOLYSIS REACTIONS

Photolysis is a process by which chemical bonds can be broken by the transfer of light energy (direct photolysis) of radiant energy (indirect photolysis) to the chemical bonds. The rate of the reaction depends on several factors including: (i) light adsorption properties; (ii) the reactivity of the chemical; and (iii) the intensity of the radiation (Lyman et al., 1982). The process is divided into three stages which are: (i) the adsorption of light which excites electrons in the target molecule; (ii) the primary photochemical processes which transform or de-excite the excited target molecule; and (iii) the secondary thermal reactions which transform the intermediates produced in the previous step [Step (ii))].

Indirect photolysis (often referred to as sensitized photolysis) occurs when the light energy absorbed by a non-target molecule (a sensitized molecule) is transferred to the target molecule (an acceptor molecule). Moreover, the probability of a sensitized molecule donating energy to target molecule is proportional to the concentration of both chemical species.

8.3.4 REARRANGEMENT REACTIONS

Rearrangement reactions typically yield products that are in accordance with the generally accepted mechanism of the reactions-in many cases, the rearrangement affords products of an isomerization, coupled with some stereochemical changes.

A rearrangement reaction falls into a class of reactions where the carbon skeleton of an molecule is rearranged to form a structural isomer of the original molecule that assumes the minimal energy content of the product, i.e., the most stable product is formed (March, 1992; Morrison et al., 1992; Moulay, 2002). As a result of the reaction, a substituent group typically moves from one atom to another atom in the same molecule to yield an isomer of the original reactant:

$$\underset{\substack{| \\ R}}{\text{—C}}\text{—C—C—} \longrightarrow \text{—C—}\underset{\substack{| \\ R}}{\text{C}}\text{—C—}$$

Carbon atom: 1 2 3 1 2 3

In the above equation, there has been movement of the substituents group (represented by R) from carbon atom number 1 to carbon atom number 2.

Thus, a rearrangement reaction is a reaction in which an atom or a group (or in some cases a bond) is caused to move or migrate to another part of the molecule. The reaction may involve several steps, but the defining feature is that the atom or a group or the bond shifts from one site of attachment to another site. The simplest (perhaps, the most common) types of rearrangement reactions are intramolecular rearrangements insofar as the reactions occur in within one reactant molecule and the product of the reaction is a structural isomer of the reactant (Moulay, 2002).

However, in some instances, reactions do not give exclusively and solely the anticipated products but may lead to other products that arise from unexpected and mechanistically different reaction paths. An energetic requirement is also observed in order for a rearrangement to take place; that is, the rearrangement usually involves an evolution of energy (typically in the form of heat, i.e., the reaction is overall an exothermic reaction) evolution to be able to yield a more stable compound (Moulay, 2002).

8.3.5 REDOX REACTIONS

Redox reactions (reduction-oxidation reactions, also referred to as oxidation-reduction reactions) are reactions in which one chemical undergoes reduction and another chemical undergoes is oxidation (Table 8.7). Therefore, the oxidation state of the species involved must change.

TABLE 8.7 Examples of Reduction and Oxidation Reactions of Organic Compounds

Reduction:
$[CH_2NOS] + [H] \rightarrow [CH_3] + H_2O + NH_3 + H_2S$
Oxidation:
Organic matter
$[CH_2NOS] + [O] \rightarrow CO_2 + H_2O + NO_x + SO_2$

In organic chemistry, the term hydrogenation could be used instead of reduction, since hydrogen is the reducing agent in a large number of reactions, especially in chemistry and biochemistry. But, unlike oxidation, hydrogenation has maintained its specific connection to reactions that add hydrogen to another substance such as the hydrogenation processes used in a crude oil refinery (Speight, 2014, 2017). On the other hand, the word oxidation originally implied reaction with oxygen to form an oxide, but the word has been expanded to encompass oxygen-like substances that accomplished parallel chemical reactions, and ultimately, the meaning was generalized to include all processes involving loss of electrons.

For example, there is the oxidation of ethyl alcohol (CH_3CH_2OH) is oxidized to acetaldehyde (CH_3CHO) and the reverse reaction in which acetaldehyde is reduces to ethyl alcohol:

$$CH_3CH_2OH \rightarrow CH_3CHO + H_2 \rightarrow CH_3CH_2OH$$

This equation can be sub-divided as follows into oxidation by loss of hydrogen and reduction by gain of hydrogen. Thus:

Oxidation by loss of hydrogen:

$$CH_3CH_2OH \rightarrow CH_3CHO \text{ (oxidation by loss of hydrogen)}$$

Reduction by gain of hydrogen:

$$CH_3CHO \rightarrow CH_3CH_2OH \text{ (reduction by gain of hydrogen)}$$

The species that is oxidized is the reducing agent and the species that is reduced is the oxidizing agent. To complicate matters even further, the oxidizing and reducing agents can be the same element or compound, as is the case when disproportionation of the reactive species occurs.

8.3.6 SUBSTITUTION REACTIONS

A substitution reaction (sometime referred to as a single displacement reaction or single replacement reaction) is a reaction in which a functional group in an organic chemical is replaced by another functional group (March, 1992; Morrison et al., 1992). Substitution reactions are of prime importance in environmental organic chemistry because of the simplicity of the reaction which is accompanied by a substantial change in the chemical and physical properties of the product vis-à-vis compared to the chemical and physical properties of the starting chemical. Substitution reactions in organic chemistry are classified either as (i) nucleophilic substitution or (ii) electrophilic substitution depending upon the reagent involved.

Thus, a nucleophilic substitution reaction (which is common in organic chemistry) is a reaction in which an electron-rich nucleophile selectively bonds with or attacks the positive or partially positive charge of an atom or a group of atoms to replace the leaving electrophile:

$$\text{(Nucleophile)} + \text{R-(Leaving Group)} \rightarrow \text{R-(Nucleophile)} + \text{(Leaving Group)}$$

On the other hand, electrophilic substitution involves electrophiles and an example is electrophilic aromatic substitution. In this type of substitution, the benzene ring (which has a π-electron cloud above and below the plane of the ring of carbon atoms) is attacked by an electrophile (shown as E^+). The π-electron cloud is disturbed and a carbocation resonating structure results after which a proton (H^+) is ejected from the intermediate and a new aromatic compound is formed:

$$C_6H_6 + E^+ \rightarrow C_6H_5E + H^+$$

In the hydrolysis reaction, water can act as an acid or a base: (i) if that water acts as an acid, the water molecule would donate a proton (H^+), also written as a hydronium ion (H_3O^+); or (ii) if the water acts as a base, the water molecule would accept a proton (H^+).

In the aqueous hydrolysis reaction, the reacting water molecules are split into hydrogen (H^+) and hydroxide (OH^-) ions, which react with and break up the other reacting compound. The term *hydrolysis* is also applied to the electrolysis of water (that is, breaking up of water molecules by an electric current) to produce hydrogen and oxygen. The hydrolysis reaction is distinct from a hydration reaction, in which water molecules attach to molecules of the other reacting compound without breaking up the latter compound.

A relevant reaction in the environment (because of the ever-presence of water in many ecosystems) is the hydrolysis reaction in which an chemical compound is decomposed by reaction with water-also the hydrolysis reaction should not be confused with the *hydrogenolysis* reaction which is a reaction of hydrogen as practiced widely in the crude oil industry to produce liquid fuels (Speight, 2014, 2017). Hydrolysis is an example of a larger class of reactions referred to as nucleophilic displacement reactions in which a nucleophile (an electron-rich species containing an unshared pair of electrons) attacks an electrophilic atom (an electron-deficient reaction center). Hydrolytic processes encompass several types of reaction mechanisms that can be defined by the type of reaction center (i.e., the atom bearing the leaving group, X) where hydrolysis occurs. The reaction mechanisms encountered most often are direct and indirect nucleophilic substitution and nucleophilic addition-elimination.

This type of reaction can be used to predict the persistency of a chemical in the environment, the physical-chemical properties of the chemical, and the reactivity of the chemical in the environment need to be known or at least estimated (Rahm et al., 2005). The chemicals that can undergo elimination reactions are rapidly transformed, as are perhalogenated chemicals that can undergo substitution reactions. These chemicals are not likely to persist in the environment, while those that do not show any observable reactivity under similar hydrolytic conditions are likely to be POPs and the fate of these chemicals is intimately linked to the cycling of chemicals in the environment (deBruyn and Gobas, 2004).

$$CH_3COOH + H_2O \rightleftharpoons H_3O^+ + CH_3COO^-$$

In the above reaction, the proton H^+ from CH_3COOH (acetic acid) is donated to water, producing the hydronium ion (H_3O^+) and an acetate ion (CH_3COO^-). The bonds between proton and the acetate ion are dissociated by the addition of water molecules. A reaction with acetic acid (CH_3COOH), a weak acid, is similar to an acid-dissociation reaction, and water forms a conjugate base and a hydronium ion. When a weak acid is hydrolyzed, a hydronium ion is produced. Thus, the products of hydrolysis depend very much upon that substrate that is to be hydrolyzed (Table 8.8).

More pertinent to the present text, the hydrolysis of a pesticide is basically a reaction with a water molecule involving specific catalysis by proton or hydroxide, and sometimes in ions such as phosphate ion, present in the aquatic environment that play a role in general acid-base catalysis (Katagi, 2002).

TABLE 8.8 Examples of Hydrolysis Reactions of Organic Chemicals*

1. The hydrolysis of a primary amide forms a carboxylic acid and ammonia:

   $RCONH_2 + H_2O \rightarrow RCOOH + NH_3$
2. The hydrolysis of a secondary amide forms a carboxylic acid and primary amine**: $R^1CONHR^2 + H_2O \rightarrow RCOOH + RNH_2$
3. The hydrolysis of an ester forms a carboxylic acid and an alcohol**:

   $R^1COOR^2 + H_2O \rightarrow RCOOH + ROH$
4. The hydrolysis of a halogenoalkane forms an alcohol:

   $RBr + H_2O \rightarrow ROH + H^+ + Br^-$

*The chemical equations presented above illustrate the hydrolysis by reaction with water.

**R^1 and R^2 can be the same or different alkyl groups.

However, the hydrolytic profiles depend on the chemical structure and functional group(s) in the pesticide molecule, which are not always consistent within a chemical class of pesticides (Stoytcheva, 2011). For example, pesticides that are composed of organophosphorus derivatives are primarily susceptible to alkaline hydrolysis with less acidic catalysis, but some of phosphorodithioate derivatives are found to be acid labile. Various instrumental techniques have been applied to the chemical identification of degraded products, leading to clarification of the reaction mechanisms involved. Moreover, pesticides are usually applied as a suitable formulation, and thus the effects of surfactants and other formulation reagents on hydrolysis should be examined in more detail. To assess the fate and impact of pesticides and their degraded products in real aquatic environments, these concerns should be further examined using the various analytical techniques together with simulation models.

8.4 CHEMISTRY IN THE ENVIRONMENT

Chemical reactions occur at a characteristic rate (the reaction rate) at a given temperature and chemical concentration. Typically, reaction rates increase with increasing temperature because there is more thermal energy available to reach the activation energy necessary for breaking bonds between atoms. The general rule of thumb (see above) is that for every 10°C (18°F) increase in temperature the rate of an chemical reaction is doubled and there is no reason to doubt that this would not be the case for chemicals discharged into the environment (Jury et al., 1987).

The chemical industry involves physical, thermal, and manufacture of chemical intermediates and end-product chemicals. The product slate is varied but includes fuels, petrochemicals, fertilizers, pesticides, paints, waxes, thinners, solvents, cleaning fluids, detergents, refrigerants, antifreeze, resins, sealants, insulations, latex, rubber compounds, hard plastics, plastic sheeting, plastic foam and synthetic fibers. The composition of the chemicals is varied, and there are very few indications of how this chemical will behave once they are discharged into the environment either as a single chemical or as a mixture. It is at this stage that a knowledge of chemical properties can bring some knowledge of predictability related to chemical behavior.

These chemicals vary from simple hydrocarbon derivatives of low-to-medium molecular weight to higher molecular weight compounds containing sulfur, oxygen, and nitrogen, as well as compounds containing metallic constituents, particularly vanadium nickel, iron, and copper and contain one or more functional groups that dictate the behavior of the chemical. However, the behavior of a chemical on the basis of a functional groups depends upon (i) the type of functional group, (ii) the number of functional groups, (iii) the position of the functional groups within the molecule, and (iv) the ecosystem into which the chemical is discharged.

Chemical contaminants can enter the environment (air, water, and soil) when they are produced, used, or disposed and the impact on the environment is determined by (i) the amount of the contaminant that is released, (ii) the type of the contaminant, (iii) the concentration of the contaminant, and (iv) any transformation that occur after the contaminant has entered the environment whether it is in the atmosphere, the aquasphere, or the terrestrial biosphere (Jury et al., 1987). Some contaminants can be harmful if released to the environment even when there is not an immediate impact. Other contaminants are of concern as they can work their way into the food chain and accumulate and/or persist in the environment for prolonged periods, including years (Jury et al., 1987).

The volatility of a chemical is of concern predominantly for surface-located chemicals and is affected by (i) temperature of the soil, (ii) the water content of the soil, (iii) the adsorptive interaction of the chemical and the soil, (iv) the concentration of the chemical in the soil, (v) the vapor pressure of the chemical, and (vi) the solubility of the chemical in water, which is the predominant liquid in the soil.

However, before delving into the realm of chemicals in the environment, it is necessary for any investigator to recognize that are chemicals that exist naturally in the environment and which must be taken into account before accurate assessment of chemicals in the environment can be made. These naturally-occurring chemicals are often grouped under the umbrella name natural matter (NOM) is an inherently complex mixture of polyfunctional molecules (Macalady and Walton-Day, 2011). Because of their universality and chemical reversibility, oxidation/reductions (redox) reactions of natural matter have an especially interesting and important role in geochemistry. Variabilities in natural matter composition and chemistry make studies of its redox chemistry particularly challenging, and details of natural matter-mediated redox reactions are only partially understood. This is in large part due to the analytical difficulties associated with natural matter characterization and the wide range of reagents and experimental systems used to study natural matter redox reactions.

When dealing with chemicals that have been released (advertently or inadvertently depending upon the circumstances), there are several types of chemical transformations of chemical transformation reactions that can occur in the environment. These reactions can be grouped into four major categories: (i) oxidation-reduction reactions, also known as redox reactions; (ii) carbon-carbon bond formation; (iii) carbon-heteroatom bond formation in which a carbon atom of one molecule forms a bond with the nitrogen atom or oxygen atom or sulfur atom of another molecule; (iv) carbon-carbon bond cleavage; (v) carbon-heteroatom bond cleavage; and (vi) intermolecular interactions.

Redox reactions would include: (i) the hydrogenation of olefin derivatives and acetylene derivatives; (ii) the loss of hydrogen through aromatization reactions; (iii) the oxidation or reduction of alcohol derivatives, aldehyde derivatives, and ketone derivatives, as well as the oxidative cleavage of olefins. Examples of chemical transformations involving bond formation are polymerization or condensation reactions, esterification, or amide ($-CONH_2$) formation, and cyclization (ring formation) reactions. Several types of bond cleavage reactions which might affect the fate or longevity of chemicals discharged into the environment are the formation of amino acids from peptides and proteins, and the hydrolysis of esters and amides to form carboxylic acids, as well as another form of chemical degradation (Wham et al., 2005).

Long-term trends of chemical species in the environment are determined by emissions from anthropogenic and natural sources as well as by transport of the chemical, physical, and chemical processes that affect the behavior of the chemical, and deposition. While continually increasing emissions of such trace species as carbon dioxide (CO_2), nitrous oxide (N_2O, and methane (CH_4) that can arise form transformation occurring during the life-cycle of chemicals are predicted to raise global temperatures via the greenhouse effect (Figure 8.1) (Speight, 1996), growing emissions of sulfur dioxide (SO_2), which forms sulfate ($-SO_4$) aerosol through oxidation most likely will have a cooling effect by reflecting solar radiation back to space. However, these postulates do not take into account the fact that the Earth is in an interglacial period during which time there will be an overall rise in climatic temperature as the natural order of climatic variation. Therefore, the extent of the anthropogenic contributions to temperature rise (climate change) cannot be accurately assessed (Speight and Islam, 2016). Complicating matters is the fact that the chemical reactions are sensitive to climatic conditions, being functions of temperature, the presence of water vapor, as well as a variety of other physical parameters.

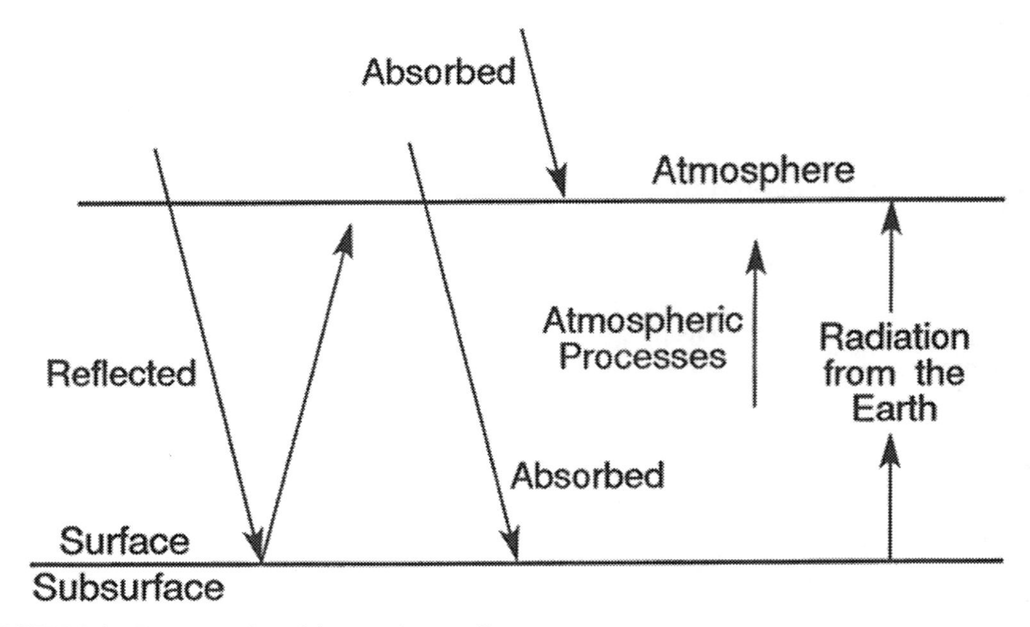

FIGURE 8.1 Representation of the greenhouse effect.

Source: Reproduced with permission from: Speight (1996). © Taylor and Francis.

Thus, environmental chemistry-the study of chemical processes occurring in the environment-is impacted by a variety of external activities (including anthropogenic activities and climatic variations) and these impacts may be felt on a local scale (through the presence of urban air pollutants or toxic substances arising from a chemical waste site) or on a global scale (through depletion of stratospheric ozone or the phenomenon that has become known as global climate change).

8.4.1 CHEMISTRY IN THE ATMOSPHERE

Chemicals can be emitted directly into the atmosphere or formed by chemical conversion or by through chemical reactions of precursors species. In these reactions, highly toxic chemicals can be converted into less toxic products, but the result of the reactions can also be products having a higher toxicity than the starting chemicals. In order to understand these reactions, it is also necessary to understand the chemical composition of the natural atmosphere, the way gases, liquids, and solids in the atmosphere interact with each other and with the surface of the Earth and associated biota, and how human activities may be changing the chemical and physical characteristics of the atmosphere.

Much of the anthropogenic (human) impact on the atmosphere is associated with the increasing use of fossil fuels as an energy source-for things such as heating, transportation, and electric power production. Photochemical smog/tropospheric ozone is a serious environmental problem that has been associated with burning such fuels and the result has been the formation and deposition of acid rain (Figure 8.2) (Speight, 1996).

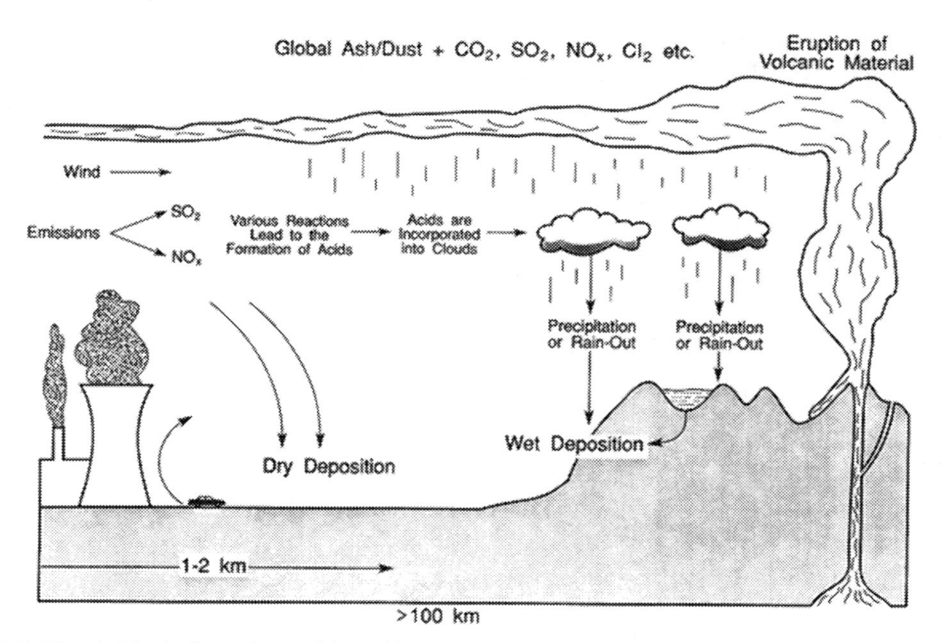

FIGURE 8.2 Acid rain formation and deposition.

Source: Reproduced with permission from: Speight (1996). © Taylor and Francis.

Acid rain is formed when sulfur oxides and nitrogen oxides react with water vapor and other chemicals in the presence of sunlight to form various acidic compounds in the atmosphere. The principle source of acid rain-causing pollutants, sulfur oxides and nitrogen oxides, is from fuel combustion-specifically from fuels that contain sulfur and nitrogen:

$$2[C]_{fuel} + O_2 \rightarrow 2CO$$

$$[C]_{fuel} + O_2 \rightarrow CO_2$$

$$2[N]_{fuel} + O_2 \rightarrow 2NO$$

$$[N]_{fuel} + O_2 \rightarrow NO_2$$

$$[S]_{fuel} + O_2 \rightarrow SO_2$$

$$2SO_2 + O_2 \rightarrow 2SO_3$$

$$SO_2 + H_2O \rightarrow H_2SO_3$$

sulfurous acid

$$SO_3 + H_2O \rightarrow H_2SO_4$$

sulfuric acid

$$NO + H_2O \rightarrow HNO_2$$

nitrous acid

$$3NO_2 + 2H_2O \rightarrow HNO_3$$

nitric acid

In addition, as a result of a variety of human activities (e.g., agriculture, transportation, industrial processes), a large number of different toxic chemical pollutants are emitted into the atmosphere. Among the chemicals that may pose a human health risk are pesticides, polychlorobiphenyl derivatives (PCBs), polycyclic aromatic hydrocarbon derivatives (PAHs), dioxin derivatives, and volatile compounds (e.g., benzene, carbon tetrachloride).

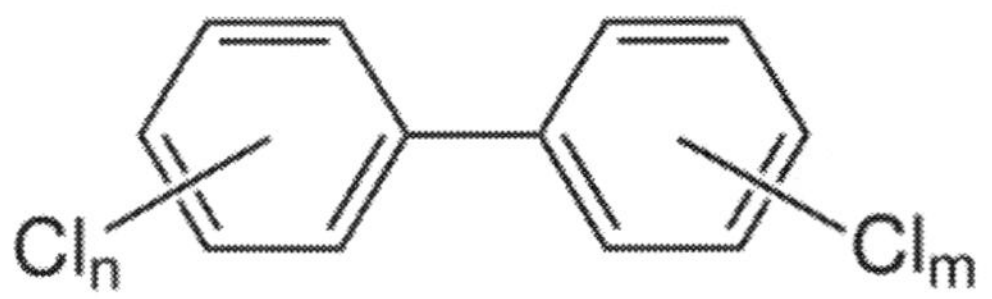

Cl_n Cl_m

Polychlorobiphenyl derivatives; *n* and *m* can vary by any number from 1 to 5.

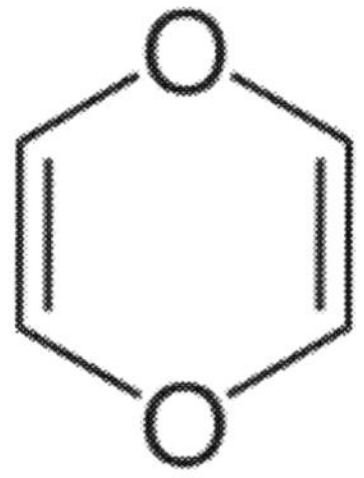

O O

1,4-Dioxin

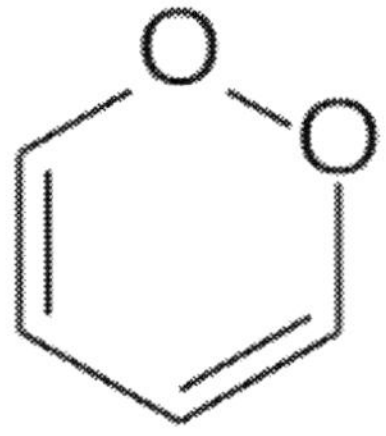

1,2-Dioxin

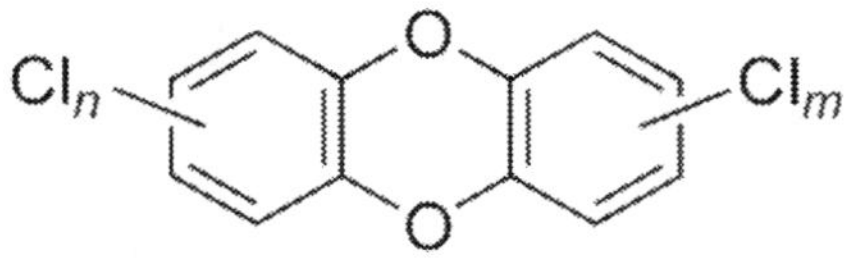

Polychlorinated dibenzo-p-dioxin derivatives; *n* and *m* can vary by any number from 1 to 5.

Many of the more environmentally persistent compounds (such as the polychlorobiphenyl derivatives) have been measured in various floral and faunal species.

8.4.2 CHEMISTRY IN THE AQUASPHERE

Water pollution has become a widespread phenomenon and has been known for centuries, particularly the pollution of rivers and groundwater (Samin and Janssen, 2012). By way of example, in ancient times up to the early part of the 20th century, many cities deposited waste into the nearby river or even into the ocean.

In addition, there are many are indications that the chemicals materials in the aquasphere (also called, when referring to the sea, the marine aquasphere) are subject to intense chemical transformations and physical recycling processes imply that a total-carbon approach is not sufficient to resolve the numerous processes occurring. The transport of anthropogenically produced or distributed compounds such as crude oil hydrocarbon derivatives and halogenated hydrocarbon derivatives, including the PCBs, the DDT family, and the Freon derivatives and the chemistry of these chemicals in water is not fully understood.

The effects of a chemical released into the marine environment (or any part of the aquasphere) depend on several factors such as (i) the toxicity of the chemical, (ii) the quantity of the chemical, (iii) the resulting concentration of the chemical in the water column, (iv) the length of time that floral and faunal organisms are exposed to that concentration, and (v) the level of tolerance of the organisms, which varies among different species. Even if the concentration of the chemical is below what would be considered as the lethal concentration, a sub-lethal concentration of a chemical can still lead to a long-term impact within the aqueous marine environment. In addition, the characteristics of some chemicals can result in an accumulation of the chemical within an organism (*bio-accumulation*) and the organism may be particularly vulnerable to this problem. Furthermore, subsequent bio-magnification may also occur if the chemical (or a toxic product produced by one or more transformation reactions) can be passed on, following the food chain up to higher flora or fauna.

In terms of a chemical spill into the environment, the complex processes of transformation start developing almost as soon as the chemical contacts the land (or the water) although the progress, duration, and result of the transformations depend on the properties and composition of the chemical, parameters of the spill, and environmental conditions. The major operative processes are (i) physical transport, (ii) dissolution, (iii) emulsification, (iv) oxidation, (v) sedimentation, (vi) microbial degradation, (vii) aggregation, and (viii) self-purification.

In terms of physical transport, the distribution of a contaminant spilled on the surface of the ocean occurs under the influence of gravitation forces and is controlled by the viscosity of the (liquid) contaminant as well as the surface tension of the water. In addition, during the first several days after a spill of a liquid chemicals or a mixture of liquid chemicals, a part of spilled chemical may be lost through evaporation and any water-soluble constituents disappear into the water. The portion of the chemical mixture that remains is the more viscous fraction. Further changes take place under the combined impact of meteorological and hydrological factors.

Many organic chemicals are not *soluble* in water, although some constituents may be water-soluble to a certain degree, especially low-molecular-weight aliphatic and aromatic hydrocarbon derivatives. Compared to the evaporation process, the dissolution of organic chemicals in water is a slow process. However, the *emulsification* of a chemical (or chemicals) in the aquasphere does occur but depends predominantly on the presence of organic functional groups in the spilled material which can increase with time due to oxidation.

Oxidation is a complex process that can ultimately results in the destruction of the crude boil constituents. The final products of oxidation (such as hydroperoxide derivatives, phenol derivatives, carboxylic acid derivatives, ketone derivatives, and aldehyde derivatives) usually have increased water solubility. This can result in the apparent disappearance of the chemicals from the surface of the water. This is due to the incorporation of oxygen-containing functional groups into the chemicals, which results in a change in density with an increase in the ability of the transformed chemicals to become miscible (or emulsify) and sink to various depths of the water system as these changes intensify. These chemical changes also result in an increase in the viscosity of the chemicals which promotes the formation of solid oil aggregates. The reactions of photo-oxidation, photolysis in particular, also initiate the transformation of the more complex (polar) chemicals.

Self-purification is a result of the processes previously described above in which a chemical in the environment rapidly loses the original properties and disintegrates into various products. These products may have different chemical composition and structure to the original chemical and exist in different migrational forms, and they undergo chemical transformations that slow after reaching thermodynamic equilibrium with the environmental parameters. Eventually, the original and intermediate compounds disappear, and carbon dioxide and water form. This form of self-purification inevitably happens in water ecosystems if the amount of toxic chemicals spilled into the system does not exceed acceptable limits.

While acid-base reactions are not the only chemical reactions important in aquatic systems, they do present a valuable starting point for understanding the basic concepts of chemical equilibria in such systems. Carbon dioxide (CO_2), a gaseous inorganic chemical that is of vital importance to a variety of environmental processes, including growth and decomposition of biological systems, climate regulation and mineral weathering, has acid-base properties that are critical to an understanding of its chemical behavior in the environment.

Furthermore, the almost unique physical and chemical properties of water as a solvent are of fundamental concern to aquatic chemical processes. For example, in the liquid state water has unusually high boiling point and melting point temperatures compared to its hydride analogs from the periodic table of the elements (Table 8.9), such as ammonia (NH_3), hydrogen fluoride (HF) and hydrogen sulfide (H_2S). Hydrogen bonding between water molecules means that there are strong intermolecular forces making it relatively difficult to melt or vaporize.

TABLE 8.9 Periodic Table of the Elements

1	2	3	4	5	6	7	8	9	10	11	12	13	14	15	16	17	18
H																	He
Li	Be											B	C	N	O	F	Ne
Na	Mg											Al	Si	P	S	Cl	Ar
K	Ca	Sc	Ti	V	Cr	Mn	Fe	Co	Ni	Cu	Zn	Ga	Ge	As	Se	Br	Kr
Rb	Sr	Y	Zr	Nb	Mo	Tc	Ru	Rh	Pd	Ag	Cd	In	Sn	Sb	Te	I	Xe
Cs	Ba	57-71	Hf	Ta	W	Re	Os	Ir	Pt	Au	Hg	Tl	Pb	Bi	Po	At	Rn
Fr	Ra	89-103	Rf	Db	Sg	Bh	Hs	Mt	Ds	Rg	Cn	Nh	Fl	Mc	Lv	Ts	Og
		La	Ce	Pr	Nd	Pm	Sm	Eu	Gd	Tb	Dy	Ho	Er	Tm	Yb	Lu	
		Ac	Th	Pa	U	Np	Pu	Am	Cm	Bk	Cf	Es	Fm	Md	No	Lr	

In many electrolyte solutions of interest, the presence of ions can alter the nature of the water structure. Ions tend to orient water molecules that are near to them. For example, cations attract the negative oxygen end of the water dipole towards them. This reorientation tends to disrupt the ice-like structure further away. This can be seen by comparing the entropy change on transferring ions from the gas phase to water with a similar species that does not form ions.

As an example, chemical transformation that can occur in a water system, the chemistry of methyl iodide (which is thermodynamically unstable in seawater) is known and its chemical fate is kinetically controlled. The equations showing the fate of methyl iodide are as follows (Gacosian and Lee, 1981):

$$CH_3I + Cl^- \rightarrow CH_3Cl + I$$

$$CH_31 + Br^- \rightarrow CH_3Br + I^-$$

$$CH_3Br + Cl^- \rightarrow CH_3Cl + Br^-$$

$$CH_3X + H_2O \rightarrow CH_3OH + X\text{-}$$

In this equation, $X = Cl^-, Br^-, I^-$

Steroids are a class of biogenic compounds which may serve as an indicator of certain processes transforming matter in seawater and sediments. The steroid hydrocarbon structure forms a relatively stable nucleus which may incorporate functional groups such as alcohols (sterol derivatives and stanol derivatives), ketone derivatives (stanone derivatives) and olefin linkages (styrene derivatives) either in the four-ring system or on the side chain originating at C-17.

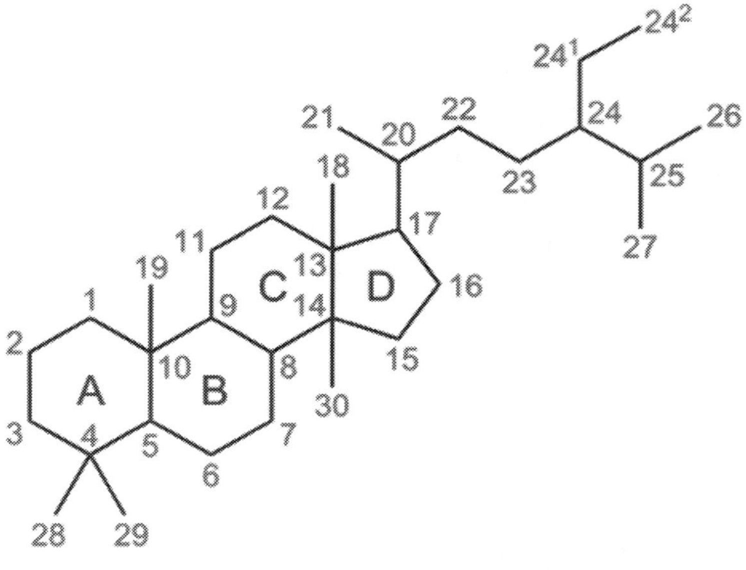

The hydrocarbon framework of the steroid system (ring lettering and atom numbering are shown). These compounds are produced by a wide variety of marine and terrestrial organisms and often have specific species sources. Diagenetic alteration of steroids by geochemical and biochemical processes can lead to the accumulation of transformed products in seawater and sediments.

Within the group of chlorinated compounds, chlorinated ethylene derivatives are the most often detected groundwater pollutants. Tetrachloroethylene (PCE) is the only chlorinated ethylene derivative that resists aerobic biodegradation. Trichloroethylene (TCE), all three isomers of dichloroethylene (CCl_2=CH_2 and the cis/trans isomers of CHCl=CHCl), and vinyl chloride (CH_2=CHCl) are mineralized in aerobic co-metabolic processes by methanotrophic or phenol-oxidizing bacteria. Oxygenase derivatives with broad substrate spectra are responsible for the co-metabolic oxidation. Vinyl chloride is furthermore utilized by certain bacteria as carbon and electron source for growth. All chlorinated ethylene derivatives are reductively dechlorinated under anaerobic conditions with possibly ethylene or ethane as harmless end-products.

Tetrachloroethylene (CCl_2=CCl_2) is dechlorinated to TCE (CCl_2=CHCl) in a co-metabolic process by methanogens, sulfate reducers, homoacetogen derivatives and others. Furthermore, tetrachloroethylene, and TCE serve in several bacteria as terminal electron acceptors in a respiration process. The majority of these isolates dechlorinate tetrachloroethylene and TCE to cis-l,2-dichloroethene although they have been isolated from systems where complete dechlorination to ethene occurred.

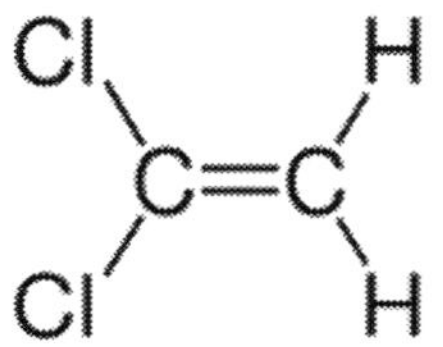

```
Cl     H
 \    /
  C=C
 /    \
Cl     H
```

1,1-Dichloroethylene

```
H      H
 \    /
  C=C
 /    \
Cl     Cl
```

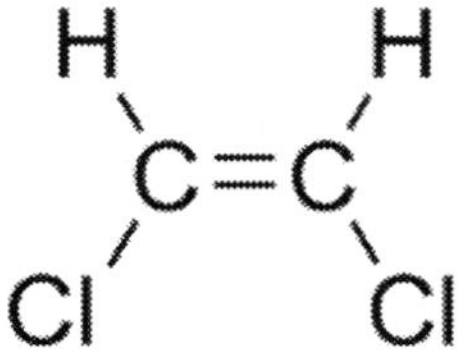

Cis-1,2-dichloroethylene

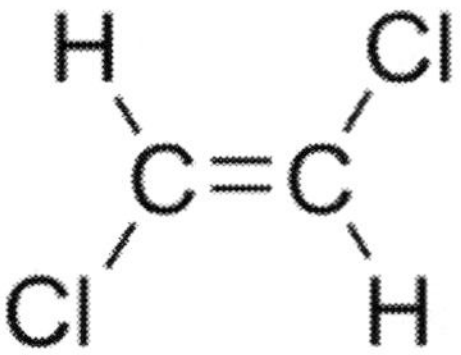

```
H      Cl
 \    /
  C=C
 /    \
Cl     H
```

Trans-1,2-dichloroethylene

If the chemical (or chemicals) have become subsurface contaminants that threaten important drinking water resources. A strategy to remediate such polluted subsurface environments is with the help of the degradative capacity of bacteria.

8.4.3 CHEMISTRY IN THE GEOSPHERE

An important aspect of the chemistry of the geosphere (Chapter 5) is the chemistry that occurs in the soil (Strawn et al., 2015). The soil itself is a complex mixture of inorganic chemicals-even before pollution occurs. Eight chemical elements comprise the majority of the mineral matter in soils. Of these eight elements, oxygen, a negatively-charged ion (anion) in crystal structures, is the most prevalent on both a weight and volume basis. The next most common elements, all positively-charged ions which, in decreasing order, are silicon, aluminum, iron, magnesium, calcium, sodium, and potassium. Ions of these elements combine in various ratios to form different minerals.

The most chemically active fraction of soils consists of colloidal clay minerals and organic matter. Colloidal particles are so small (< 0.0002 mm) that they remain suspended in water and exhibit a large surface area per unit weight. These materials also generally exhibit net negative charge and high adsorptive capacity. Several different silicate clay minerals exist in soils, but all have a layered structure. It is these clay minerals that act as adsorbents for inorganic pollutants (as well as organic pollutants).

The soil transports and moves water, provides refuge for bacteria and other fauna, and has many different arrangements of weathered rock and minerals. When soil and minerals weather over time, the chemical composition of soil also changes. However, many of the problems of soil chemistry problems are concerned with environmental sciences such as the accidental or deliberate disposal of chemicals. This can involve several physical-chemical interactions that can cause changes to the soil: (i) ion-exchange; (ii) soil pH; (iii) sorption and precipitation; and (iv) oxidation-reduction reactions.

Ion exchange involves the movement of cations (positively charged elements such as calcium, magnesium, and sodium) and anions (negatively charged elements such as chloride, and compounds like nitrate) through the soils. More specifically, *cation exchange* is the interchanging between a cation in the solution of water around the soil particle and another cation that is on the surface of the clay mineral.

A cation is a positively charged ion and most inorganic contaminants are cations, such as: ammonium (NH_4^+), calcium (Ca^{2+}), copper (Cu^{2+}), magnesium (Mg^{2+}), manganese (Mn^{2+}), potassium (K^+), and zinc (Zn^{2+}). These cations are in the soil solution and are in dynamic equilibrium with the cations adsorbed on the surface of clay and organic matter. It is the cation exchange interaction of ions with clay minerals in the soil that gave spoil the ability of soil to absorb and exchange cations with those in soil solution (water in soil pore space). The adsorption capacity of clay minerals can cause a mixture of inorganic pollutants to physically transform when one or more constituents of the mixture adsorbs onto a clay mineral. In addition, the total amount of positive charges that the clay can absorb (the cation exchange capacity, CEC) impacts the rate of movement of the pollutants through the soil and the physical and chemical changes that can occur to the pollutants. For example, soil (i.e., clay) with a low CEC is much less likely to retain pollutants and, furthermore, the soil (i.e., the clay) is less able to retain inorganic chemicals with the potential that the chemicals are released into the groundwater.

Soil pH is a commonly measured soil chemical property and is also one of the more informative properties since the data imply certain characteristics that might be associated with soil. The other principal variables affecting life in soil in the environment include moisture, temperature, and aeration and the balance of these factors controls the abundance and activities of the floral and faunal inhabitants in the soil which in turn have a marked influence on the critical processes of soil aggregation retention of pollutants.

Although soil is an inorganic matrix of metals and minerals, the main reason why the soil becomes contaminated is due to the presence of anthropogenic waste. This waste is typically: (i) chemicals in one form or another that are not indigenous to the soil; or (ii) chemicals that are indigenous chemicals that are deposited into the soil in amounts that exceed the natural abundance. The typical actions that lead to pollution of the soil by inorganic chemicals are: (i) industrial activities; (ii) agricultural activities; (iii) waste disposal; and (iv) acid rain.

Waste disposal has also been a major cause for concern insofar as the manner of disposal is subject to scrutiny. While industrial waste is sure to cause contamination. This type of waste typically contains toxins and chemicals which are now seeping into the land and causing pollution of soil. Finally, acid rain is caused when sulfur-containing and nitrogen-containing chemicals present in the air mixes up with the rain and falls back to the soil. The polluted (acidified) water has the ability to dissolve away some of the soil constituents and, been more drastic, change the structure of the soil. This latter action could release pollutants that were adsorbed (and may have even remained adsorbed) and once released can find their way into groundwater.

KEYWORDS

- **natural organic matter**
- **persistent inorganic pollutant**
- **persistent pollutants**
- **polychlorobiphenyl derivatives**
- **polycyclic aromatic hydrocarbon**
- **trichloroethylene**

REFERENCES

Calvet, R., (1989). Adsorption of organic chemicals in soils. *Environmental Health Perspectives, 83*, 145–177.

De Bruyn, A. M. H., & Gobas, F. A. P. C., (2004). Modeling the diagenetic fate of persistent organic pollutants in organically enriched sediments. *Ecological Modelling, 179*, 405–416.

Gacosian, R. B., & Lee, C., (1981). Processes controlling the distribution of biogenic organic compounds in seawater. In: Duursma E. K., &. Dawson, R., (eds.), *Marine Organic Chemistry: Evolution, Composition, Interactions, and Chemistry of Organic Matter in Seawater* (pp. 91–123). Amsterdam, Netherlands. Chapter 5.

Habashi, F., (1994). Conversion reactions in inorganic chemistry. *J. Chem. Educ., 71*(2), 130.

Jury, W. A., Winer, A. M., Spencer, W. F., & Focht, D. D., (1987). Transport and transformations of organic chemicals in the soil-air-water ecosystem. In: Ware, G. W., (ed.), *Reviews of Environmental Contamination and Toxicology* (Vol. 99, pp. 119–164). Springer, New York.

Katagi, T., (2002). Abiotic hydrolysis of pesticides in the aquatic environment. *Rev. Environ. Contam. Toxicol., 175*, 79–261.

Lyman, W. J., Reehl, W. F., & Rosenblatt, D. H., (1982). *Handbook of Chemical Property Estimation Methods*. McGraw-Hill, New York.

Macalady, D. L., & Walton-Day, K., (2011). Redox Chemistry and natural organic matter (NOM): Geochemists dream, analytical chemists nightmare. In: Tratnyek, P. G., Grundl, T. J., & Haderlein, S. B., (eds.), *Aquatic Redox Chemistry* (pp. 85–111). ACS Symposium Series No. 1071. American Chemical Society, Washington, DC. Chapter 5.

Manzetti, S., Van, D. S. E. R., & Van, D. S. D., (2014). Chemical properties, environmental fate, and degradation of seven classes of pollutants. *Chem. Res. Toxicology, 27*(5), 713–737.

March, J., (1992). *Advanced Organic Chemistry: Reactions, Mechanisms, and Structure* (4th edn.). John Wiley & Sons Inc., Hoboken, New Jersey.

Morrison, R. T., Boyd, R. N., & Boyd, R. K., (1992). *Organic Chemistry* (6th edn.). Benjamin Cummings, Pearson Publishing, San Francisco, California.

Moulay, S., (2002). The most well-known rearrangements in organic chemistry at hand. *Chemistry Education: Research and Practice in Europe, 3*(1), 33–64.

Neilson, A. H., & Allard, A. S., (2012). *Organic Chemicals in the Environment: Mechanisms of Degradation and Transformation* (2nd edn.). CRC Press, Taylor & Francis Group, Boca Raton, Florida.

Petrucci, R. H., Herning, G. E., Madura, J., & Bissonnette, C., (2010). *General Chemistry: Principles and Modern Application* (11th edn.). Prentice-Hall, Upper Saddle River, New Jersey.

Rahm, S., Green, N., Norrgran, J., & Bergman, A., (2005). Hydrolysis of environmental contaminants as an experimental tool for indication of their persistency. *Environ. Sci. Technol., 39*(9), 3128–3133.

Samin, G., & Janssen, D. B., (2012). Transformation and biodegradation of 1,2,3-trichloropropane (TCP). *Environ. Sci. Pollut. Res. Int., 19*(8), 3067–3078.

Speight, J. G., & Islam, M. R., (2016). *Peak Energy: Myth or Reality*. Scrivener Publishing, Salem, Massachusetts.

Speight, J. G., (1996). *Environmental Technology Handbook*. Taylor & Francis, Washington, DC.

Speight, J. G., (2014). *The Chemistry and Technology of Petroleum* (5th edn.). CRC Press, Taylor & Francis Group, Boca Raton, Florida.

Speight, J. G., (2017). *Handbook of Petroleum Refining*. CRC Press, Taylor & Francis Group, Boca Raton, Florida.

Strawn, D. G., Bohn, H. L., & O'Connor, G. A., (2015). *Soil Chemistry* (4th edn.). John Wiley & Sons Inc., Hoboken, New Jersey.

Wayne, C. E., & Wayne, R. P., (2005). *Photochemistry*. Oxford University Press, Oxford, United Kingdom.

Wayne, R. P., (2000). *Chemistry of Atmospheres* (3rd edn.). Oxford University Press, Oxford, United Kingdom.

Wham, R. M., Fisher, J. F., Forrester, R. C. III., Irvine, A. R., Salmon, R., Singh, S. P. N., Ulrich, et al., (2005). Synthesis and hydrolytic degradation of aliphatic polyesteramides. *Polymer Degradation and Stability, 88*(2), 309–316.

Zhou, Y. F., & Haynes, R. J., (2010). Sorption of heavy metals by inorganic and organic components of solid wastes: significance to use of wastes as low-cost adsorbents and immobilizing agents. *Critical Reviews in Environmental Science and Technology, 40*(11), 909–977.

CHAPTER 9

Management of Chemical Waste

9.1 INTRODUCTION

Chemical waste management refers to an organized system for waste handling and disposal in which chemicals pass through appropriate pathways leading to the elimination of the waste or disposal in ways that protect the environment (Kocurek and Woodside, 1994; Syed, 2006). While a waste can be a sludge discarded from a process, the contaminant can also be in the form of volatile organic compounds (often referred to as VOCs), semi-volatile organic compounds (SVOCs) (often referred to as SVOCs), metals, radioactively contaminated chemicals, or a mixture of any or all of these types. Thus, the choice of an appropriate technology to treat chemical waste is dictated by the nature of the waste, such as whether the waste is nonhazardous or hazardous (Chapter 6). Moreover, the definition of a hazardous waste and nonhazardous waste is determined by the relevant local, regional, or national legislation (Chapter 10).

The migration of chemical waste is very much a part of waste management as well as treatment. In fact, the migration of chemical waste through the land, water, and the atmosphere is largely a function of the physical properties of the waste as well as the physical properties of the surrounding matrix and the physical conditions to which the waste is subjected. In addition, chemical factors also play an important role, and the potential for reaction of the chemical waste with an outside agent (i.e., a chemical not originally in the waste) is worthy of consideration. The potential for a chemical waste to react with an outside agent and be converted to a more hazardous chemical is real.

The distribution of chemical waste constituents between the atmosphere, the aquasphere, and the geosphere is largely a function of the volatility of the chemical as well as any transformation in the environment that occurs before treatment begins. Highly volatile chemicals are more amenable to transportation through the atmosphere and more soluble chemicals are more amenable to transportation by water. Volatile chemicals are more mobile under hot, windy conditions and soluble chemicals can become mobile during (and after) periods of heavy rainfall. Chemicals (especially nonabsorbable chemicals) will migrate further in porous formations (such as sandstone) than they will in dense formations (such as soil, especially soil containing clay minerals), even if adsorptive effects were absent from all media. And chemicals that are more reactive (in the chemical and biochemical

The Science and Technology of the Environment. James G. Speight (Author)

sense) will not migrate to any great extent before they are converted to other products (Chapter 6, Chapter 7, Chapter 8).

On the other hand, in water and in the soil, the tendency of water and the soil to retain the chemical is a factor in its mobility. Physical interactions such as hydrogen bonding (in water) and adsorption (on the soil) can retain the chemical at temperatures above the boiling point. The physical interactions with the soil constituents may also be weaker than hydrogen bonding forces in the water resulting in a more ready release from the soil than from water (Chapter 8). Furthermore, soil that has been exposed to chemical waste can be severely damaged by alteration of the physical and chemical properties of the minerals in the soils as well as the ability to support plants and the soil may also become susceptible to erosion.

A consideration in selecting a treatment technology is the location where the wastes are to be treated (Wise and Trantolo, 1994). For example, wastes may be treated in place (in situ), within the confines of the site, or at an off-site facility (ex-situ). However, most treatment processes yield chemical waste, such as sludge from wastewater treatment or incinerator ash, which requires disposal and which may be hazardous to some extent. In addition, more emphasis in treatment is being placed on the recovery of recyclable chemicals and production of innocuous byproducts.

The technologies to manage chemical waste that are covered in this chapter are subdivided into four general categories which are: (i) chemical methods; (ii) physical methods; (iii) thermal methods; (iv) solidification and stabilization; and (v) biodegradation. The effectiveness of the application of each of these methods to chemical waste varies depending on: (i) the type of waste, including the concentration of the individual chemicals in the waste; (ii) the physical phase-solid or liquid-of the waste material; (iii) the desired level of treatment; and (iv) the method of disposal of any residue remaining after the treatment. Another consideration in selecting a treatment technology is where the wastes are to be treated. Wastes may be treated in place (*in situ*), within the confines of the site, or at an off-site facility (*ex-situ*).

9.2 CHEMICAL METHODS

As a first consideration unless proven otherwise by the application of suitable test methods, all chemical wastes should be considered as being harmful to humans, even poisonous, depending upon the amount of the water ingested.

The first step in selecting a technology for the treatment of wastes is to determine whether the waste is hazardous or nonhazardous (Chapter 6). And the definition of a hazardous/nonhazardous waste is determined by the relevant legislation (Chapter 12). Also, chemical waste is not always in the form of a sludge, but the waste can also be in the form of volatile organic compounds (VOCs), SVOCs, metals, radioactive chemicals, radioactively contaminated materials, or a mixture of any or all of these types.

The ideal treatment process should reduce the quantity of chemical waste to an appropriate level by converting the hazardous waste to a nonhazardous form (Chapter 6). There are various technologies, such as chemical treatment and biological treatment that can be applied to the waste (US EPA, 1994). Chemical treatment processes and physical treatment processes as well as thermal methods of treatment (including incineration) as well as solidification or stabilization are the subject of this chapter. These processes are used to: (i) destroy the waste by complete elimination of the chemical; or (ii) by conversion of the chemical to a benign form and produce a final residual chemical that is suitable for disposal. Another consideration in selecting a treatment technology is the location where the wastes are to be treated (Wise and Trantolo, 1994).

The effectiveness of the application of any waste treatment technology varies depending on: (i) the type of waste; (ii) the concentration of the individual components in the waste; (iii) the physical phase of the chemical; (iv) the desired level of treatment; and (v) the final method of disposal of any remaining residue. The waste characteristics can include such properties as volatility (gases, volatile chemicals in water, gases or volatile chemicals adsorbed on solids, such as catalysts), liquid phase chemicals (such as organic solvents), dissolved or soluble chemicals (water-soluble inorganic species, water-soluble organic species, compounds soluble in organic solvents), semisolid chemicals (sludge, grease), and solid chemicals (dry solids, including granular solids with a significant water content, such as dewatered sludge, as well as solids suspended in liquids). Furthermore, waste treatment may occur at three major levels: (i) primary waste treatment; (ii) secondary waste treatment; and (iii) tertiary waste treatment, sometimes referred to as polishing.

In many cases, primary waste treatment is often considered to be the preparation of the waste for further treatment, although the primary process can result in a reduction in the quantity of the waste. Following the primary treatment process, the secondary treatment process is often used to detoxify, destroy, and remove any hazardous constituents in the waste. Finally, the tertiary treatment process (often referred to as polishing) usually refers to treatment of the waste for safe disposal charge. An example is the treatment of recycled water to remove any chemicals so that it may be safely discharged or repurposed (Benefield et al., 1982; Cheremisinoff, 1995).

The applicability of chemical treatment processes to waste management depends upon the chemical properties of the constituents of the waste and the products of the treatment process which alter the chemical structure of the constituents of the waste to produce either an innocuous or a less hazardous by-product. Particularly relevant properties that are the focus of the treatment processes are: (i) acid-base character of the waste; (ii) oxidation and/or reduction behavior of the waste; (iii) precipitation and/or the tendency of the waste to form complexes; (iv) flammability and combustibility of the waste; (v) reactivity and corrosivity of the waste; and (vi) compatibility of the waste with other chemical waste. As a consequence of these considerations, there is a variety of chemical-based operations commonly used in treating wastes (Long, 1995).

9.2.1 NEUTRALIZATION

The neutralization process involves a chemical reaction in which an acid and a base are reacted. In the reaction in water, neutralization results in there being no excess of hydrogen or hydroxide ions present in the solution:

$$\text{Acid} + \text{alkali} \rightarrow \text{salt} + \text{water}$$

$$HCl + NaOH \rightarrow NaCl + H_2O$$

The goal of the process (sometimes referred to as the pH adjustment process) is to reduce the acidity or alkalinity of a waste stream by mixing acids and bases to produce a neutral solution. Also, the neutralization process can be used as a pretreatment option prior to the application of other waste treatment processes. As examples, lime (calcium oxide, CaO) or slaked lime [calcium hydroxide, $Ca(OH)_2$] are regularly used as bases for treating acidic wastes. Sulfuric acid (H_2SO_4) is a relatively inexpensive acid that is used for treating alkaline wastes. For some applications, acetic acid (CH_3COOH) is preferable to sulfuric acid since acetic acid is a weak acid and it is also a natural product and biodegradable. Hydrolysis may also be considered to be a form of neutralization and one of the methods to dispose of water-reactive waste is to react the waste with water under controlled conditions. Inorganic chemicals that can be treated by hydrolysis include metals that react with water (example are: (i) metal carbide derivatives; (ii) hydride derivatives; (iii) amide derivatives; (iv) alkoxide derivatives; (v) halogen derivatives; (vi) nonmetal oxyhalide derivatives; and (vii) sulfide derivatives (Table 9.1).

TABLE 9.1 Hydrolysis Reactions of Selected Chemicals*

Substrate	Equation
Active metal	$Ca + 2H_2O \rightarrow Ca(OH)_2 + H_2$
Alkoxide derivatives	$C_2H_5ONa + H_2O \rightarrow C_2H_5OH + NaOH$
Amide derivatives	$NaNH_2 + H_2O \rightarrow NH_3 + NaOH$
Carbide derivatives	$CaC_2 + H_2O \rightarrow Ca(OH)_2 + CH{\equiv}CH$
Halide derivatives	$SiCl_4 + H_2O \rightarrow SiO_2 + 4HCl$
Hydride derivatives	$NaAlH_4 + H_2O \rightarrow Al(OH)_3 + NaOH$

*Arranged alphabetically.

9.2.2 PRECIPITATION

Precipitation is a major component of the hydrological cycle (the water cycle), and is responsible for depositing most of the fresh water on the planet (Figure 9.1) (Speight, 1996). The term may also be applied to any product of the condensation of atmospheric water vapor that falls to the surface of the earth under the force of gravity from the clouds. The main forms of precipitation include drizzle, rain, sleet, snow, ice, and hail while fog and mist are not precipitation in the true sense of the word but remain as colloidal systems because the water vapor does not condense sufficiently to precipitate.

In the current context, precipitation is the separation (by chemical reaction) of a solid from a solution. However, without sufficient coagulation of the solid particles or settling, the precipitate can remain in suspension-an undesirable outcome if the suspension cannot be separated from the solution. If the suspension settles as a solid mass (sedimentation), separation can be achieved by any one of several physical methods, such as filtration of centrifugation.

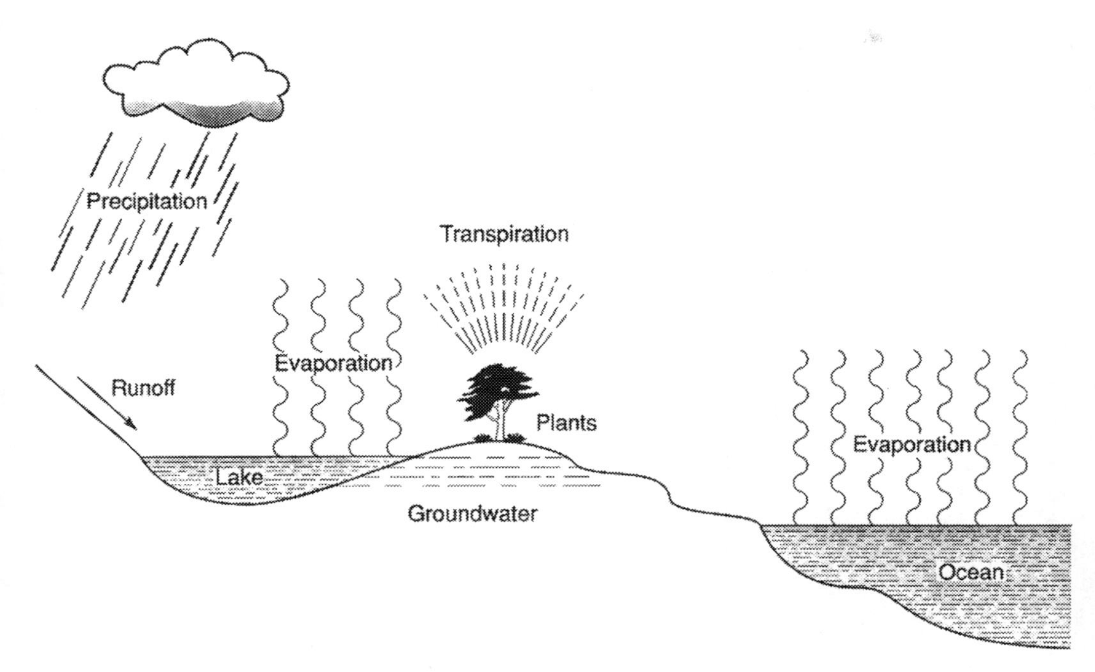

FIGURE 9.1 The hydrological cycle (also known as the water cycle).
Source: Reproduced with permission from: Speight (1996). © Taylor and Francis.

The formation of a precipitate often indicate the occurrence of a chemical reaction such as when a solution of silver nitrate ($AgNO_3$) is added to a solution of sodium chloride (NaCl), a chemical reaction occurs forming a white precipitate of silver chloride or when potassium iodide (KI) solution reacts with lead nitrate ($PbNO_3$) solution, a yellow precipitate of lead iodide (PbI_2)is formed.

$$AgNO_3 + NaCl \rightarrow AgCl\downarrow + NaNO_3$$
$$Pb(NO_3)_2 + 2KI \rightarrow PbI_2\downarrow + 2KNO_3$$

Note: The down-pointing arrow (↓) indicates the product that is precipitated.

An important stage of the precipitation process is the addition of a specific chemical is added to coagulate the offending species and form a precipitate (Table 9.2). The process is used primarily for the removal of the ions of heavy (high density, high atomic number) metal ions from water-the most widely used method of precipitating metal ions is by the formation of the hydroxide derivative by the addition of chemicals such as lime [$Ca(OH)_2$], sodium hydroxide (NaOH), or sodium carbonate (Na_2CO_3).

Heavy metal ions in soil that has been contaminated by chemical wastes may be present in a co-precipitated form with insoluble oxides of iron (Fe^{3+}) and manganese (Mn^{4+}) (i.e., FeO, Fe_2O_3 and MnO_2). These oxides can be dissolved by reducing agents, such as solutions of sodium dithionate/citrate or hydroxylamine. This results in the production of soluble iron and manganese and the release of heavy metal ions which are removed with the water.

TABLE 9.2 Coagulation Reactions of Selected Chemicals*

$Al_2(SO_4)_3 + 3Ca(HCO_3)_2 \rightarrow 2Al(OH)_3 + 3CaSO_4 + 6CO_2$
$Al_2(SO_4)_3 + K_2SO_4 + 3Ca(HCO_3)_2 \rightarrow 2Al(OH)_3 + K_2SO_4 + 3CaSO_4 + 6CO_2$
$Al_2(SO_4)_3 + 6NaOH \rightarrow 2Al(OH)_3 + 3Na_2SO_4$
$Al_2(SO_4)_3 + 3Na_2CO_3 + 3H_2O \rightarrow 2Al(OH)_3 + 3Na_2SO_4 + 3CO_2$
$Al_2(SO_4)_3 + (NH_4)_2SO_4 + 3Ca(HCO_3)_2 \rightarrow 2Al(OH)_3 + (NH_4)_2SO_4 + 3CaSO_4 + 6CO_2$
$4Fe(OH)_2 + O_2 + 2H_2O \rightarrow 4Fe(OH)_3$
$FeSO_4 + Ca(OH)_2 \rightarrow Fe(OH)_2 + CaSO_4$
$Fe_2(SO_4)_3 + 3Ca(HCO_3)_2 \rightarrow 2Fe(OH)_3 + 3CaSO_4 + 6CO_2$
$Na_2Al_2O_4 + Ca(HCO_3) + 2H_2O \rightarrow 2Al(OH)_3 + CaCO_3 + Na_2CO_3$

*Arranged alphabetically by first chemcial.

9.2.3 ION EXCHANGE

Ion exchange processes are used for the purification of aqueous solutions and employ a solid polymeric ion exchange resin (Wachinski, 2016; Gupta, 2017). The ion exchange process is one of the most utilized techniques in water treatment and wastewater treatment as well as in separation processes and offers a method of removing cations or anions from solution onto a solid resin, which can be regenerated by treatment with acids, bases, or salts (Long, 1995).

Typical ion-exchange resins (functionalized porous or gel polymer polymer), zeolites, montmorillonite, clay minerals, and soil humus which are either cation media (that exchange positively charged ions or anion media (exchange negatively charged ions. There are also amphoteric ion exchanger media that are able to exchange both cations and anions simultaneously. However, the simultaneous exchange of cations and anions can be more efficiently performed (i) in mixed beds, which contain a mixture of an anion-exchange resin and a cation-exchange resin, or (ii) by passing the solution through several different ion-exchange chemicals.

Zeolites are relatively soft minerals and are not very abrasion resistant, and the frameworks of zeolites are less open and more rigid than those of most other ion exchangers. They, therefore, have minimal swelling and the counterions in their pores are not very mobile. Cation exchangers include the zeolite minerals, which have a porous structure that can accommodate a wide variety of cations such as sodium (Na^+), potassium (K^+), calcium (Ca^{2+}), and magnesium (Mg^{2+}). These positive ions are rather loosely held and can readily be exchanged for others in a contact solution. Some of the more common mineral zeolites are analcime ($NaAlSi_2O_6 \cdot H_2O$), chabazite ($Al_2Si_4O_{12} \cdot 6H_2O$), and natrolite ($Na_2Al_2Si_3O_{10} \cdot 2H_2O$).

Many naturally-occurring chemicals are able to capture certain ions and retain them in an exchangeable state-this ability is referred to as the exchange capacity of the chemical. Most natural ion-exchange chemicals are crystalline aluminosilicates (i.e., clay minerals) with cation exchange properties insofar as they remove positively charged cations from solution), although certain aluminosilicates can also act as anion exchangers.

9.2.4 *OXIDATION-REDUCTION*

An oxidation-reduction reaction (or a reduction-oxidation reaction, a redox reaction) is a type of chemical reaction that involves a transfer of electrons between two species and, therefore, can be used for the removal of chemicals from an ecosystem (Table 9.3). Chemically, the oxidation-reduction reaction is a chemical reaction in which the oxidation number of a molecule, atom, or ion is changed by gaining an electron or by losing an electron.

TABLE 9.3 Examples of Oxidation-Reduction Reactions for the Treatment of Chemical Waste

Inorganic Chemical Waste
Reduction
Chromate
$2CrO_4^- + 3SO_2 + 4H+ \rightarrow Cr_2(SO_4)_3 + 2H_2O$
Permanganate
$MnO_4^- + 3Fe^{2+} + 7H_2O \rightarrow MnO_2 + 3Fe(OH)_3 + 5H^+$
Oxidation
Cyanide
$2CN^- + 5OCl^- + H_2 \rightarrow N_2 + 2HCO_3^- + 5Cl^-$
Iron (Fe^2+)
$4Fe^{2+} + O_2 + 10H_2O \rightarrow 4Fe(OH)_3 + 8H^+$
Sulfur dioxide
$2SO_2 + 2O_2 + H_2O \rightarrow 2H_2SO_4$
Organic Chemical Waste
Reduction
$[CH_2NOS] + [H] \rightarrow [CH_3] + H_2O + NH_3 + H_2S$
Oxidation
$[CH_2NOS] + [O] \rightarrow CO_2 + H_2O + NO_x + SO_2$

The oxidation of sulfur dioxide in acid rain is an example of a redox reaction that occurs in the environment, Sulfur dioxide (SO_2) is released into the atmosphere naturally through volcanic activity as well as through industrial sources and the sulfur dioxide released is carried in the atmosphere by the prevailing winds. Sulfur dioxide is oxidized to form sulfuric acid (H_2SO_4) from a variety of different chemical reactions. The oxidation state of sulfur in SO_2 is +4 and is often referred to as S^{IV}. The oxidation state of sulfur in sulfuric acid and sulfate is +6 referred to as S^{VI}. Sulfate derivatives (SO_4^{2-}) or sulfuric acid occur as aerosols in the atmosphere.

Sulfur dioxide gas dissolves in water according to the following equations in which three species are produced (i) hydrated sulfur dioxide (SO_2. H_2O), (ii) the bisulfite ion (HSO_3^-), and (iii) the sulfite ion (SO_3^{2-}):

$$SO_2 + H_2O \rightarrow SO_2.\ H_2O$$

$$H_2O \rightarrow HSO_3^- + H^+$$

$$2HSO_3^- \rightarrow SO_3^{2-} + 2H^+$$

The predominant product depends on the acidity of the solution-the pH range of atmospheric droplets is on the order of pH 2 to pH 6 (acidic) and most of the dissolved sulfur dioxide is converted to the bisulfite ion (HSO_3^-) which is then oxidized by hydrogen peroxide to the sulfate ion (SO_4^{2-}):

$$HSO_3^- + H_2O_2(aq) + 2H^+ \rightarrow H_2SO_4$$

$$H_2SO_4 \rightarrow SO_4^{2-} + 2H^+ + H_2O$$

Acid rain acidifies soils and rivers and lakes and the change in acid balance affects the community of organisms capable in these systems.

Finally, electrolysis is a process in which a chemical species in solution (usually a metal ion) is reduced by electrons at the cathode. In waste disposal applications, electrolysis is most widely used in the recovery of cadmium (Cd), copper (Cu), gold (Au), lead (Pb), silver (Ag), and zinc (Zn). However, the recovery of metals by electrolysis is complicated when cyanide ions (CN^-) are present. This ion stabilizes the metals in solution as the cyanide complexes-an example is nickel tetra cyanide [$Ni(CN)_4$] which requires careful treatment and disposal.

9.2.5 EXTRACTION AND LEACHING

Whilst not strictly a chemical-based method for waste treatment, the extraction and leaching processes are included here since the procedures depend to a large extent on the chemical properties of the released contaminants. More simply, extraction and leaching are processes for the selective solubilization and removal of a chemical (or chemicals) from the surrounding medium. Common examples of the extraction process include liquid-liquid extraction and solid-phase extraction. The distribution of a solute between two phases is an equilibrium condition described by partition theory which is based on the manner in which the analyte moves from the initial solvent into the extracting solvent.

On the other hand, leaching is the process of a solute is extracted from its carrier by use of a solvent. Specific extraction methods depend on the solubility characteristics (or properties) of the leached material relative to the characteristics (or properties) of the sorbent. Leaching can be applied to enhance water quality and contaminant removal, as well as for disposal of hazardous waste products.

In addition, chemical extraction or chemical leaching is the removal of a contaminant by chemical reaction. Low-solubility heavy metal salts can be extracted by reaction of the salt anions with acid (H+). Acids also dissolve nitrogen-containing basic organic compounds such as amine derivatives. However, as a note of caution, extraction with acids should be avoided if cyanide derivatives (X-CN) or sulfide derivatives (X-S) are present. The extraction of these derivatives can lead to the formation of toxic hydrogen cyanide (HCN) and toxic hydrogen sulfide (H_2S):

$$\text{X-CN} + \text{acid} \rightarrow HCN$$

$$\text{X-S} + \text{acid} \rightarrow H_2S$$

The use of the non-toxic weak acids, such as include acetic acid (CH_3COOH) and monosodium phosphate (NaH_2PO_4) are usually safer process options. Chelating agents are also employed to remove heavy metal contaminants from soil by forming soluble species with the metal salts. Chelating ion-exchange resins can also be employed for removal of heavy metals from liquid streams.

9.3 PHYSICAL METHODS

Physical processes for the treatment of contaminants include processes that separate components of a contaminant stream or change the physical form of the contaminant without altering the chemical structure of the contaminant(s) (Long, 1995).

These processes, which are based upon one or more of the physical properties of the released contaminant, are particularly useful for separating hazardous contaminants from otherwise nonhazardous contaminants so that the hazardous contaminants can be treated and reduced to a more manageable (less-hazardous) form. In this way, any blended contaminant streams can be separated into the various contaminant streams for different treatment processes, and preparing the contaminant(s) for ultimate destruction in biological processes or thermal treatment processes (Long, 1995). These processes include: (i) phase separation, such as filtration; (ii) phase transfer, such as extraction and sorption; (iii) phase transition, such as distillation, evaporation, and precipitation; and (iv) membrane separation, such as reverse osmosis, hyperfiltration, and ultrafiltration.

9.3.1 PHASE SEPARATION

Phase separation involves separation of the components of a mixture that are already in two different phases. Typically, the separation is aided by mechanical means such as: (i) sedimentation; (ii) the use of screens; (iii) centrifugation; (iv) flotation; and (v) filtration. These processes will divide a waste into two or more distinct streams based upon size, density, or chemical type. In fact, these phase separation processes allow for a more efficient operation of any subsequent-technology applied to the waste while also reducing the quantities of waste chemicals to be treated.

Separation by sedimentation is usually accomplished by providing the necessary sufficient time and space for settling in a storage tank or in a holding pond. In addition, coagulating agents are often added to assist in the settling of fine particles. Following from this, the screening process is a process for removing particles from waste streams that is used to protect downstream processes. Four general categories of screens are available: (i) grizzly screens, which are sets of parallel bars used for the removal of coarse chemical, (ii) revolving screens, which are cylindrical frames covered with wire cloth, (iii) vibrating screens, which often used when higher capacities are required, and (iv) oscillating screens, which are used at lower speeds than vibrating screens and are used for separating particles by grain size.

Flotation is a process for removing solids from liquids and by moving the particles to the surface using air bubbles as flotation media. The process is useful for removing particles that are too small and not of sufficient density to be removed by sedimentation. In the process known as dissolved air flotation (DAF), air is dissolved in the suspending medium under pressure and when the pressure is released, the air comes out of solution as minute air bubbles with attached to suspended particles that float to the surface. As an example, a contaminant might be removed by a relatively simple (and perhaps standard) flushing technique or by a somewhat more innovative technique that involves the principles of laminar flow and which can be applied not only to lighter-than-water liquids but also to heavier-than-water liquids.

Centrifugation is a process for separating solid and liquid components of a contaminant stream by rapid rotation of a mixture of solids and liquids inside a vessel. The process is used frequently to dewater sludge.

Emulsion breaking is an important but often a difficult treatment step in which emulsions are caused to aggregate and settle from the liquid phase. Agitation, heat, acidification, and the addition of coagulants (flocculating agents)-typically organic polyelectrolytes or inorganic chemicals, such as an aluminum salts-may be used for this purpose.

Filtration is a physical, chemical, or biological operation that separates solid matter (such as insoluble contaminants) and fluid from a mixture using a filter medium. In the process, oversize solid particles form a filter cake on the filter. The size of the largest particles that can successfully pass through a filter is or controlled by called the effective pore size of that filter.

9.3.2 PHASE TRANSFER

Adsorption is a process for removing organic and inorganic contaminants from waste streams using the surface of a porous chemical, such as activated carbon, as the adsorbent (ASTM D3922). The adsorbent may need frequent replacement and regenerated with heat or a suitable solvent when the capacity of the adsorbent to attract the contaminants is reduced.

Activated carbon sorption is the most effective process for removing contaminants from water that are poorly water-soluble, such as xylene, naphthalene, cyclohexane, chlorinated hydrocarbon derivatives, phenol, aniline, dyes, and surfactants. Also, adsorbents other than activated carbon (including synthetic resins composed of organic polymers and mineral substances can be used for sorption of contaminants from liquid wastes. For example, clay minerals and regenerable adsorbents are employed to remove impurities from waste lubricating oils in some used oil recycling processes (Speight and Exall, 2014).

Solvent extraction is a process for separating liquids by mixing the stream with an immiscible solvent which will extract specific contaminants of the waste stream, depending upon the properties of the contaminants and the solvent (Long, 1995). The extracted contaminants are then removed from the solvent for repurposing or for disposal. One approach to solvent extraction and leaching of chemical waste constituents is the use of supercritical fluids as extraction solvents. After a chemical (either a contaminant or a desired chemical) substance has been extracted from a waste into a supercritical fluid at high pressure, the pressure can be released, resulting in separation of the extracted chemical.

By way of clarification, a supercritical fluid is a fluid that has characteristics of both liquid and gas and consists of a chemical that is above its supercritical temperature and pressure (31.1°C/88°F and 73.8 atmospheres/1,085 psi, respectively, for carbon dioxide) (Kiran and Brennecke, 1993).

9.3.3 PHASE TRANSITION

Another class of physical separation processes involves the use of phase transition processes in which a contaminant changes from one physical phase to another-an example is the distillation process for separating liquids with different boiling points (Long, 1995). In the process, the mixed stream is exposed to increasing amounts of heat by which the various components of the mixture are vaporized and recovered. The vapor may be recovered and reboiled several times to effect a complete separation of components.

On the other hand, the evaporation process is used to concentrate nonvolatile solids in a solution by boiling off the liquid (volatile) portion of the waste stream (Long, 1995). Evaporation units are often operated under partial vacuum or under full vacuum to lower the thermal energy required to boil the solution. The process is typically employed to remove water from an aqueous waste to concentrate the contaminants in the non-volatile portion of the waste. A special case of this technique is thin-film evaporation in which volatile constituents are removed by heating a thin layer of liquid waste or sludge waste that has been spread on a heated surface.

Drying (sometimes referred to as dewatering), involves the removal of a solvent or water from a solid or semisolid (sludge) or the removal of solvent from a liquid mixture or from a suspension. This is an important operation because water is often the major constituent of waste products, such as sludge obtained from emulsion breaking. For example, in the freeze-drying process, the solvent, usually water, is sublimed from a frozen contaminant.

Also, solids and sludge are dried to reduce the quantity of waste, to remove the solvent (or water) that might interfere with subsequent treatment processes, and to remove volatile constituents. Also, dewatering can often be achieved thermally and can be improved with the addition of a filter aid, such as diatomaceous earth, during the filtration step.

Stripping is a form of distillation that has been modified as a redistillation process to separate volatile components from less volatile components in a liquid mixture by the partitioning of the more volatile chemicals to a gas phase of air or steam (air stripping or steam stripping) (Long, 1995). The gas phase is introduced into the aqueous solution or suspension containing the contaminant in a stripping tower that is equipped with trays or packed to provide maximum turbulence and contact between the liquid phase and the gas phase (Billet, 1995; Long, 1995). The two major products are condensed vapors and a stripped residue that is non-volatile (under the conditions of the process).

Another form of the phase transition processes involves the precipitation process, which is used to separate a solid from a solute in solution as a result of a physical change in the solution (Long, 1995). The major changes that can cause physical precipitation are (i) cooling the solution, (ii) evaporation of the solvent, or (iii) alteration of the composition of the solvent.

The most common type of physical precipitation is by alteration of the composition of the solvent, which occurs when a water-miscible organic solvent is added to an aqueous solution, so that the solubility of a salt is lowered below its concentration in the solution.

9.3.4 PHASE CONVERSION

Vitrification (also referred to as glassification) is a phase conversion process in which a chemical waste (or constituents of the waste) are melted at high temperature to form an impermeable capsule around the remainder of the waste. The principles behind vitrification are the same as those applied to the production of glass. High temperature electrodes are used to melt the wastes, and organic constituents are transformed by pyrolysis and either collected as product or destroyed in secondary processes. The inorganic components are immobilized in the resulting glass matrix. In ex situ applications, the waste is introduced into the furnace along with the silica, soda, and lime. The organic chemicals are driven off, captured, and treated while the inorganic chemicals are incorporated into the glass. The plasma arc has also been shown to be suitable for ex-situ vitrification.

The in-situ process involves the insertion of suitably-sized electrodes into the soil and graphite can be spread between the electrodes on the surface of the soil to complete the circuit. A negatively pressurized hood is placed over the site to collect any off-gases for later treatment. High voltage (on the order of 4,160 volts, a 3,000 kW electrical source is required) is applied across the electrodes to produce temperatures reaching 3,600°C (6,510°F). The use of in situ vitrification is limited by (i) high groundwater tables, (ii) buried metal objects, and (iii) the need for sufficient quantities of glass-forming chemical in the waste or in the soil.

The process can be used to treat radioactive waste and mixed wastes that are immobilized in the glass matrix (US DOE, 1995). In the case of mixed wastes, the vitrification process drives off the non-radioactive contaminants thereby allowing these contaminants to be treated as non-radioactive waste while immobilizing the radioactive contaminants.

9.3.5 MEMBRANE SEPARATION

Another major class of physical separation involves the use of membranes in which dissolved contaminants or solvent pass through a size-selective membrane under pressure (Huang, 1991; Long, 1995; Noble and Stern, 1995). The products are relatively pure solvent phase (usually water) and a concentrate that has been enriched in the contaminants. Thus, in the process, water, and lower molecular mass solutes (i.e., solutes of lower molecular size) under pressure pass through the membrane as a stream of purified permeate, leaving behind a stream of concentrate containing impurities in solution or suspension. The membrane used in the process is a generally non-porous layer, so there will not be a severe leakage of gas through the membrane and the performance of the membrane depends on permeability and selectivity (Speight, 2019a).

A similar membrane-based process (dialysis) is used for separating components in a liquid stream. The process can be used to separate mixed waste streams into relatively pure solutions and relatively impure solutions. The electrodialysis process is an extension of the dialysis process and is used to separate the components of an ionic solution by applying an electric current to the solution, which causes ions to move through the dialysis membrane.

The related process of electrolysis has been reported as a method for the dechlorination of polychlorinated biphenyl derivatives (PCBs) (Rusling and Zhang, 1994). In the process, polychlorinated biphenyl derivatives in mixtures have been reduced to biphenyl derivatives and hydro-biphenyl derivatives.

Reverse osmosis is a process that separates components in a liquid stream by applying external pressure to one side of the membrane so that solvent will flow in the opposite direction. The process operates on the principle that the membrane is selectively permeable to water and excludes ionic solutes. On the other hand, the process known as ultrafiltration is more applicable to higher molecular weight materials. The result is that dissolved components with low molecular weights will pass through the membrane with the bulk liquid while higher molecular-weight components become concentrated through the loss of solvent.

Finally, in the sense of ultrafiltration and hyperfiltration, the fate and behavior of nanoparticles (i.e., particles with at least one dimension between 1 and 100 nanometers in length, 1 nanometer = 1×10^{-9} meter = 3.9370×10^{-8} inch) which are released to the environment must also be addressed. Nanoparticle wastes should be (i) collected in the same way as hazardous chemical waste-in tightly closed containers free of leaks and cracks, (ii) labeled as a hazardous chemical waste, including, and (iii) removed from the site by scheduling a pickup through the hazardous chemical waste program.

9.4 THERMAL METHODS

Application of thermal processes is a common form of waste treatment and suitable and efficient gas cleaning processes are available to ensure the minimum amount of contaminants in the gaseous emission streams (Speight, 2019a). Except for the vitrification process, thermal technologies are ex situ processes requiring the wastes to be transported to the processing unit. However, the use of a thermal destruction technology may be (depending upon the nature of the emissions) superior to wet scrubbing technologies for emissions cleaning.

9.4.1 INCINERATION

Incineration is the controlled combustion of chemical contaminants in an enclosed area. For example, the incineration of sewage sludge is a process that involves exposure of the waste to oxidizing conditions at a high temperature, usually in excess of 900°C (1,650°F) whereby the oxides of carbon (dioxide) and hydrogen (water) are, under the precise conditions, the predominant products. In addition, the destruction of polychlorobiphenyl derivatives (PCBs) commences at a temperature on the order of 800°C (1,470°F) and commercial incinerators operate at temperatures in excess of 1,000°C (1,830°F).

This, of course, is a convenient point at which to note only one of the many ways in which environmental events in the air (the atmosphere, Chapter 3), in the water (the aquasphere, Chapter 4), and on land (the geosphere, Chapter 5) are interconnected. Moreover, the connection between the air, the water, and the land systems is not guaranteed to be sequential and interaction between any two of the systems can occur at any time. In summary, a pollutant in any one of the three Earth system can also spread to at least one of the other two systems. Thus:

$$\text{Atmosphere} \leftrightarrow \text{Aquasphere} \leftrightarrow \text{Geosphere}$$
$$\text{Atmosphere} \leftrightarrow \text{Geosphere} \leftrightarrow \text{Aquasphere}$$
$$\text{Aquasphere} \leftrightarrow \text{Atmosphere} \leftrightarrow \text{Geosphere}$$
$$\text{Aquasphere} \leftrightarrow \text{Geosphere} \leftrightarrow \text{Atmosphere}$$
$$\text{Geosphere} \leftrightarrow \text{Atmosphere} \leftrightarrow \text{Aquasphere}$$
$$\text{Geosphere} \leftrightarrow \text{Aquasphere} \leftrightarrow \text{Atmosphere}$$

Briefly, emissions from the incinerators may be gaseous (sulfur oxides, SOx, nitrogen oxides, NOx, and PM) which can pollute the atmosphere or, through the formation of the constituents of acid deposition, the water systems and the land systems. In addition, solid waste (mineral ash) from an incinerator can pollute the land and, as a result of leaching by rain (acidic or otherwise) pollute the water systems and, therefore, attention must also be focused on the disposal of the ash (Goodwin, 1993). In addition, sulfur-containing fuels emit sulfur dioxide and other sulfur containing gases as a result of combustion and these sulfur-containing gases have the potential for conversion to sulfates (sulfuric acid in the atmosphere) which can be deposited on the land as sulfate.

Nitrogen oxides (also produced during fuel combustion) are converted to nitrate derivatives ($-NO_3$) in the atmosphere and the nitrate derivatives eventually are deposited on soil. However, the soil also adsorbs nitric oxide (NO) and nitrogen dioxide (NO_2) readily, and these gases are also oxidized to nitrate in the soil. Carbon monoxide is converted to carbon dioxide and possibly to biomass by soil bacteria and fungi. Elevated levels of heavy metals (such as lead) are also found in the soil near facilities such as mines and smelters.

Thermal treatment of chemical waste can be used to accomplish the common objectives of waste treatment which are (i) volume reduction, and (ii) removal of volatile, combustible, mobile organic matter, and destruction of toxic and pathogenic chemicals. Incineration utilizes high temperatures, an oxidizing atmosphere, and often turbulent combustion conditions to destroy wastes. However, the effective incineration of wastes depends upon the combustion conditions, which are (i) a sufficient supply of oxygen in the combustion zone, (ii) thorough mixing of the waste, the oxidant, and any supplemental fuel, (iii) combustion temperatures above 900°C (>1,650°F), and (iv) sufficient residence time to allow reactions to occur. Usually, the heat required for incineration is provided by the oxidation of organically-bound carbon and organically-bound hydrogen either in the waste or in the supplemental fuel or both:

$$C_{organic} + O_2 \rightarrow CO_2 + \text{heat}$$
$$H_{organic} + O_2 \rightarrow H_2O$$

These equations represent the destruction of the organic matter and the generation of the heat required for endothermic reactions, such as the rupture of carbon-chlorine (C-Cl) bonds in organo-chlorine compounds in the waste.

9.4.2 THERMAL DESORPTION

Thermal desorption involves heating a waste in a controlled environment in order to volatilize any organic constituents. The process is well-suited for VOCs but can also be employed for semivolatile organic compounds (SVOCs).

Prior to entering the thermal desorption unit, the waste is screened to eliminate coarse pieces also for waste with a high moisture content, it is advisable to remove the excess moisture. The waste is then passed to a furnace which operates at temperatures in the range 300 to 600°C (570 to 1,110°F). VOCs become gaseous in the process and are either (i) collected on an adsorbent, such as activated carbon, for further treatment or (ii) passed through an incinerator connected inline with the thermal desorption unit for complete destruction of the waste.

9.4.3 PYROLYSIS

Pyrolysis is a process by which a chemical change that can be achieved by the action of heat. The process differs from incineration (which is the destruction of a waste in the presence of oxygen) insofar as in the pyrolysis process, the amount of oxygen in the system is extremely limited (preferably zero).

In the process, the organic constituents of the waste are converted to combustible gas, charcoal, organic liquids, and ash/metal residues. In some instances, the organic liquid fraction produced during pyrolysis has the potential to produce constituents for synthetic crude oil or petrochemical manufacture (Speight, 2019b). As with many thermal processes, the effectiveness of waste destruction by pyrolysis depends upon (i) the residence time within the retort, (ii) the rate of temperature increase, (iii) the final temperature, and (iv) the composition of the feed material. Pyrolysis units, which operate at temperatures on the order of 500 to 800°C (930 to 1,470°F) have achieved 99.9999% destruction/removal efficiencies.

Plasma torch processes have been applied to the principles of pyrolysis at temperatures in the range 5,000 to 15,000°C (9,030 to 27,030°F). In the process, the waste is fed into the thermal plasma where it is dissociated into the basic atomic components which recombine in the reaction chamber to form carbon monoxide, nitrogen, hydrogen, as well as methane and ethane. Acid gases that are removed from the emissions by scrubbers and any solid products are either incorporated into the molten bath at the bottom of the chamber or removed from exhaust gases by particulate scrubbers or filters.

9.5 SOLIDIFICATION AND STABILIZATION

Solidification and stabilization are treatment processes that are designed (i) to improve the physical characteristic of waste thereby making the waste more amenable to handling, or (ii) or to decrease the surface area across which transfer or loss of contained pollutants can occur, (iii) or to limit the solubility of, or detoxify, any chemical constituents in the waste.

Stabilization/solidification processes are used to minimize the potential for groundwater pollution from land disposal of hazardous wastes by transforming the waste into anon-mobile product by stabilizing and solidifying the waste. As an example, Portland cement alone or in combination with fly ash, cement kiln dust lime, or other ingredients is the principal solidifying agent used. The processes are effective for immobilizing any heavy metals present in sludge, contaminated soils and other wastes but may not be totally effective for the immobilization of toxic organic chemicals. Organically modified clay minerals are one additive for the stabilization and/or solidification processes in order to absorb and retain the organic pollutants in the solidified waste form.

Typically, stabilization involves the addition of materials (especially chemicals) to ensure that any chemical contaminants are maintained in the least soluble form and/or the least toxic form. On the other hand, in the solidification a monolithic block of treated waste with high structural integrity is created. The process limits the solubility of, or detoxify, the waste contaminants even though the physical characteristics of the waste may not be changed.

Encapsulation, in which is the action of enclosing something in a solid matrix-as if in a capsule-is another perhaps more specific form of waste stabilization in which a mixed waste, i.e., waste containing radioactive chemicals and higher-level radioactive waste can be rendered less harmful to the environment The use of polyethylene, chosen for its durability, has been used as a suitable encapsulating agent.

9.6 BIODEGRADATION

Biodegradation is a term that is applied to processes in which microorganisms are employed to decompose the constituents of organic waste either into water, carbon dioxide, and simple inorganic products, or into simpler organic substances, such as aldehyde derivatives (-CH=O) and organic acid derivatives ($-CO_2H$). The complete bioconversion of a substance to inorganic species such as carbon dioxide, ammonia, and phosphate (often referred to as mineralization).

Aerobic processes for the treatment of waste utilize aerobic bacteria and fungi that require molecular oxygen and are often favored by microorganisms. Aerobic and anaerobic digestion processes have been adapted to the use of an activated sludge process. These treatments can be applied to wastes such as chemical process wastes and landfill leachates and some processes use powdered activated carbon as an additive to remove any non-biodegradable organic wastes by adsorption.

On the other hand, anaerobic processes for the treatment of waste are those processes in which microorganisms degrade wastes in the absence of oxygen. Compared to the aerated activated sludge process, anaerobic digestion (i) yields a lesser amount of the sludge byproduct, (ii) generates hydrogen sulfide which precipitates toxic heavy metal ions, and (iii) produces methane which can be used as an energy source.

The biodegradability of a compound is influenced by its physical characteristics of the compound and examples are (i) solubility in water, (ii) vapor pressure, (iii) chemical properties, including molecular mass, molecular structure, and (iv) the presence of various kinds of functional groups, some of which provide a point for the initiation of biodegradation.

In addition, the properties of chemical contaminants can be changed to increase biodegradability which can be accomplished by adjustment of conditions to (i) the optimum temperature, (ii) the acidity or alkalinity with the pH usually in the range of 6 to 9), (iii) stirring or agitation, (iv) oxygen level, and (v) the amount of the material. Also, the biodegradation process can be aided by the removal of toxic organic chemicals and toxic inorganic chemicals, such as ions of the heavy metals.

Activated sludge is the biologically active sediment produced by the repeated aeration and settling of sewage and/or organic wastes. The dissolved organic matter acts as food for the growth of aerobic flora. These species produce a biologically active sludge which is usually brown in color and which destroys the polluting organic matter in the sewage and waste. The process is known as the activated sludge process. The activated sludge process (Figure 9.2) (Speight, 1996) is an effective waste treatment process in which microorganisms in the aeration tank (the reactor) convert organic contaminants in wastewater to microbial biomass and carbon dioxide. Organic nitrogen is converted to ammonium ion (NH_4^+)or nitrate ($-NO_3^-$) and organic phosphorus is converted to orthophosphate ($-PO_4^{3-}$).

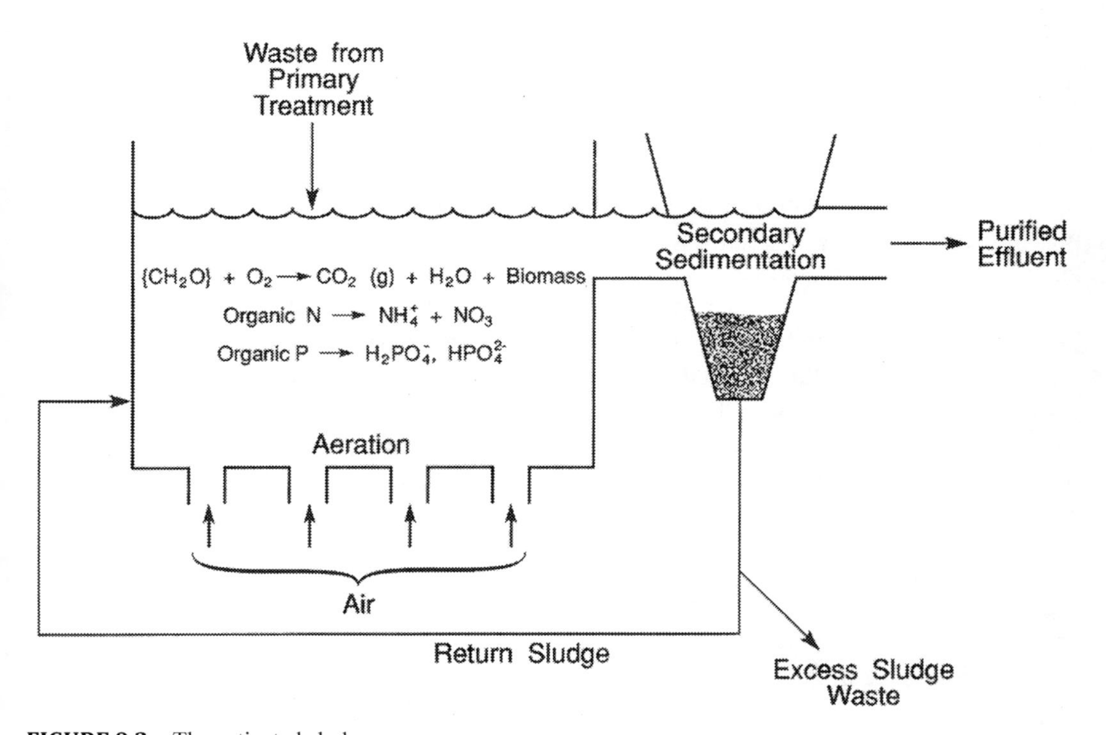

FIGURE 9.2 The activated sludge process.

Source: Reproduced with permission from: Speight (1996). © Taylor and Francis.

9.6.1 SOLID PHASE BIODEGRADATION

Solid-phase biodegradation (also referred to as land farming) is a process in which conventional soil management practices are used to enhance the microbial degradation of the waste. The goal of the process is to stimulate indigenous biodegradative microorganisms and facilitate the aerobic degradation of contaminants. Land farming remains a viable disposal alternative. In the process, nutrients and minerals are also added to promote the growth of the indigenous species.

Typically, the process requires excavation of contaminated soil to facilitate microbial degradation and, depending on the type and state (liquid or solid) of the contaminant to be removed, ex-situ biodegradation is classified into two process types which are (i) a solid phase system, which includes land treatment and soil piles or (ii) a slurry phase system, which includes solid-liquid suspensions in bioreactors. A particular advantage of the ex-situ biodegradation process is that the process requires less time than the in situ biodegradation process.

Solid-phase biodegradation is a process in which the contaminated soil is excavated and placed into piles (often referred to as biopiles). Bacterial growth is stimulated through a network off pipes that are distributed throughout the piles and by providing air through the pipes the necessary ventilation is provided for microbial respiration and the necessary moisture is introduced by spraying the soil with water. The biopiles provide a favorable environment for sustaining the activity of the indigenous aerobic and anaerobic microorganisms.

9.6.2 SLURRY PHASE BIODEGRADATION

Slurry phase biodegradation (also known as bioreactor remediation) is a controlled treatment process that involves the excavation of the contaminated soil, mixing it with water, and placing it in a bioreactor.

In the process, contaminated soil is combined with water and other additives in a large tank (the bioreactor) and mixed to maintain the contact between microorganisms (which are already present in the soil) and the contaminants in the soil. Nutrients and oxygen are added, and conditions in the bioreactor are controlled to create the optimum environment for the microorganisms to degrade the contaminants. When the treatment is complete, the water is removed from the solids, which are sent for disposal or treated further if they still contain pollutants.

9.6.3 IN SITU BIODEGRADATION

Aerobic in situ techniques can vary in the way they supply oxygen to the organisms that degrade the contaminants such as (i) bioventing and (ii) injection of hydrogen peroxide. In the bioventing, oxygen is provided by pumping air into the soil above the water table or by delivering the oxygen as hydrogen peroxide (H_2O_2).

Typically, the process involves the use of multiple air injection points and multiple soil vapor extraction points to extract vapor phase contaminants above the water table. A blower is attached to wells, usually through a manifold, below the water table creating pressure, and the pressurized air forms small bubbles that travel through the contamination in and above the water column and volatilize the contaminants, thereby carrying the contaminants to the unsaturated soils above.

9.7 CONTROL OF GASEOUS POLLUTANTS

Gas cleaning (Mokhatab et al., 2006; Speight, 2019a) consists of separating all of the various obnoxious constituents from a gas stream since modern environmental regulation (Chapter 10) impose restrictions on the make-up of the gas that is allowed into the atmosphere or for sale to a consumer and there are several methods that can be employed to assure that the gas meets the specifications for sales and use (Table 9.4) (Speight, 2019a).

TABLE 9.4 Examples of Methods Used for Cleaning Gas Streams

Process Type*	Description
Absorption	Gaseous contaminants that are soluble in aqueous liquids can be removed in absorbers.
	Mainly used for the removal of acid gases (such as sulfur dioxide, hydrogen chloride, and hydrogen fluoride) as well as water-soluble organic compounds (such as alcohol derivatives, aldehyde derivatives, and organic acid derivatives).
	The contaminant is absorbed from the gas stream when the stream is in contact with the liquid.
Adsorption	Gaseous contaminants can be removed by the use of a solid adsorbent.
	Adsorbent can be in a wide variety of physical forms: (i) pellets in a thick bed; (ii) small beads in a fluidized bed; or (iii) fibers pressed onto a flat surface.
	Two types of adsorption mechanisms: (i) physical; and (ii) chemical which differ in the manner by which the contaminant is held to the adsorbent surface.
Condensation	Used for the recovery of organic compounds present at moderate-to-high concentrations in industrial process effluent gas streams.
	Three main categories of condensation systems: (i) water-based direct and indirect condensers in the temperature range (4 to 26°C/40 to 80°F); (ii) refrigeration condensers in the temperature range –45 to –100°C/–50 to –150°F; and (iii) cryogenic condensers in the temperature range –73 to –195°C/–100 to –320°F).
	Refrigeration and cryogenic systems are used primarily for the high efficiency recovery of high value contaminants.
Oxidation	Oxidizers can be used for the destruction of a wide variety of organic compounds to produce
	There are two main categories: (i) thermal oxidizers; and (ii) catalytic oxidizers.
Reduction	Used primarily for the destruction of NOx compounds emitted from combustion processes.
	These systems include selective noncatalytic reduction systems (SNCR) and selective catalytic reduction systems (SCR).
	In both systems, a chemically-reduced form of nitrogen is injected into the gas stream to react with the oxidized nitrogen compounds, namely nitric oxide (NO) and nitrogen dioxide (NO_2).
	The reactions between the reduced and oxidized forms of nitrogen result in molecular nitrogen, the major constituent of clean air.

*Listed alphabetically.

However, the selection and design of a gaseous contaminant control system must be based on specific information such as: (i) the composition of the gas stream; (ii) the flow rate of the gas stream; (iii) the temperature of the gas stream; as well as (iv) the amount and characteristics of the PM in the gas stream.

Raw (unprocessed) gas steams contain a variety of constituents and can include (i) acid gases such as carbon dioxide (CO_2), hydrogen sulfide (H_2S) and mercaptan derivatives (RSH) and other gases such as nitrogen and helium. Water vapor and liquid water may also be present as well as mercury (primarily in elemental form, Hg°) but chlorides and other species are possibly present as well as higher molecular weight hydrocarbon derivatives (VOCs, SVOCs) and PM. The raw gas must be purified to meet the standards specified by the local, state, or national regulations before it is sent to a consumer.

PM is particularly worthy of mention since this type of constituents entrained in a gas stream with the gaseous contaminants can have a severe impact on the efficiency and reliability of the collector. If not controlled, PM can gas flow and the impact of PM is especially severe if it is relatively large (such as >3 micrometers, $>3 \times 10^{-6}$ meter)-if the contaminant control system is subject to issues arising from the presence of the PM, a pre-collector might be needed.

KEYWORDS

- **chemical waste management**
- **geosphere**
- **polychlorinated biphenyl derivatives**
- **semi-volatile organic compounds**
- **volatile organic compounds**

REFERENCES

ASTM. D3922, (2020). *Estimating the Operating Performance of Virgin/Reactivated Granular Activated Carbon (GAC) for Removal of Soluble Pollutants from Water*. Annual Book of Standards. ASTM International, West Conshohocken, Pennsylvania.

Benefield, L. D., Judkins, J. F., & Weand, B. L., (1982). *Process Chemistry for Water and Wastewater Treatment.* Prentice-Hall, Englewood Cliffs, New Jersey.

Billet, R., (1995). *Packed Towers: In Processing and Environmental Technology*. VCH Publishers, Deerfield Beach, Florida.

Goodwin, R. W., (1993). *Combustion/Ash Residue Management: An Engineering Perspective.* Noyes Data Corp., Park Ridge, New Jersey.

Gupta, A. K., (2017). *Ion Exchange in Environmental Processes: Fundamentals, Applications, and Sustainable Technology*. John Wiley & Sons Inc., Hoboken, New Jersey.

Huang, R. Y. M., (1991). *Pervaporation Membrane Separation.* Elsevier, Amsterdam, Netherlands.

Kiran, E., & Brennecke, J. F., (1993). *Supercritical Fluid Engineering Science*. Symposium Series No. 514. American Chemical Society, Washington, D.C.

Kocurek, D. S., & Woodside, G., (1994). *Resources and References: Hazardous Waste and Hazardous Materials Management*. Noyes Data Corp., Park Ridge, New Jersey.

Ladisch, M. R., & Bose, A., (1992). *Harnessing Biotechnology for the 21st Century*. American Chemical Society, Washington, D.C.

Long, R. B., (1995). *Separation Processes in Waste Minimization*. Marcel Dekker Inc., New York.

Mokhatab, S., Poe, W. A., & Speight, J. G., (2006). *Handbook of Natural Gas Transmission and Processing*. Elsevier, Amsterdam, Netherlands.

Noble, R. S., & Stern, A., (1995). *Membrane Separations Technology: Principles and Applications*. Elsevier, New York.

Speight, J. G., (1996). *Environmental Technology Handbook*. Taylor & Francis, Washington, DC. Speight, J. G., & Exall, D. I., (2014). *Refining Used Lubricating Oils*. CRC Press, Taylor & Francis Group, Boca Raton.

Speight, J. G., (2019a). *Natural Gas: A Basic Handbook* (2nd edn). Gulf Publishing Company, Elsevier, Cambridge, Massachusetts.

Speight, J. G., (2019b). *Handbook of Petrochemical Processes*. CRC Press, Taylor & Francis Group, Boca Raton, Florida.

Syed, S., (2006). Solid and liquid waste management. *Emirates Journal for Engineering Research*, *11*(2), 19–36.

US DOE, (1995). *Mixed Waste Characterization, Treatment, and Disposal Focus Area*. Report No. DOE/EM-0252. Office of Environmental Management. United States Department of Energy, Washington, DC.

US EPA, (1994). *The Superfund Innovative Technology Program*. Annual Report to Congress. Report No. EPA/540/R-94/518. United States Environmental Protection Agency, Washington, DC.

Wachinski, A. M., (2016). *Environmental Ion Exchange: Principles and Design*. CRC Press, Taylor & Francis Group, Boca Raton, Florida.

Wise, D. L., & Trantolo, D. J., (1994). *Remediation of Hazardous Waste Contaminated Soils*. Marcel Dekker Inc., New York.

CHAPTER 10

Environmental Regulations

10.1 INTRODUCTION

The latter part of the 20th century started with the important realization that all chemicals can act as environmental pollutants depending upon the amount of chemical discharged and the ecosystem into which the chemical is released. In addition, there came the realization, in the current context, that emissions of compounds such as carbon dioxide (CO_2), methane (CH_4), and nitrous oxide (N_2O) to the atmosphere had either a direct impact or even an indirect impact on the global climate as well as on depletion of the ozone layer. As a result, unprecedented efforts were then made to reduce all global emissions of all chemicals through that has resulted in the evolution of environmental-oriented thinking followed by the formulation and passage of various environmental regulations (Table 10.1).

TABLE 10.1 Examples of Environmental Events and Regulations in the United States)

1906	The Pure Food and Drug Act established the Food and Drug Administration (FDA) that now oversees the manufacture and use of all foods, food additives, and drugs; amendments (1938, 1958, and 1962) strengthened the law considerably.
1924	The Oil Pollution Act was enacted.
1935	The Chemical Manufacturers Association (CMA), a private group of people working in the chemical industry and especially involved in the manufacture and selling of chemicals, established a Water Resources Committee to study the effects of their products on water quality.
1948	The Chemical Manufacturers Association established an Air Quality Committee to study methods of improving the air that could be implemented by chemical manufacturers.
1948	Clean Water Act enacted; often referred to as the Water Pollution Control Act.
1953	The Delaney Amendment to the Food and Drug Act defined and controlled food additives; any additives showing an increase in cancer tumors in rats, even if extremely large doses were used in the animal studies, had to be outlawed in food; recent debates have focused on a number of additives, including the artificial sweetener cyclamate.
1957	The Windscale fire, the worst nuclear accident in the history of Great Britain, released substantial amounts of radioactive contamination into the surrounding area at Windscale, Cumberland (now Sellafield, Cumbria).
1959	Just before Thanksgiving, the government announced that it had destroyed cranberries contaminated with a chemical, aminotriazole, that produced cancer in rats; the cranberries were from a lot frozen from two years earlier when the chemical was still an approved weed killer.

TABLE 10.1 *(Continued)*

1960	Diethylstilbestrol (DES), taken in the late 1950s and early 1960s to prevent miscarriages and also used as an animal fattener, was reported to cause vaginal cancer in the daughters of these women as well as premature deliveries, miscarriages, and infertility.
1962	Thalidomide, a prescription drug used as a tranquilizer and flu medicine for pregnant women in Europe to replace dangerous barbiturates that cause 2000–3000 deaths per year by overdoses, was found to cause birth defects. Thalidomide had been kept off the market in America because of the insistence that more safety data be produced for the drug.
1962	The Kefauver-Harris Amendment to the Food and Drug Act began required that drugs be proven safe before being placed on the market.
1962	The publication of Silent Spring (authored by Rachel Carson) that outlined many environmental problems associated with chlorinated pesticides, especially DDT and its use was banned in 972.
1963	The Clean Air Act was enacted. with subsequent amendments in 1970, 1977, and 1990.
1965	Non-linear, non-biodegradable synthetic detergents made from propylene tetramer were banned after these materials were found in large amounts in rivers, so much as to cause soapy foam in many locations. Phosphates in detergents were banned in detergents by many states in the 1970s.
1965	Mercury poisoning from concentration in the food chain recognized.
1966	Polychlorinated biphenyls (PCBs) were first found in the environment and in contaminated fish; banned in 1978 except in closed systems.
1968	TCDD (a dioxin derivative) tested positive as a teratogen in rats.
1969	The artificial sweetener cyclamate was banned because of its link to bladder cancer in rats fed with large doses; many 20 subsequent studies have failed to confirm this result but cyclamate remains banned.
1970	Occupational Safety and Health Act enacted.
1970	Earth day recognized because of concern related to the effects of many chemicals on the environment.
1970	The Clean Air Act amended.
1971	TCDD (see above) was outlawed by the Environmental Protection Agency.
1971	The Chemical Manufacturers Association established the Chemical Emergency Transportation System (CHEMTREC) to provide immediate information on chemical transportation emergencies.
1972	The clean water act enacted.
1972	Federal Water Pollution Control Act enacted.
1974	Safe Drinking Water Act enacted.
1974	Hazardous Materials Transportation Act enacted.
1974	Vinyl chloride was investigated as a possible carcinogen.
1976	The Toxic Substances Control Act (TSCA or TOSCA); Environmental Protection Agency developed rules to limit manufacture and use of PCBs.
1976	The Resource Conservation and Recovery Act (RCRA).
1977	Dibromochloropropane (DBCP) investigated for causes leading to sterility; now banned.
1977	Benzene was linked to an abnormally high rate of leukemia, there has been increased concern with benzene use in industry.
1977	Saccharin found to cause cancer in rats; banned by the FDA temporarily but Congress placed a moratorium on this ban because of public pressure; saccharin is still available.
1977	Clean Air Act amended 1977

TABLE 10.1 *(Continued)*

1978	Ban on chlorofluorocarbons (CFCs) as aerosol propellants; react with ozone in the stratosphere causing an increase the penetration of ultraviolet sunlight and increase the risk of skin cancer.
1978	The Amoco Cadiz sank near the northwest coast of France, resulting in the spilling of 68,684,000 gallons (US gallons of crude oil) (approximately 1,635,000 US barrels).
1978	Love Canal, Niagara Falls, New York.
1979	Three Mile Island accident-a partial nuclear meltdown.
1979	The Ixtoc oil spill.
1980	CHEMTREC is recognized by the Department of Transportation as the central service to provide immediate information on chemical transportation emergencies.
1980	The Comprehensive Environmental Response, Compensation, Liability Act enacted.
1980	The Resource Conservation and Recovery Act (RCRA) amended.
1984	Bhopal disaster in India.
1984	Toxic Substances Control Act amended.
1986	The Safe Drinking Water Act amended.
1986	The Emergency Planning and Community-Right-to-Know Act enacted; states that companies must report inventories of specific chemicals kept in the workplace and annual release of hazardous materials into the environment.
1986	The Superfund Amendments and Reauthorization Act enacted.
1986	The Comprehensive Environmental Response, Compensation, Liability Act amended.
1986	Chernobyl meltdown and explosion, contaminating surrounding area.
1986	The Sandoz disaster in Schweizerhalle (Switzerland) released tons of toxic agrochemicals into the Rhine.
1989	Pasadena, TX: explosion caused by leakage of exploded when ethylene and iso-butane leaked from a pipeline.
1989	The Montreal Protocol comes into effect, phasing out chlorofluorocarbons (CFCs) and other chemicals responsible for ozone depletion.
1989	The Exxon Valdez oil spill.
1990	The Clean Air Act amended.
1990	The Oil Pollution Act emended.
1990	Hazardous Materials Transportation Act amended.
1990	Channelview, TX: explosion in a petrochemical treatment tank of wastewater and chemicals.
1991	Sterlington, LA: explosion at a nitro-paraffin plant.
1991	Charleston, SC: explosion at a plant manufacturing Antiblaze 19®, a phosphonate ester and flame retardant used in textiles and polyurethane foam; manufactured from trimethyl phosphite, dimethyl methylphosphonate, and trimethyl phosphate.
1991	MT Haven explosion and oil spill of the coast of Italy.
1992	The Earth Summit is held in Rio-attended by representatives of 192 nations.
1992	Mingbulak oil spill in Uzbekistan.
1997	The Kyoto Protocol is signed, committing nations to reducing greenhouse gases emissions.
2005	Texas City refinery explosion.
2005	Hurricanes Katrin, Rita, and Wilma cause widespread destruction and environmental harm to coastal communities in the Gulf Coast region of the United States, especially the New Orleans area.

TABLE 10.1 *(Continued)*

2010	The Deepwater Horizon oil spill in the Gulf of Mexico causes millions of barrels of oil to pollute the Gulf.
2010	ExxonMobil oil spill in the Niger Delta (Nigeria).
2010	Xingang Port (China) oil spill into the Yellow Sea.
2011	Earthquake and tsunami in Japan caused release of radiation from damaged nuclear power plant.
2013	Minamata Convention on Mercury is signed, committing nations to reducing mercury poisoning.
2013	Lac-Mégantic, Quebec, Canada; derailment of an oil shipment train.
2015	A global climate change pact is agreed at the COP 21 summit, committing all countries to reduce carbon emissions for the first time.
2016	Representatives of 150 nations meet at the UNEP summit in Rwanda agree to phase out hydrofluorocarbons (HFCs), as an extension to the Montreal Protocol.
2019	Philadelphia refinery explosion.
2020	MV Wakashio oil spill in south Mauritius.

Notwithstanding early efforts at environmental protection and control, the concept of laws and regulations that were oriented to the protection of the environment but as a separate and distinct from the typical civil laws is a 20th century development (Lazarus, 2004). However, the collective recognition of the public and lawmakers alike that the environment is a fragile organism that is need of legal protection did not occur until late in the 1960s (Chapter 1). At that time, numerous influences including: (i) a growing awareness of the unity and fragility of the various ecosystems; (ii) increased public concern over the impact of industrial cycle (Figure 10.1) (Speight, 1996) on natural resources and the various flora and fauna-including human health; (iii) the increasing strength of the western government as well as the regulatory ability of these governments; and (iv) the success of various civil movements to protect the environment led to a collections of laws in a relatively short period of time (Tables 10.1 and 10.2). While the modern history of environmental law is the story of continuing (political) discussion and evolution, a collection of environmental laws had been established in the United States and many other countries by the end of the 20th century and formed a legal landscape which has been followed by similar regulatory movements in many developing countries.

By way of introduction to the term, environmental law includes regulation of pollutants and natural resource conservation and allocation. The regulations focus on energy development and use, agriculture, real estate, and land use, and has been expanded to include international environmental governance, international trade, environmental justice and climate change. The practice and application of environmental law typically requires extensive knowledge of administrative law and aspects of tort law, property, legislation, constitutional law, and land use law. Also, the broad category of environmental regulations may be broken down into a number of more specific regulatory subjects and while there is no single agreed-upon division or sub-division into specific categories, the core environmental regulations that are of interest in the context of this book can be employed address the environmental impact of the production, use, and disposal of chemicals on the environment.

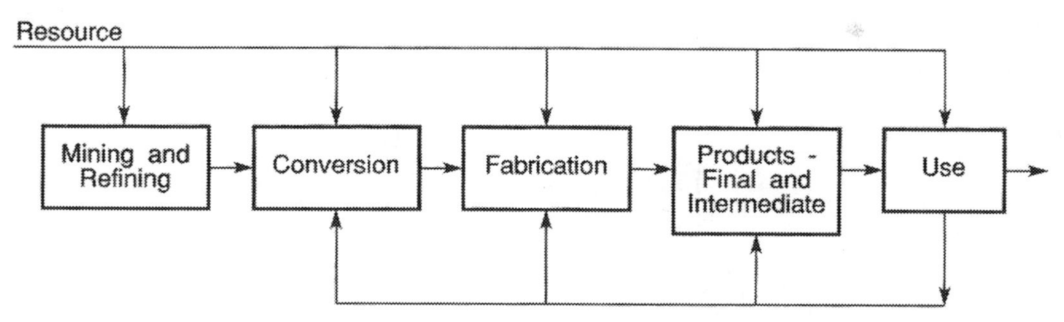

FIGURE 10.1 Representation of the industrial cycle.

Source: Reproduced with permission from: Speight (1996). © Taylor and Francis.

In concert with the evolution and institution of laws and regulations that are designed to the protect the environment, commercial processes have been designed to reduce the direct emissions of various chemicals and the related products as well as emissions of the by-products of chemical manufacturing processes into the air (the atmosphere), the water (the aquasphere), and on to the land (the geosphere).

Furthermore, in order to ensure that the movement to a clean environment continues, satellite-based instruments are employed to detect, quantify, and to monitor areas where there is a potential to discharge e a wide range of chemical pollutants into the various ecosystems. In addition, through an increased knowledge and behavior of the properties of various chemicals (Chapter 6) there is an accompanying increased understanding of the fate and consequences of the discharge of these chemicals into the environment and any potential changes to the chemicals as they reside in an ecosystem (Chapters 6 and 7) (Figure 10.2) (Speight, 1996). Not surprisingly, there has also been the on the effect of chemical properties and chemical behavior in the environment (Chapter 8) has increased dramatically and there are now available the means of predicting, with much greater precision, many of the environmental, ecological, and biochemical consequences of the inadvertent introduction of chemicals into the environment.

It is the purpose of this chapter to introduce the reader to an overview of a selection of the many and varied regulations instituted in the United States that regulate the disposition of chemicals into the environment.

10.2 ENVIRONMENTAL IMPACT OF PRODUCTION PROCESSES

Chemicals can enter the air, water, and soil during production, usage, or disposal. The impact of these chemicals on the environment is determined by (i) the type of the chemical, (ii) the concentration of the chemical, which is related to the amount of the chemical that is released, and (iii) the location where the chemical is introduced into the environment. In addition, the adverse effects of the chemical on an ecosystem may not be evident for some time after the introduction of the chemical. Nevertheless, the inadvertent or deliberate disposal of chemicals into the environment must also be of concern as any chemical has the ability to become part of the floral and faunal food chain (including humans) after which the chemical can also accumulate and/or persist in the environment for many years, even for decades.

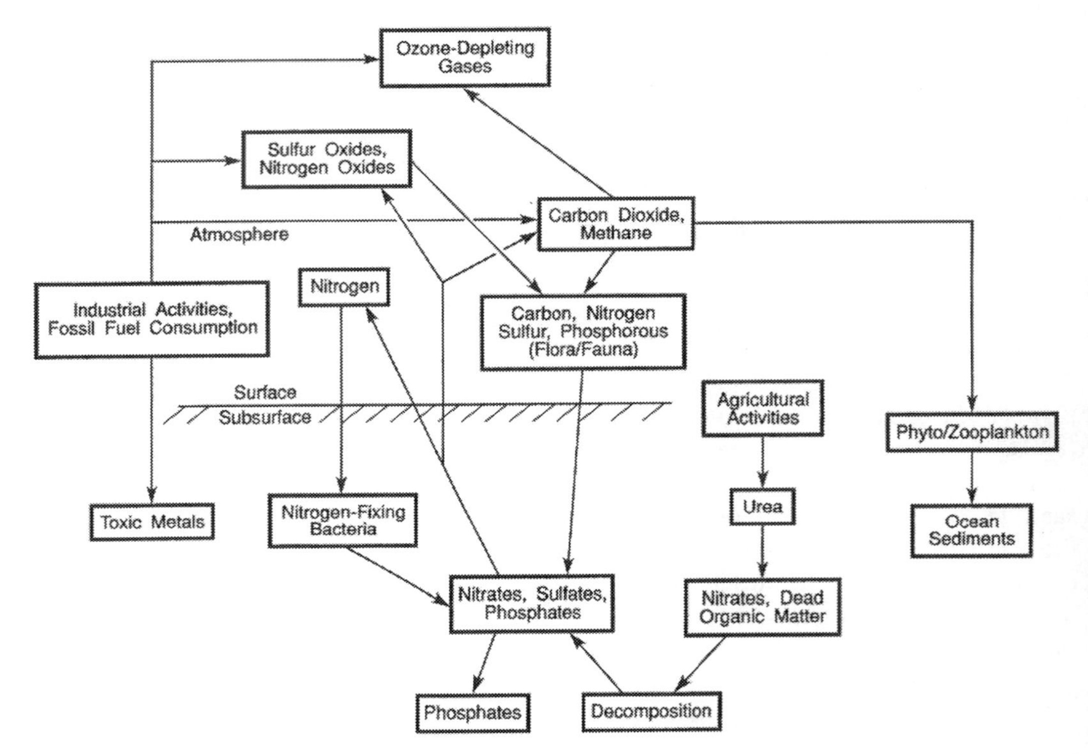

FIGURE 10.2 Chemical entry into the environment.
Source: Reproduced with permission from: Speight (1996). © Taylor and Francis.

In a more localized context, the chemicals industry is one of the largest industries in the United States and potential environmental hazards have caused the increased need for environmental. Briefly, production of chemicals involves a series of steps that includes separation and blending of chemicals. Production facilities are generally considered a major source of pollutants in areas where they are located and are regulated by a number of environmental laws related to air, land, and water (Table 10.2).

10.2.1 AIR POLLUTION

Air pollution can arise from several sources within the chemicals industry including, such as from: (i) equipment leaks from valves or other devices; (ii) high-temperature combustion processes in the actual burning of fuels for electricity generation; (iii) the heating of steam and process fluids; and (iv) the transfer of products. These pollutants are typically emitted into the environment over the course of a year through normal emissions, fugitive releases, accidental releases, or plant upsets. The combination of oxides of nitrogen, oxides of sulfur, and volatile hydrocarbon derivatives also contribute to ozone formation, which is one of the most important problems related to air pollution.

TABLE 10.2 Federal Regulations Relevant to Effects of Chemicals in the Environment*

- Atomic Energy Act (AEA)
- Beaches Environmental Assessment and Coastal Health (BEACH) Act
- Chemical Safety Information, Site Security, and Fuels Regulatory Relief Act
- Clean Air Act (CAA)
- Clean Water Act (CWA; original title: Federal Water Pollution Control Amendments of 1972
- Comprehensive Environmental Response, Compensation, and Liability Act (CERCLA, also known as Superfund or the Superfund Act); also contains the Superfund Amendments and Reauthorization Act (SARA)
- Emergency Planning and Community Right-to-Know Act (EPCRA)
- Endangered Species Act (ESA)
- Energy Independence and Security Act (EISA)
- Energy Policy Act
- Federal Food, Drug, and Cosmetic Act (FFDCA)
- Federal Insecticide, Fungicide, and Rodenticide Act (FIFRA)
- Food Quality Protection Act (FQPA)
- Marine Protection, Research, and Sanctuaries Act (MPRSA, also known as the Ocean Dumping Act)
- National Environmental Policy Act (NEPA)
- National Technology Transfer and Advancement Act (NTTAA)
- Nuclear Waste Policy Act (NWPA)
- Occupational Safety and Health (OSHA)
- Oil Pollution Act (OPA)
- Pesticide Registration Improvement Act (PRIA)
- Pollution Prevention Act (PPA)
- Resource Conservation and Recovery Act (RCRA)
- Safe Drinking Water Act (SDWA)
- Shore Protection Act (SPA)
- Toxic Substances Control Act (TSCA)

*The various Acts are listed alphabetically and not in the order of the date of enactment or the importance to the chemical industries. Executive Orders signed by the President are not included in this list.

Air quality laws govern the emission of air pollutants into the atmosphere and air quality laws are designed specifically to limit or completely eliminate the concentration of airborne pollutants as well as the complete elimination of some airborne pollutants. Other regulations are designed to address broader ecological problems, such as the limitation on chemicals that affect the ozone layer, and emissions trading programs to address acid rain or to address climate change. Regulatory efforts include: (i) identifying and categorizing air pollutants; (ii) setting limits on acceptable emissions levels, if any; and (iii) dictating or suggesting or recommending the necessary or appropriate mitigation technologies.

Chemicals are a source of hazardous and toxic air pollutants: they are also a major source of criteria air pollutants: particulate matter (PM), nitrogen oxides (NOx), carbon monoxide (CO), hydrogen sulfide (H_2S), and sulfur oxides (SOx). However, it is worthy of note that at the current time and because of improved process options for gas-cleaning operations, chemicals production (including petrochemical operations) release less toxic chemicals than in prior decades.

10.2.2 WATER POLLUTION

Chemicals found in water supplies and maybe natural due to the geology (mineralogy) of the area within and surrounding the water system or caused by activities of man through mining, industry, or agriculture. It is not uncommon to have trace amounts of many chemicals in water supplies but, in terms of pollution, chemicals (either as ionic species or non-dissociated compounds) are the greatest proportion of chemical contaminants in drinking water and are usually present in natural water systems. Many of the chemicals such as calcium carbonate ($CaCO_3$) and calcium bicarbonate, [$Ca(HCO_3)_2$] in hard water-are naturally occurring and should be considered as an integral part of that type of water rather than as contaminants.

Sometimes the most startling differences in the chemical content of drinking water arise as a consequence of the difference between groundwater and surface water. Such differences are usually a reflection of the solution of minerals as water percolates through the ground. However, some are due to relatively low groundwater flow compared to the flow of surface water, and the subsequent build-up of pollutants such as nitrate. Moreover, an assessment of the potential health effects of chemical contaminants in drinking water may be complicated by the limited database on the toxicity of these chemicals by the oral route and the fact that many of these elements are essential for human nutrition.

Water quality laws govern the release of the various types of pollutants into water resources, including surface water, groundwater, and stored drinking water. Some water quality laws, such as drinking water regulations, may be designed solely with particular reference to human health. Many other regulations, including restrictions on the disposal of chemicals into water resources, may also reflect broader efforts to protect the aqua sphere (i.e., aquatic ecosystems).

Typically, the regulatory efforts include: (i) identifying and categorizing water pollutants; (ii) specifying acceptable pollutant concentrations in water resources, and, above all; (iii) limiting pollutant discharges from effluent sources. Regulatory areas include disposal of industrial waste, sewage treatment and disposal, management of agricultural chemical waste and agricultural wastewater, as well as control of surface runoff from construction sites and other urban environments.

Chemical production facilities are also potential contributors to groundwater and surface water contamination. In some cases, companies have (with the appropriate permits) continued to use deep-injection wells to dispose of wastewater generated by the various processes, but this method of disposal may allow some of these wastes end up in aquifers and groundwater. Industrial wastewater may be highly contaminated and may arise from various processes, such as wastewaters from desalting processes, water from cooling towers, stormwater, distillation processes, or thermal processes. This water is recycled through many stages during the production process and goes through several treatment processes, including a wastewater treatment plant (WWTP), before being released (through government-issued permits) into surface waters.

Many wastewater issues (such as chemicals in waste process waters) face the chemicals industry (which also includes the fossil fuel industries as well as other process industries). However, efforts by the industry are being continued to eliminate any water contamination that may occur, whether it be from inadvertent leakage of crude oil or crude oil products or leakage of contaminated water from any one of several processes. In addition to monitoring the more complex salts (ionized chemicals) in the water, metals concentration must be continually monitored since heavy metals tend to concentrate in the tissues of floral and, in particular, faunal species (such as fish and animals) and increase in concentration as they go higher in the food chain. In addition, general sewage problems related to the disposal of sewage face every municipal sewage treatment facility, regardless of size.

Primary treatment (solid settling and removal) is required and secondary treatment (use of bacteria and aeration to enhance degradation of chemicals) is becoming more routine, tertiary treatment (filtration through activated carbon, applications of ozone, and chlorination) have been, or are being, implemented by many chemical production companies. Wastewater pre-treaters that discharge wastewater into sewer systems have more stringent requirements, and a variety of pollutant standards for sewage sludge have been enacted. Toxic chemicals in the water must be identified and plans must be developed and accepted by the various levels of governmental authority to alleviate any potential problems. In addition, regulators have established, and continue to establish, water-quality standards for priority toxic pollutants.

Many of the wastes are regulated under the safe drinking water act (SDWA) while the wastes discharged into surface waters are subject to state discharge regulations and are regulated under the clean water act (CWA). As examples of the regulations, the discharge guidelines typically limit the amounts of sulfides, ammonia, suspended solids (particulate matter or sediment) and other chemical constituents that may be present in the wastewater. Although these guidelines are in place, contamination from past unmanaged discharge of chemicals may be persistent in surface water bodies.

10.2.3 SOIL POLLUTION

Contamination of soils from the various processes is also an issue. Past (pre-1970) production practices may have led to spills on company property that now need to be cleaned up. In some cases, the natural bacteria that may use the chemicals as food are often effective at cleaning up chemicals spills and leaks compared to many other pollutant chemicals. Many waste materials are produced during the production of chemicals, and some of them are recycled through other stages in the process. Other wastes are collected and disposed of in landfills, or they may be recovered by other facilities. Soil contamination, including some hazardous wastes, spent catalysts or coke dust, tank bottoms, and sludge from the treatment processes can occur from leaks as well as accidents or spills on or off-site during the transport process.

10.3 ENVIRONMENTAL REGULATIONS IN THE UNITED STATES

The term environmental law (or environmental regulations) is a term that is commonly used to signify a collection of laws that work together and often overlap in areas and the terms covers the laws that regulate the discharge of chemicals into the environment Moreover, the broad category of environmental law may be broken down into a number of more specific regulatory subjects (Table 10.2). While there is no single agreed-upon classification of these regulations, the core environmental regulation addresses environmental pollution and many do specify pollution by chemical or by chemicals such as pesticides. A related but distinct set of regulatory regulations focus on the management of specific natural resources, such as forests, minerals, or fisheries. Other areas, such as environmental impact assessment (EIA), may not fit neatly into either category, but are nonetheless important components of the protection of the environmental.

In the United States, there are also common law protections that allow a land-owner who has land that is being polluted can seek a judgment (by filing a lawsuit) against the polluter. A landowner may sue under a theory of trespass (a physical invasion of the property) or nuisance (an interference with the enjoyment of the property by the landowner). However, each of these theories must include an element of reasonableness, and there can be no recovery (of financial damages) if the neighbor is making a reasonable use of the land. Additionally, the degree of *reasonableness* depends on the facts of the specific case. Also, an action may be brought under public nuisance where the suit is brought by a public entity if it is the public that is harmed (rather than a uniquely harmed individual).

State laws also reflect the same concerns and common law actions which allow adversely affected property owners to seek a judicial remedy for environmental harms. Although laws on the state level vary from state to state, many of use the federal laws as a base, thereby allowing an additional forum for aggrieved landowners to be heard. Finally, state laws may require a higher level of protection than federal law.

The toxic chemicals found within the chemicals industry are not necessarily unique, and although general air pollution, water pollution, and land pollution controls are affected by the chemicals defining, these problems and solutions are not unique to the industry. In fact, because the issues are so diverse, the chemicals industry (and because a segment of the industry-the crude oil refining industry-is an industrial complex consisting of many integrated unit processes) may be looked upon as a series of complex pollution-prevention issues and each issue is unique to the unit process from which the effluent originates. Therefore, there may be many examples of laws and controls that have been enacted by governments with input from the producers of chemicals that address pollution prevention and control and discharge of hazardous chemicals into the environment.

Thus, air quality laws govern the emission of chemical pollutants into the atmosphere (Chapter 6). Other initiatives are designed to address broader ecological problems, such as limitations on chemicals that affect the ozone layer, and emissions trading programs that address the formation and disposition of acid rain. Regulatory efforts include identifying and categorizing air pollutants, setting limits on acceptable emissions levels, and dictating necessary or appropriate mitigation technologies.

Similarly, water quality laws govern the release of chemical pollutants into water resources, including surface water, groundwater, and drinking water (Chapter 6). Some of the water quality laws, such as regulations that influence the quality of drinking water regulations, may be designed solely with reference to human health but many other water quality laws, including restrictions on the alteration of the chemical, physical, radiological, and biological characteristics of water resources are designed to protect the aquatic ecosystems and are often broader in their respective application to ecosystems. Regulatory efforts may also include identifying and categorizing water pollutants, dictating acceptable pollutant concentrations in water resources, and limiting pollutant discharges from effluent sources. Regulatory areas include sewage treatment and disposal, industrial wastewater management and agricultural wastewater management, and control of surface runoff from construction sites and urban environments, especially where there is a high likelihood that chemicals are dissolved in the waste water.

Waste management regulations govern the transport, treatment, storage, and disposal of all manner of waste, including municipal solid waste, hazardous waste, and nuclear waste, among many other types of chemical-containing waste. These laws are typically designed to minimize or eliminate the uncontrolled dispersal of waste (chemical) materials into the environment in a manner that may cause ecological or biological harm, and the regulations also include laws that are designed to reduce the generation of waste and promote or mandate waste recycling. Regulatory efforts include identifying and categorizing waste types and mandating transport, treatment, storage, and disposal practices.

Environmental cleanup laws govern the removal of (chemical) pollutants or (chemical) contaminants from environmental ecosystems such as soil, sediment in aqueous ecosystems, surface water, or groundwater. Unlike pollution control laws (which are designed to be followed before-the-fact), cleanup laws are designed to respond after-the-fact to environmental contamination, and consequently must often define not only the necessary response actions, but also the parties who may be responsible for undertaking the actual cleanup as well as bearing the cost of the cleanup actions. Regulatory requirements may include rules for emergency response, liability allocation, site assessment, remedial investigation, feasibility studies, remedial action, post-remedial monitoring, and site reuse.

Thus, pollution prevention and control of hazardous chemical materials is an issue not only for the chemicals industry but also for many industries and has been an issue for decades (Table 10.1) (Noyes, 1993). In this context, there are specific definitions for terms such as *hazardous* substances, toxic substances, and hazardous waste (Chapter 1). These are all terms of art and must be fully understood in the context of their statutory or regulatory meanings and not merely limited to their plain English or dictionary meanings. It is absolutely imperative from a legal sense that each statute or regulation promulgated be read in conjunction with terms defined in that specific statute or regulation (Majumdar, 1993).

In order to combat any threat to the environment, it is necessary to understand the nature and magnitude of the problems involved. It is in such situations that environmental technology has a major role to play. Environmental issues even arise when outdated laws are taken to task. Thus, the concept of what seemed to be a good idea at the time the action occurred may no longer be satisfactory when the law influences the environment.

Finally, it is worth of note that regulatory disincentives to voluntary reductions of emissions from chemicals also exist. Many environmental statutes define a baseline period and measure progress in pollution reductions from that baseline. Any reduction in emissions before it is required could lower the baseline emissions of a facility. Consequently, future regulations requiring a specified reduction from the baseline could be more difficult (and, consequently, have a much greater effect on the economic bottom line) to achieve because the most easily-applied and, hence, the most cost-effective reductions would already have been made and establishing the environmental base case may by no longer realistic. With no credit given for voluntary reductions, those facilities that do the minimum may be in fact be rewarded when emissions reductions are required.

As a start to the passage of various federal laws in the United States, the National Environmental Policy Act (NEPA) was passed in 1970 along with the Environmental Quality Improvement Act, the National Environmental Education Act, and the Environmental Protection Agency (EPA). The NEPA has been described as a relevant piece of environmental legislation passed by the Congress of the United States. The basic purpose of NEPA is to ensure that governmental agencies to consider the effects of their decisions on the environment. Thus, the main objective of these federal enactments was to assure that the environment be protected against both public and private actions that failed to take account of costs or harms inflicted on the eco-system. It is the duty of the United States EPA to monitor and analyze the environment, conduct research, and work closely with state and local governments to devise pollution control policies.

Thus, over time, the Congress of the United States (as has also happened in many other countries) has enacted a variety of protections with the goal of protecting the environment. These laws are summarized in the following sections.

10.3.1 CLEAN AIR ACT

The formulators of the Clean Air Act (CAA) recognized that virtually all metals are present in the atmosphere at low levels. PM emitted from combustion of fossil fuels contains trace metals that were present in the original fuel sample. The greatest health hazard is from aerosols that are smaller than 2.5 micrometers (μm) ion diameter and contain lead, beryllium, mercury, cadmium, and chromium. On the upside, these particles can settle out of the atmosphere over time or by precipitation events. On the downside, the particles must go somewhere, usually into the oceans and onto the land. Therefore, removal from the atmosphere by natural processes is not the answer to pollution. Furthermore, there is also the issues of mercury as a pollutant-mercury is the only metal that can exist as a gas (it is liquid at room temperature) and is therefore the only metal that can exist in a steady-state concentration in air. All other metals are emitted to the atmosphere from natural or anthropogenic sources, and then removed by settling or precipitation events.

However gaseous chemicals in various forms (Table 10.3) remain in the atmosphere for long residence times and would continue to build up much higher levels than observed currently if it was not for the fact that the gases undergo reactions in the gas phase reactions that convert these chemicals to other chemical products, such as acid rain (Chapters 3 and 8).

TABLE 10.3 Gaseous Chemical Pollutants

Element	Atmospheric Forms
Carbon	CO, CO_2
Nitrogen	NH_3, N_2O, NO, NO_2, N_2O_5*
Oxygen	O_3
Sulfur	H_2S, SO_2, SO_3

*Dinitrogen pentoxide can be created in the gas phase by reacting nitrogen dioxide (NO_2) with ozone (O_3). Thus: $2NO_2 + O_3 \rightarrow N_2O5 + O_2$

Furthermore, although ozone (O_3) is a desirable chemical in the stratosphere, it is a major environmental hazard at ground level. Ozone is a by-product of photochemical smog and reacts with hydrocarbon derivatives to form peroxy-nitrate derivatives that can cause severe environmental damage. Excessive ozone levels in the troposphere have been blamed for causing floral destruction through reactions with chlorophyll. Ozone is formed naturally when oxygen molecules are photochemically dissociated into oxygen atoms that can then react with a second oxygen molecule to make ozone. The presence of nitrogen oxides (NO, NO_2) leads to higher than normal background levels of ozone through several well-understood photochemical reactions.

Carbon monoxide is a product of the incomplete combustion of hydrocarbon fuels and as much as 20% v/v of the carbon monoxide released into the atmosphere each year comes from natural sources, but the greatest health problem is in metropolitan areas near high densities of vehicular traffic. It has been estimated that carbon monoxide has a 4-month lifetime in the atmosphere, where it reacts with the hydroxyl radical to form carbon dioxide:

$$CO + HO\cdot \rightarrow CO_2 + H\cdot$$
$$H\cdot + O_2 \rightarrow HOO\cdot$$
$$2HOO\cdot \rightarrow H_2O_2 + O_2$$
$$H_2O_2 + h\nu \rightarrow 2HO\cdot$$

Atmospheric carbon dioxide levels are determined by a long-term equilibrium between carbon dioxide in the air: (i) carbon dioxide dissolved in the aquasphere; (ii) releases of carbon dioxide from natural and anthropogenic sources; and (iii) losses by plant growth. Elevated levels of atmospheric carbon dioxide may have a major impact on the climate of the Earth.

Also, sulfur dioxide that is generated by coal-fired electric power generating plants is also an issue. Gas cleaning technologies (scrubbing technologies) have been used since the early 1970s to remove sulfur dioxide from power plant emissions (flue gas desulfurization, FGD) (Mokhatab et al., 2006; Kidnay et al., 2011; Speight, 2013, 2014, 2017, 2019; Bahadori, 2014). All current technologies involve exposing the combustion gases to a chemical that will absorb most of the sulfur dioxide. The FGD technologies are categorized into wet technologies that expose the flue gases to an aqueous solution and dry technologies that expose the flue gas to solid absorbents. Thus:

Lime scrubbing:

$$Ca(OH)_2 + SO_2 \rightarrow CaSO_3 + H_2O$$

Limestone slurry scrubbing:

$$CaCO_3 + SO_2 \rightarrow CaSO_3 + CO_2(g)$$

Magnesium oxide scrubbing:

$$Mg(OH)_2 + SO_2 \rightarrow MgSO_3 + H_2O$$

Sodium base scrubbing:

$$Na_2SO_3 + H_2O + SO_2 \rightarrow 2NaHSO_3$$

$$2NaHSO_3 \rightarrow Na_2SO_3 + H_2O + SO_2 \text{ (regeneration)}$$

Double alkali scrubbing:

$$2NaOH + SO_2 \rightarrow Na_2SO_3 + H_2O$$

$$Ca(OH)_2 + Na_2SO_3 \rightarrow 2NaOH + CaSO_3(s) \text{ (regeneration)}$$

There is also air pollution caused by the presence of particulate (PM) which is a term used for a mixture of solid particles and liquid droplets found in the air. Some particles, such as dust, dirt, soot, or smoke, are large or dark enough to be seen with the naked eye. Others are so small they can only be detected using an electron microscope. Pollution by such material includes: (i) PM_{10}, which refers to inhalable particles with a diameter on the order of 10 micrometers (10 μm) and smaller, and (ii) $PM_{2.5}$, which refers to fine inhalable particles with a diameter on the order of 2.5 micrometers (2.5 μm) and smaller. These particles have various sizes and shapes and can be made up of hundreds of different chemicals. Some are emitted directly from a source, such as construction sites, unpaved roads, fields, smokestacks or fires. Most particles form in the atmosphere as a result of complex reactions of chemicals such as sulfur dioxide (SO_2) and nitrogen oxides (NO_x), which are pollutants emitted from power plants, industries, and automobiles.

Because of these air pollution issues; the CAA was enacted in 1970 and is designed to protect air quality by regulating stationary and mobile sources of pollution. The Amendments of 1990 to the CAA have made significant changes in the basic CAA. The Clean Air allowed the establishment of air quality standards and provisions for their implementation and enforcement. This law was strengthened in 1977 and the CAA Amendments of 1990 imposed many new standards that included controls for industrial pollutants.

The CAA of 1970 and the 1977 amendments that followed consist of three titles (i) Title I deals with stationary air emission sources, (ii) Title II deals with mobile air emission sources, and (iii) Title III includes definitions of appropriate terms, provisions for citizen suits, and applicable standards for judicial review. However, in contrast to the previous clean air statutes, the 1990 Amendments contained extensive provisions for control of the accidental release of air toxins from storage or transportation as well as the formation of acid rain (Chapters 3 and 8). At the same time, the 1990 Amendments provided new and added requirements for such original ideas as state implementation plans (SIP) for attainment of the national ambient air quality standards and permitting requirements for the attainment and non-attainment areas. Title III now calls for a vastly expanded program to regulate hazardous air pollutants (HAPs) or the so-called air toxics.

Under the CAA Amendments of 1990, the mandate is to establish, during the first phase, technology-based maximum achievable control technology (MACT) emission standards that apply to the major categories or subcategories of sources of the listed HAPs. In addition, Title III provides for health-based standards that address the issue of residual risks due to air toxic emissions from the sources equipped with MACT and to determine whether the MACT standards can protect health with an ample margin of safety.

Section 112 of the original CAA that dealt with HAPs has been greatly expanded by the 1990 Amendments. The list of HAPs has been increased many fold. In addition, the standards for emission control have been tightened and raised to a high level, referred to as the best of the best, in order to reduce the risk of exposure to various HAPs.

Thus, the 1990 CAA Amendments aimed to encourage voluntary reductions above the regulatory requirements by allowing facilities to obtain emission credits for voluntary reductions in emissions. These credits would serve as offsets against any potential future facility modifications resulting in an increase in emissions. Other regulations established by the amendments, however, will require the construction of major new units within existing chemicals producers to reduce emissions even further and these new operations will require emission offsets in order to be permitted. This will consume many of the credits available for existing facility modifications. A shortage of credits for facility modifications will make it difficult to receive credits for emission reductions through pollution prevention projects.

Thus, under this CAA, the EPA sets limits on how much of a pollutant can be in the air anywhere in the United States. The law does allow individual states to have stronger pollution controls, but states are not allowed to have weaker pollution controls than those set for the whole country. The law recognizes that it makes sense for states to take the lead in carrying out the CAA because pollution control problems often require a special understanding of local industries and geography as well as housing developments near to industrial sites.

In addition, Title IV of the 1990 Amendments to the CAA (Title IV-Acid Deposition Control) mandates requirements for the control of acid deposition acid rain). The purpose of this title is to reduce the adverse effects of acid deposition through reductions in annual emissions of sulfur dioxide and also, in combination with other provisions of this act, though reductions of nitrogen oxides emissions in the 48 contiguous States and the District of Columbia. It is the intent of this title to effectuate such reductions by requiring compliance by affected sources with prescribed emission limitations by specified deadlines, which limitations may be met through alternative methods of compliance provided by an emission allocation and transfer system. It is also the purpose of this title to encourage energy conservation, use of renewable and clean alternative technologies, and pollution prevention as a long-range strategy, consistent with the provisions of this title, for reducing air pollution and other adverse impacts of energy production and use. Furthermore, individual states are required to develop a SIP that is a collection of the regulations a state will use to clean up polluted areas. The states must involve the public, through hearings and opportunities to comment, in the development of each state implementation plan. The EPA must approve each plan and if a state implementation is not acceptable, the EPA can take over enforcing the CAA in that state.

Air pollution often travels from its source in one state to another state. In many metropolitan areas, people live in one state and work or make purchases in another state thereby allowing (in some cases unknowingly allowing) air pollution from automobiles and trucks to spread throughout the interstate area. The 1990 CAA Amendments provide for interstate commissions on air pollution control, which are to develop regional strategies for cleaning up air pollution. The 1990 Amendments also cover pollution that originates in nearby countries, such as Mexico and Canada, and drifts into the United States, as well as pollution from the United States that reaches Canada and Mexico.

In the current context, the 1990 Amendments provide economic incentives for cleaning up pollution. For instance, producers of chemicals can get credits if they produce cleaner products than required, and use the credits when the product falls short of the requirements. Furthermore, chemicals (like many industrial products) can be extremely toxic to the environment. As an example, to combat such effects refiners have started to reformulate gasoline sold in the formerly smog-prone areas. This form of gasoline contains less volatile chemicals (of which benzene is an example and which is also a HAP that causes cancer and aplastic anemia, a potentially fatal blood disease). The reformulated gasoline also contains detergents, which, by preventing build-up of engine deposits, keep engines working smoothly and burning fuel cleanly.

10.3.2 CLEAN WATER ACT

The CWA came into law because of the contamination of the aquasphere (groundwater, aquifers, lakes, rivers, and oceans), which occurs when pollutants are directly or indirectly discharged into water bodies without prior adequate treatment to remove the pollutants. Water pollution affects the entire biosphere-floral and faunal species that live in the various bodies of water. In almost all cases, the effect is damaging not only to the natural biological communities, including users (human users) of the water.

Water is typically referred to as polluted when it is impaired by contaminants from anthropogenic sources: (i) either does not support a human use, such as drinking water; or (ii) undergoes a marked shift in its ability to support its constituent biotic communities, such as fish. Natural phenomena such as volcanoes, algal blooms, storms, and earthquakes also cause major changes in water quality and the ecological status of water.

The specific contaminants leading to pollution in water include a wide spectrum of chemicals and physical changes such as elevated temperature and discoloration. While many of the chemicals that are regulated may be naturally occurring (such as derivatives of calcium, sodium, iron, and manganese), the concentration (relative to the indigenous concentration) is often the key in determining what is a natural component of water and what is a contaminant. High concentrations of naturally occurring chemicals can have negative impacts on aquatic flora and fauna. It is safest to assume that many of the anthropogenic chemicals are toxic, unless proven otherwise. Even if the chemicals are non-toxic, there is no reason why anthropogenic chemicals should be in the water system. The chemicals (non-toxic or toxic) can cause alteration of the physical chemistry of the water body, which includes acidity (a change in the pH), electrical conductivity, temperature, and eutrophication. Eutrophication is an increase in the concentration of chemical nutrients in an ecosystem to an extent that increases in the primary productivity of the ecosystem. Depending on the degree of eutrophication, subsequent negative environmental effects such as anoxia (oxygen depletion) and severe reduction in water quality may occur, affecting the floral population and the animal population of the water body.

Thus, the CWA is designed to protect water by preventing discharge of pollutants into navigable waters from point sources. The CWA started life as a regulatory act. The Federal Water Pollution Control Act of 1948 was the first major U.S. law to address water pollution. Growing public awareness and concern for controlling water pollution led to amendments in 1972 and, as amended in 1972, the law became commonly known as the CWA. The 1972 amendments: (i) established the basic structure for regulating pollutant discharges into the waters of the United States; (ii) gave the EPA the authority to implement pollution control programs such as setting wastewater standards for industry; (iii) maintained existing requirements to set water quality standards for all contaminants in surface waters; (iv) made it unlawful for any person to discharge any pollutant from a point source into navigable waters, unless a permit was obtained under its provisions; (v) funded the construction of sewage treatment plants under the construction grants program; and (vi) recognized the need for planning to address the critical problems posed by nonpoint source pollution.

Subsequent amendments modified some of the earlier provisions of the CWA. For example, revisions in 1981 streamlined the municipal construction grants process, improving the capabilities of treatment plants built under the program. Further changes to the act in 1987 phased out the construction grants program, replacing it with the State Water Pollution Control Revolving Fund (the Clean Water State Revolving Fund) which addressed water quality needs by building on EPA-state partnerships.

Over the years, many other laws have changed parts of the CWA. Title I of the Great Lakes Critical Programs Act of 1990, for example, put into place parts of the Great Lakes Water Quality Agreement of 1978, signed by the U.S. and Canada, where the two nations agreed to reduce certain toxic pollutants in the Great Lakes. That law required the EPA to establish water quality criteria for the Great Lakes addressing 29 toxic pollutants with maximum levels that are safe for humans, wildlife, and aquatic life. It also required the EPA to help the States implement the criteria on a specific schedule.

Thus, the Clean Water Act (CWA or the Water Pollution Control Act) is the cornerstone of surface water quality protection in the United States and employs a variety of regulatory and non-regulatory tools to sharply reduce direct pollutant discharges into waterways and manage polluted runoff. The objective of the CWA is to restore and maintain the chemical, physical, and biological integrity of water systems. The act established the basic structure for regulating discharges of pollutants into the waters of the United States and regulating quality standards for discharge of pollutants into the waters of the United States and gave the EPA the authority to implement pollution control programs such as setting wastewater standards for industry. The CWA also continued requirements to set water quality standards for all contaminants in surface waters. The act is credited with the first comprehensive program for controlling and abating water pollution. In addition, the act made it unlawful to discharge any pollutant from a point source into navigable waters, unless a permit was obtained. Point sources are discrete conveyances such as pipes or man-made ditches. Individual homes that are connected to a municipal system, use a septic system, or do not have a surface discharge do not need an NPDES permit; however, industrial, municipal, and other facilities must obtain permits if their discharges go directly to surface waters.

The statute makes a distinction between conventional and toxic pollutants. As a result, two standards of treatment are required prior to their discharge into the navigable waters of the nation. For conventional pollutants that generally include degradable nontoxic chemicals, the applicable treatment standard is best conventional technology (BCT). For toxic pollutants, on the other hand, the required treatment standard is best available technology (BAT), which is a higher standard than BCT.

The statutory provisions of CWA have five major sections that deal with specific issues: (i) nationwide water-quality standards; (ii) effluent standards for the chemicals industry; (iii) permit programs for discharges into receiving water bodies based on the National Pollutant Discharge Elimination System (NPDES); (iv) discharge of toxic chemicals, including oil spills; and (v) construction grant program for publicly owned treatment works (POTW). In addition, section 311 of CWA includes elaborate provisions for regulating intentional or accidental discharges of oil and hazardous chemicals. Included there are response actions required for oil spills and the release or discharge of toxic and hazardous chemicals. Pursuant to this, certain elements and compounds are designated as hazardous chemicals and an appropriate list has been developed (40 CFR 116.4). The person in charge of a vessel or an onshore or offshore facility from which any designated hazardous chemicals is discharged, in quantities equal to or exceeding its reportable quality, must notify the appropriate federal agency as soon as such knowledge is obtained. Such notice should be provided in accordance with the designated procedures (33 CFR 153.203).

Under the CWA, discharge of water-borne pollutants is limited by NPDES permits. Chemical-producing companies that easily meet their permit requirements may find that the permit limits will be changed to lower values and be less stringent. However, because occasional system upsets do occur resulting in significant excursions above the normal performance values, many companies may feel that they must maintain a large operating margin below the permit limits to ensure continuous compliance. Those companies that can significantly reduce water-borne emissions may find the risk of having their permit limits lowered to be a substantial disincentive.

10.3.3 COMPREHENSIVE ENVIRONMENTAL RESPONSE, COMPENSATION, AND LIABILITY ACT

The cleanup of environmental pollution involves a variety of techniques, ranging from simple biological processes to advanced engineering technologies. Cleanup activities may address a wide range of contaminants, from common industrial chemicals such as agricultural chemicals and metals to radionuclides. Cleanup technologies may be specific to the contaminant (or contaminant class) and to the site (Chapter 9).

Cleanup costs can vary dramatically depending on the contaminants, the media affected, and the size of the contaminated area. Much of the remediation has been in response to such historical chemical management practices as dumping, poor storage, and uncontrolled release or spillage of chemicals. Greater effort in recent years has been directed toward pollution prevention, which is more cost-effective than remediation.

However, in most cases, it is financially or physically impractical to completely remove all traces of contamination. In such cases, it is necessary to set an acceptable level of residual contamination. As a result, evolution of cleanup technologies has yielded four general categories of remediation approaches: (i) physical removal, with or without treatment; (ii) *in situ* conversion by physical or chemical means to less toxic or less mobile forms; (iii) containment; and (iv) passive cleanup, or natural attenuation (Chapter 9). Combinations of the four technology types may be used at some contaminated sites.

Thus, the Comprehensive Environmental Response, Compensation, and Liability Act (CERCLA), commonly known as Superfund, 1980, is aimed at cleaning up already polluted areas. This statute assigns liability to almost anyone associated with the improper disposal of hazardous waste, and is designed to provide funding for cleanup. To achieve this goal, a tax was created on the chemical and chemicals industries and provided broad Federal authority to respond directly to releases or threatened releases of hazardous chemicals that may endanger public health or the environment.

The act was amended by the Superfund Amendments and Reauthorization Act (SARA) in 1986 and stressed the importance of permanent remedies and innovative treatment technologies in cleaning up hazardous waste sites. Thus, the act provides a federal superfund to clean up uncontrolled or abandoned hazardous-waste sites as well as accidents, spills, and other emergency releases of pollutants and contaminants into the environment. Through CERCLA, the EPA was given power to seek out those parties responsible for any release and assure their cooperation in the cleanup.

A CERCLA response or liability will be triggered by an actual release or the threat of a hazardous substance or pollutant or contaminant being released into the environment. A hazardous substance [CERCLA 101(14)] is any substance requiring special consideration due to its toxic nature under the CAA, the CWA, or the Toxic Substances Control Act (TSCA) and as defined under RCRA. Additionally, a pollutant or contaminant can be any other chemical not necessarily designated or listed but that (quote) will or may reasonably (end quote) be anticipated to cause any adverse effect in organisms and/or their offspring [CERCLA 101(33)].

The central purpose of CERCLA is to provide a response mechanism for cleanup of any hazardous chemical released, such as an accidental spill, or of a threatened release of a hazardous chemical (Nordin et al., 1995). Section 102 of CERCLA is a catchall provision because it requires regulations to establish that quantity of any hazardous substance the release of which shall be reported pursuant to section 103 of CERCLA. Thus, under CERCLA, the list of potentially responsible parties (PRPs) can include all direct and indirect culpable parties who have either released a hazardous chemical or violated any statutory provision. In addition, responsible private parties are liable for cleanup actions and/or costs as well as for reporting requirements for an actual or potential release of a hazardous chemical (hazardous pollutant, or hazardous contaminant).

CERCLA (Superfund) legislation deals with actual or potential releases of hazardous materials that have the potential to endanger people or the surrounding environment at uncontrolled or abandoned hazardous waste sites. The act requires responsible parties or the government to clean up waste sites. Among major purposes of CERCLA are the following: (i) site identification, (ii) evaluation of danger from sites where waste chemicals have been deposited, (iii) evaluation of damages to natural resources, (iv) monitoring of release of hazardous chemicals from sites, and (v) removal or cleanup of wastes by responsible parties or government.

The SARA addresses closed hazardous waste disposal sites that may release hazardous chemicals into any environmental medium. Title III of SARA also requires regular review of emergency systems for monitoring, detecting, and preventing releases of extremely hazardous chemicals at facilities that produce, use, or store such chemicals.

The most revolutionary part of SARA is the Emergency Planning and Community Right-to-Know Act (EPCRA) that is covered under Title III of SARA. EPCRA includes three subtitles and four major parts: emergency planning, emergency release notification, hazardous chemical reporting, and toxic chemical release reporting. Subtitle A is a framework for emergency planning and release notification. Subtitle B deals with various reporting requirements for hazardous chemicals and toxic chemicals. Subtitle C provides various dimensions of civil, criminal, and administrative penalties for violations of specific statutory requirements.

Other provisions of SARA basically reinforce and/or broaden the basic statutory program dealing with the releases of hazardous chemicals (CERCLA Section 313). It requires owners and operators of certain facilities that manufacture, process, or otherwise use one of the listed chemicals to report (on an annual basis) all environmental releases of these chemicals. This information related to the total annual release of chemicals from an industrial facility can be made available to the public.

The act also requires (under Section 4) testing of chemicals by manufacturers, importers, and processors where risks or exposures of concern are found and (under Section 5) issuance of significant new use rules (SNURs) when a significant new use is identified for a chemical that that could result in exposures to, or releases of, a chemical of concern. In addition, companies or persons importing or exporting chemicals are required to comply with certification reporting and/or other requirements which also requires (under Section 8) reporting and record-keeping by persons who manufacture, import, process, and/or distribute chemical chemicals in commerce as well as that any person who manufactures (including imports), processes, or distributes in commerce a chemical or mixture of chemicals and who obtains information which reasonably supports the conclusion that such chemical or mixture presents a substantial risk of injury to health or the environment to immediately inform the EPA, except where the Agency has been adequately informed of such information.

The Toxic Chemicals Control Acts is also authorized (under Section 8) to maintain the TSCA Inventory, which contains more than 83,000 chemicals and, as new chemicals are commercially manufactured or imported, they are placed on the list.

10.3.4 HAZARDOUS MATERIALS TRANSPORTATION ACT

A hazardous material is any material that is a hazardous material and a hazardous waste is any material that is subject to the hazardous waste manifest requirements of the United States EPA. It was recognized that the transportation of such materials can cause environmental problems when (for whatever reason) spillage occurs and there was the need to establish a means of protecting the environment against the risks to the environment that are inherent in the transportation of hazardous material in intrastate, interstate, and foreign commerce.

In the 1960s and 1970s, the high cost of disposal of hazardous material and hazardous waste led to increased dumping of materials that were increasingly being deemed hazardous by the public and by the government. Illegal dumping took place on vacant lots, along highways, or on the actual highways themselves. At the same time, increased accidents and incidents with hazardous materials during transportation was a growing problem, causing damage to the environment. The increasing frequency of illegal midnight dumping and spills, along with the already existing inconsistent regulations and fragmented enforcement, led to the passing of legislation to mitigate such activities.

Thus, the Hazardous Materials Transportation Act, passed in 1975, is the law governing transportation of chemicals and hazardous materials. It is the principal federal law in the United States that regulates the transportation of hazardous materials (such as, in the current context, chemicals). The purpose of the act is to protect against the risks to life, property, and the environment that are inherent in the transportation of hazardous material in intrastate, interstate, and foreign commerce under the authority of the United States Secretary of Transportation.

The act was passed as a means to improve the uniformity of existing regulations for transporting hazardous materials and to prevent spills and illegal dumping endangering the public and the environment, a problem exacerbated by uncoordinated and fragmented regulations. Regulations are enforced through four key provisions that encompass federal standards under Title 49 of the United States Code (a code that regards the role of transportation in the United States of America): (i) procedures and policies, (ii) material designations and labeling, (iii) packaging requirements, and (iv) operational rules. Violation of the Act regulations can result in civil or criminal penalties unless a special permit is granted under the discretion of the Secretary of Transportation.

Thus, the basic purpose of the Hazardous Materials Transportation Act is to ensure safe transportation of hazardous materials through the highways, railways, and waterways of the United States. The basic theme of the act is to prevent any person from offering or accepting a hazardous material for transportation anywhere within this nation if that material is not properly classified, described, packaged, marked, labeled, and properly authorized for shipment pursuant to the regulatory requirements. In addition, the act includes a comprehensive assessment of the regulations, information systems, container safety, and training for emergency response and enforcement. The regulations apply to "any person who transports, or causes to be transported or shipped, a hazardous material; or who manufactures, fabricates, marks, maintains, reconditions, repairs, or tests a package or container which is represented, marked, certified, or sold by such person for use in the transportation in commerce of certain hazardous materials."

Under this statutory authority, the Secretary of Transportation has broad authority to determine what is a hazardous material, using the dual tools of quantity and type. By this two-part approach, any material that may pose an unreasonable risk to human health or the environment may be declared a hazardous material. Such a designated hazardous material obviously includes both the quantity and the form that make the material hazardous. Furthermore, under the Department of Transportation (DOT) regulations, a hazardous material is any substance or material, including a hazardous substance and hazardous waste that is capable of posing an unreasonable risk to health, safety, and property when transported in commerce. DOT thus has broad authority to regulate the transportation of hazardous materials that, by definition, include hazardous chemicals.

10.3.5 OCCUPATIONAL SAFETY AND HEALTH ACT

All employees have the right to a workplace that is reasonably free of safety and health hazards, and there was a need to assure the safety and health of workers in the United States by setting and enforcing workplace safety standards.

Thus, the objective of the OSHA Hazard Communication Standard is to inform workers of potentially dangerous chemicals in the workplace and to train them on how to protect themselves against potential dangers. The act is formulated to (quote) "to assure safe and healthful working conditions for working men and women; by authorizing enforcement of the standards developed under the Act; by assisting and encouraging the States in their efforts to assure safe and healthful working conditions; by providing for research, information, education, and training in the field of occupational safety and health; and for other purposes."

The goal of OSHA is to ensure that "no employee will suffer material impairment of health or functional capacity" due to a lifetime occupational exposure to chemicals and hazardous chemicals. The statute imposes a duty on the employers to provide employees with a safe workplace environment, free of known hazards that may cause death or serious bodily injury. Thus, the act is entrusted with the major responsibility for safety in the workplace and health of the employees (Wang, 1994). It is responsible for the means by which chemicals are contained through the inspection of workplaces to ensure compliance and enforcement of applicable standards under OSHA. It is also the means by which guidelines have evolved for the destruction of chemicals used in chemical laboratories.

The statute covers all employers and their employees in all the states and federal territories with certain exceptions (Lunn and Sanstone, 1994). Generally, the statute does not cover self-employed persons, farms solely employing family members, and those workplaces covered under other federal statutes. Chemicals producers must evaluate whether the chemicals that the company manufactures and sell are hazardous. Under the General Duty Clause of OSHA, employers are required to provide an environment that is free from recognized hazards that could cause physical harm or death.

All employers are required to develop, implement, and maintain at the workplace a written hazard communication program. The program must include the following components: (i) a list of hazardous chemicals in the work place, (ii) the methods the employer will use to inform employees of the hazards associated with these chemicals, and (iii) a description of how the labeling, material safety data sheet (MSDS), and employee training requirements will be met.

The following information must be included in the program for employers who produce, use, or store hazardous chemicals in the workplace: (i) the means by which the manufacturer safety data sheets (MSDS) will be made available to the outside contractor for each hazardous chemical, (ii) the means by which the employer will inform the outside contractor of precautions necessary to protect the employees of the contractor both during normal operating conditions and in the case of emergencies, and (iii) the methods that the employer will use to inform contractors of the labeling system used in the workplace.

10.3.6 RESOURCE CONSERVATION AND RECOVERY ACT

Ads the 20th Century evolved, there was an increased need to the environment from (i) the potential hazards of waste disposal, (ii) to conserve energy and natural resources, (iii) to reduce the amount of waste generated, and (iv) to ensure that wastes were managed in an environmentally sound manner. As a result, there was the need for regulations that required corrective actions to address the investigation and cleanup of releases of hazardous chemicals in accordance with state and federal requirements. The degree of investigation and subsequent corrective action necessary to protect the environment varies significantly among facilities. Cleanup progress at these facilities is measured, in part, by interim cleanup milestones known as environmental indicators. The hazardous waste regulatory program began with the Resource Conservation and Recovery Act (RCRA) in 1976 addressed these issues and provided a *cradle-to-grave* system of preventing pollution by use of a manifest system to ensure that waste is properly disposed of, and thus not dumped into the environment.

Since the enactment of RCRA, the act has been amended several times, to promote safer solid and hazardous waste management programs (Dennison, 1993). The Used Oil Recycling Act of 1980 and the Hazardous and Solid Waste Amendments of 1984 (HSWA) were the major amendments to the original law. The 1984 amendments also brought the owners and operators of underground storage tanks under the Resource Conservation Recovery Act umbrella. This can have a significant effect on refineries that store chemicals in underground tanks. Now, in addition to the hazardous waste being controlled, the Resource Conservation Recovery Act Subtitle I regulates the handling and storage of chemicals.

The Resource Conservation Recovery Act controls disposal of solid waste and requires that all wastes destined for land disposal be evaluated for their potential hazard to the environment. Solid waste includes liquids, solids, and containerized gases and is divided into non-hazardous waste and hazardous waste. The various amendments are aimed at preventing the disposal problems that lead to a need for the Comprehensive Environmental Response Compensation and Liability Act (CERCLA), or Superfund, as it is known.

Subtitle C of the original Resource Conservation Recovery Act lists the requirements for the management of hazardous waste. This includes criteria for identifying hazardous waste, and the standards for generators, transporters, and companies that treat, store, or dispose of the waste. The Resource Conservation Recovery Act regulations also provide standards for design and operation of such facilities. However, before any action under the act is planned, it is essential to understand what constitutes a solid waste and what constitutes a hazardous waste. The first step to be taken by a generator of waste is to determine whether that waste is hazardous. Waste may be hazardous by being listed in the regulations, or by meeting any of the four characteristics: (i) ignitability, (ii) corrosivity, (iii) reactivity, and (iv) toxicity (Chapter 6).

Section 1004(27) of the Resource Conservation Recovery Act defines solid waste as garbage, refuse, sludge, from a waste treatment plant, water supply treatment plant, or air pollution control facility and other discarded material, including solid, liquid, semisolid, or contained gaseous material resulting from industrial, commercial, mining, and agricultural operations and from community activities, but does not include solid or dissolved materials in domestic sewage, or solid or dissolved materials in irrigation return flows or industrial discharges which are point sources subject to permits under section 402 of the Federal Water Pollution Control Act, as amended (86 Stat. 880), or source, special nuclear, or byproduct material as defined by the Atomic Energy Act of 1954, as amended (68 Stat. 923).

This statutory definition of solid waste is pursuant to the regulations of the EPA insofar as a solid waste is a hazardous waste if it exhibits any one of four specific characteristics: (i) ignitability, (ii) reactivity, (iii) corrosivity, and (iv) toxicity. However, a waste chemical listed solely for the characteristic of ignitability, reactivity, and/or corrosivity is excluded from regulation as a hazardous waste once it no longer exhibits a characteristic of hazardous waste.

In terms of ignitability, a waste is an ignitable hazardous waste if it has a flashpoint of less than 140° (40 CFR 261.21) Fahrenheit as determined by the Pensky-Martens closed cup flash point test; readily causes fires and burns so vigorously as to create a hazard; or is an ignitable compressed gas or an oxidizer (as defined by the Department of Transport regulations). A simple method of determining the flashpoint of a waste is to review the material. Ignitable wastes (of which naphtha is an example) carry the waste code D1001.

On the other hand, a corrosive waste is a liquid waste that has a pH of less than or equal to 2 (a highly acidic waste) or greater than or equal to 12.5 (a highly alkaline waste) is considered to be a corrosive hazardous waste (40 CFR 261.22). Also, a corrosive waste may be a liquid and corrodes steel (SAE 1020) at a rate greater than 6.35 mm (0.250 inch) per year at a test temperature of 55 C (130 F). Corrosivity testing is conducted using the standard test method as formulated by the National Association of Corrosion Engineers (NACE)-Standard TM-01-69 or, in place of this method, an EPA-approved equivalent test method.

For example, sodium hydroxide (caustic soda), a with a high pH, is often used by the chemicals production industry (especially by the crude oil refining industry and the natural gas industry) in the form of a caustic wash to remove sulfur compounds or acid gases. When these caustic solutions become contaminated and must be disposed of, the waste would be a corrosive hazardous waste. Corrosive wastes carry the waste code D002. Acid solutions also fall under this category.

A chemical waste material is considered to be a *reactive* hazardous waste if it is normally unstable, reacts violently with water, generates toxic gases when exposed to water or corrosive materials, or if it is capable of detonation or explosion when exposed to heat or a flame (40 CFR 261.23). Materials that are defined as forbidden explosives or Class A or B explosives by the DOT are also considered reactive hazardous waste.

Typically, reactive wastes are solid wastes that exhibit any of the following properties as defined at 40 CFR 61.23(a): (i) it is normally unstable and readily undergoes violent change without detonating; (ii) it reacts violently with water; (iii) it forms potentially explosive mixtures with water; (iv) when mixed with water, it generates toxic gases, vapors or fumes in a quantity sufficient to present a danger to human health or the environment; (v) it is a cyanide or sulfide bearing waste which, when exposed to pH conditions between 2 and 12.5, can generate toxic gases, vapors or fumes in a quantity sufficient to present a danger to human health or the environment; (vi) it is capable of detonation or explosive reaction if it is subjected to a strong initiating source or if heated under confinement; (vii) it is readily capable of detonation or explosive decomposition or reaction at standard temperature and pressure; (viii) it is a forbidden explosive as defined in 49 CFR 173.51, or a Class A explosive as defined in 49 CFR 173.53 or a Class B explosive as defined in 49 CFR 173.88.

The EPA has assigned the Hazardous Waste Number D003 to reactive characteristic waste. The fourth characteristic that could make a waste a hazardous waste is *toxicity* (40 CFR 261.24). To determine if a waste is a toxic hazardous waste, a representative sample of the material must be subjected to a test conducted in a certified laboratory using a test procedure (Toxicity Characteristic Leaching Procedure, TCLP). Under federal rules (40 CFR 261), all generators are required to use the TCLP test when evaluating wastes.

Wastes that fail a toxicity characteristic test are considered hazardous under the RCRA. There is less incentive for a company to attempt to reduce the toxicity of such waste below the toxicity characteristic levels because, even though such toxicity reductions may render the waste non-hazardous, it may still have to comply with new Land Disposal treatment standards under subtitle C of the RCRA before disposal. Similarly, there is little positive incentive to reduce the toxicity of listed hazardous wastes because, once listed, the waste is subject to subtitle C regulations without regard to how much the toxicity levels are reduced.

Besides the four characteristics of hazardous wastes, the EPA has established three hazardous waste lists: (i) hazardous wastes from nonspecific sources, such as spent non-halogenated solvents; (ii) hazardous wastes from specific sources, such as bottom sediment sludge from the treatment of wastewaters from wood preserving); and (iii) discarded commercial chemical products and off-specification species, containers, and spill residues. However, under regulations of the EPA, certain types of solid wastes (e.g., household waste) are not considered to be hazardous wastes irrespective of their characteristics. Additionally, the EPA has provided certain regulatory exemptions based on specific criteria. For example, hazardous waste generated in a product or raw material storage tank, transport vehicle, or manufacturing processes and samples collected for monitoring and testing purposes are exempt from the regulations.

Finally, in terms of waste classification, the EPA has also designated certain wastes as incompatible wastes, which are hazardous wastes that, if placed together, could result in potentially dangerous consequences. As defined at 40 CFR 260.10, an incompatible waste is a hazardous waste which is unsuitable for: (i) placement in a particular device or facility because it may cause corrosion or decay of containment materials, such as container inner liners or tank walls; or (ii) comingling with another waste or material under uncontrolled conditions because the commingling might produce heat or pressure, fire or explosion, violent reaction, toxic dusts, mists, fumes, or gases, or flammable fumes or gases.

Once the physical and chemical properties of a hazardous waste have been adequately characterized, hazardous waste compatibility charts can be consulted to identify other types of wastes with which it is potentially incompatible. For example, Appendix V to both 40 CFR 264 and 265 presents examples of potentially incompatible wastes and the potential consequences of their mixture.

Under the Resource Conservation Recovery Act, the hazardous waste management program is based on a cradle-to-grave concept so that all hazardous wastes can be traced and fully accounted for. Section 3010(a) of the Act requires all generators and transporters of hazardous wastes as well as owners and operators of all treatment, storage, and disposal facilities (commonly known as TSD facilities) to file a notification with the EPA within 90 days after the promulgation of the regulations. The notification should state the location of the facility and include a general description of the activities as well as the identified and listed hazardous wastes being handled.

Submission of the Part A permit application for existing facilities prior to November 19, 1980, qualified a refinery for interim status. This meant that the refinery was allowed to continue operation according to certain regulations during the permitting process. The HSWA represented a strong bias against land disposal of hazardous waste. Some of the provisions that affect the companies (especially refineries) involved in the production of chemicals are:

- A ban on the disposal of bulk or non-containerized liquids in landfills. The prohibition also bans solidification of liquids using absorbent material including absorbents used for spill cleanup.
- Five hazardous wastes from specific sources come under scheduled for disposal prohibition and/or treatment standards. These five are: dissolved air flotation (DAF) float, slop oil emulsion solids, heat exchanger bundle cleaning sludge. API separator sludge, and leaded tank bottoms (EPA waste numbers: K047–K051).
- Producers of chemicals must retrofit surface impoundments that are used for hazardous waste management. Retrofitting must involve the use of double liners and leak detection systems.

Under the Resource Conservation Recovery Act, the EPA has the authority to require a company to clean up releases of hazardous waste or waste constituents. The regulation provides for the cleanup of hazardous waste released from active treatment, storage, and disposal facilities. Superfund was expected to handle contamination that had occurred before that date.

10.3.7 SAFE DRINKING WATER ACT

In terms of drinking water, the term contaminant means any physical, chemical, biological, or radiological chemical or matter in water which is, in actual fact, broadly defined as any molecular species other than water molecules. Drinking water may reasonably be expected to contain at least small amounts of some contaminants, but some drinking water contaminants may be harmful if consumed at certain levels in drinking water while others may be harmless.

These are four general categories of drinking water contaminants and examples of each: (i) physical contaminants; (ii) chemical contaminants; (iii) radiological contaminants; and (iv) biological contaminants.

Physical contaminants primarily impact the physical appearance or other physical properties of water. Examples of physical contaminants are sediment or material suspended in the water of lakes, rivers, and streams from soil erosion. Chemical contaminants are elements or compounds and are often of anthropogenic origin-examples of chemical contaminants include nitrate ($-NO_3$) derivatives, bleach (NaOCl), salts, pesticides, and metals. Radiological contaminants are chemical elements with an unbalanced number of protons and neutrons resulting in unstable atoms that can emit ionizing radiation and examples of radiological contaminants include cesium (Cs), plutonium (Pu), and uranium (U). Finally, although out of the current context of this text but still worthy of mention because of possible interactions with the other three contaminants, biological contaminants are organisms in the water and are also referred to as microbes or microbiological contaminants. Examples of biological or microbial contaminants include bacteria, viruses, parasites, and protozoans. The latter organisms (protozoans) are single-celled eukaryotes (organisms whose cells have nuclei) that commonly show characteristics usually associated with animals, most notably mobility and ate heterotrophic insofar as the organism that cannot fix carbon from sources (such as carbon dioxide) but uses organic carbon for sustenance and growth.

The SDWA was enacted in 1974 to assure high-quality water supplies through public water system. In the United States, it is the federal law that protects public drinking water supplies throughout the nation. Under the SDWA, the EPA is authorized to set the standards for drinking water quality and with its partners implements various technical and financial programs to ensure drinking water safety.

The act is truly the first federal intervention to set the limits of contaminants in drinking water. The 1986 amendments came 2 years after the sage of the HSWA or the so-called Resource Conservation Recovery Act amendments of 1984. As a result, certain statutory provisions were added to these 1986 amendments to reflect the changes made in the underground injection control (UIC) systems. In addition, the SARA of 1986 set the groundwater standards the same as the drinking water standards for the purpose of necessary cleanup and remediation of an inactive hazardous waste disposal site.

The law was amended in 1986 and 1996 and requires many actions to protect drinking water and its sources—rivers, lakes, reservoirs, springs, and groundwater wells. (the SDWA does not regulate private wells which serve fewer than 25 individuals.). For example, the 1986 amendments of SDWA included additional elements to establish maximum contaminant level goals (MCLGs) and national primary drinking water standards. The MCLGs must be set at a level at which no known or anticipated adverse effects on human health occur, thus providing an adequate margin of safety. Establishment of a specific MCLG depends on the evidence of carcinogenicity in drinking water or a reference dose that is individually calculated for each specific contaminant. The MCLGs, an enforceable standard, however, must be set to operate as the nation primary drinking water standard (NPDWS). The 1996 amendments greatly enhanced the existing law by recognizing source water protection, operator training, funding for water system improvements, and public information as important components of safe drinking water. This approach ensures the quality of drinking water by protecting it from source to tap.

The SDWA calls for regulations that: (i) apply to public water systems; (ii) specify contaminants that may have an adverse effect on the health of persons; and (iii) specify contaminant levels. The difference between primary and secondary drinking water regulations are defined, as well as other applicable terms. Information concerning national drinking water regulations and protection of underground sources of drinking water is given.

In the context of the chemicals industry, the priority list of drinking water contaminants is important since it includes the contaminants known for their adverse effect on public health. Furthermore, most if not all are known or suspected to have hazardous or toxic characteristics that can compromise human health.

10.3.8 TOXIC SUBSTANCES CONTROL ACT

A toxic substance is (in the current context since there are toxic substances other than chemicals) any chemical that may cause harm to an individual (floral or faunal) organism if it enters the organism. Toxic chemicals may enter the organism in different ways referred to as routes of exposure (Table 10.4). A chemical can be toxic, or hazardous, or both. In fact, any chemical can be toxic or harmful under certain conditions. Some chemicals are hazardous because of their physical properties: they can explode, burn or react easily with other chemicals. Some chemicals are more toxic than others. The toxicity of a chemical is described by the types of effects it causes and its potency (Table 10.4).

The degree of hazard associated with any toxic chemical is related to: (i) the individual chemical; (ii) concentration of the material; (iii) the route into the organism; and (iv) the amount absorbed by the organism-the dose. Individual susceptibility of the organism also plays a role. Once a chemical enter an organism, the effects may occur immediately, or the effects may be delayed-acute effects or chronic effects (Table 10.4). Generally, acute effects are caused by a single, relatively high exposure while chronic effects tend to occur over a longer period of time and involve lower exposures (e.g., exposure to a smaller amount over time). Some toxic chemicals can have both acute and chronic health effects.

TABLE 10.4 General Terminology Related to Toxic Chemicals

Types of Effect*	Comments
Exposure	A chemical can cause health effects only when it contacts or enters the body.
Exposure medium	Exposure occurs when exposed to any medium (air, water, or land) that contains chemicals.
Length of exposure	Acute (short-term) exposure or chronic (long-term exposure) may cause serious effects.
Potency	A measure of the toxicity of a chemical.
Routes of exposure	Inhalation, ingestion, or direct contact.

*Depending on the chemical, the effect may be quick-acting or slow acting.

The Toxic Substances Control Act (TSCA), enacted in 1976, was designed to understand the use or development of chemicals and to provide controls, if necessary, for those chemicals that may threaten human health or the environment (Ingle, 1983; Sittig, 1991). The act provides the EPA with authority to require reporting, record-keeping, and testing requirements, and restrictions relating to chemical chemicals and/or mixtures thereof.

This act has probably had more effect on the producers of chemicals, and the reining industry, than any other Act. It has caused many changes in the industry and may even create further modifications in the future. The basic purpose of the act is (i) to develop data on the effects of chemicals on the environment and on human health, (ii) to grant authority to the EPA to regulate chemicals presenting an unreasonable risk, and (3) to assure that this authority is exercised so as not to impede technological innovation. In short, the act calls for regulation of chemical substances and chemical mixtures that present an unreasonable risk or injury to health or the environment. Furthermore, the introduction and evolution of this act has led to a central bank of information on existing commercial chemical substances and chemical mixtures, procedures for further testing of hazardous chemicals, and detailed permit requirements for submission of proposed new commercial chemical substances and chemical mixtures.

As used in the act, the term chemical substance means any chemical of a particular molecular identity, including any combination of such chemicals occurring in whole or in part as a result of a chemical reaction or occurring in nature, and any element or uncombined radical. Items not considered chemical substances are listed in the definition section of the act. The term mixture means any combination of two or more chemical substances if the combination does not occur in nature and is not, in whole or in part, the result of a chemical reaction; except that such term does include any combination which occurs, in whole or in part, as a result of a chemical reaction if none of the chemical substances comprising the combination is a new chemical substance and if the combination could have been manufactured for commercial purposes without a chemical reaction at the time the chemical substances comprising the combination were combined.

For many, familiarity with the TSCA generally stems from the specific reference to polychlorinated biphenyl derivatives, which raise a vivid, deadly characterization of the harm caused by them. But the act is not a statute that deals with a single chemical or chemical mixture or product. In fact, under the TSCA, the EPA is authorized to institute testing programs for various chemical substances that may enter the environment. Under the broad authorization of the TSCA, data on the production and use of various chemical substances and mixtures may be obtained to protect public health and the environment from the effects of harmful chemicals. In actuality, the act supplements the appropriate sections dealing with toxic chemicals in other federal statutes such as the CWA (Section 307) and the Occupational Safety and Health Act (Section 6).

At the heart of the TSCA is a premanufacture notification (PMN) requirement under which a manufacturer must notify the EPA at least 90 days prior to the production of a new chemical. In this context, a new chemical is a chemical that is not listed in the act-based inventory of chemical substances or is an unlisted reaction product of two or more chemicals. For chemicals already on this list, a notification is required if there is a new use that could significantly increase human or environmental exposure. No notification is required for chemicals that are manufactured in small quantities solely for scientific research and experimentation.

The Chemical Substances Inventory of the TSCA is a comprehensive list of the names of all existing chemical substances and currently contains over 70,000 existing chemicals. Information in the inventory is updated every four years. A facility must submit a premanufacture notice (PMN) prior to manufacturing or importation for any chemical substances not on the list and not excluded by the act. Examples of regulated chemicals include lubricants, paints, inks, fuels, plastics, and solvents.

10.4 OUTLOOK

The chemicals production industry will increasingly feel the effects of the land bans on their hazardous waste management practices. Current practices of land disposal must change along with management attitudes for waste handling. The way companies handle waste products in the future depends largely on the ever-changing regulations. Waste management is the focus and reuse/recycle options must be explored to maintain a balanced waste management program. This requires that waste be recognized as either non-hazardous or hazardous.

A good deal is already known related to the effects of chemical substances on man during their production, transport, and use, from the study of occupational medicine, toxicology, pharmacology, etc., and to a lesser extent their effects on resource-organisms, from veterinary science and plant pathology. Much less is known related to the effects of a chemical upon wildlife species, following its disposal in the environment. This last area of knowledge needs improving because ecological cycles and food chains may deliver the potentially hazardous chemical from affected wildlife back to humans. Moreover, wildlife can often be used as indicators of environmental states and trends for a potentially harmful chemical, giving an early warning of future risks to man. We are also frequently ignorant of how far wildlife may be supportive to human well-being: as a food-base for an important resource-species, e.g., fisheries or grazing animals; as a key species maintaining the stability of economically valuable ecosystems; as predators of crop or livestock pests; as a species involved in mineral recycling or biodegradation; as an important amenity.

Failure to recognize the mutually interactive roles of man, resource-species, wildlife organisms and climate in the biosphere and their different tolerances to chemical chemicals has hindered the development of a unitary environmental management policy embracing all four biosphere components. Although a good deal is already known related to the influence of molecular structure on the toxicity to human beings of drugs and certain other chemicals, much less is known related to the influence of molecular structure on the environmental persistence of a chemical. For wildlife, persistence is probably the most important criterion for predicting potential harm because there is inevitably some wild species or other which is sensitive to any compound and any persistent chemical, apparently harmless to a limited number of toxicity-test organisms, will eventually be delivered by biogeochemical cycles to a sensitive target-species in nature. This means that highly toxic, readily biodegradable chemicals may pose much less of an environmental problem, than a relatively harmless persistent chemical which may well damage a critical wild species. The study of chemical effects in the environment resolves itself into a study of (a) the levels of a chemical accumulating in air, water, soils (including sediments) and biota (including man), and (b) when the threshold action-level has been reached, effects produced in biota which constitute a significant adverse response (i.e., environmental dose-response curve). In order to predict trends in levels of a chemical, much more information is needed related to the rates of injection and flow as well as the partitioning between air, water, soils, and biota and the loss by means of degradation (environmental balance-sheets).

These dynamic phenomena are governed by the physicochemical properties of the molecule. Fluid mechanics and meteorology may in the future provide the conceptual and technical tools for producing predictive models of such systems. Most of the knowledge of effects derives from the accumulation of data relating to acute exposure (short contact with a chemical that may last a few seconds or a few hours) and medical studies on humans, but since environmental effects are usually associated with chronic exposure (continuous or repetitive exposure to a chemical), studies are being increasingly made of long-term continuous exposure to minute amounts of a chemical. The well-known difficulty of recognizing such effects when they occur in the field is aggravated by the fact that many of the effects are non-specific and are frequently swamped by similar effects deriving from exposure to such natural phenomena as famines, droughts, cold spells, etc. Even when a genuine effect is recognized, a candidate causal agent must be found and correlated with it. This process must be followed by experimental studies, unequivocally linking chemical cause and adverse biological effect. All three stages are difficult and costly, and it is not surprising that long delays are often experienced between the recognition of a significant adverse effect and a generally agreed chemical cause. There is often ample uncertainty to allow under-reaction as well as over-reaction to potential hazards, both are supported by scientific evidence.

In the mid-twentieth century, solid waste management issues rose to new heights of public concern in many areas of the United States because of increasing solid waste generation, shrinking disposal capacity, rising disposal costs, and public opposition to the siting of new disposal facilities. These solid waste management challenges continue as many communities are struggling to develop cost-effective, environmentally protective solutions. The growing amount of waste generated has made it increasingly important for solid waste management officials to develop strategies to manage wastes safely and cost-effectively.

10.4.1 HAZARDOUS WASTE REGULATIONS

By way of introduction, a hazardous waste is a waste with properties that make it dangerous or capable of having a harmful effect on floral and faunal species (including human species) or the environment. Hazardous waste is generated from many sources, ranging from industrial manufacturing process wastes to batteries and may come in many forms, including liquids, solids, gases, and sludge. The hazardous waste regulatory program, as is currently practiced, began with the RCRA in 1976. The Used Oil Recycling Act of 1980 and Hazardous and Solid Waste Amendments of 1984 (HSWA) were the major amendments to the original law.

The Resource Conservation Recovery Act provides for the tracking of hazardous waste from the time it is generated, through storage and transportation, to the treatment or disposal sites. The Act and the various amendments are aimed at preventing the disposal problems that lead to a need for the Comprehensive Environmental Response Compensation and Liability Act (CERCLA), or Superfund, as it is known. Subtitle C of the original Resource Conservation Recovery Act lists the requirements for the management of hazardous waste. This includes the EPA criteria for identifying hazardous waste, and the standards for generators, transporters, and companies that treat, store, or dispose of the waste. The Resource Conservation Recovery Act regulations also provide standards for design and operation of such facilities.

New regulations are becoming even more stringent, and they encompass a broader range of chemical constituents and processes. Continued pressure from the U. S. Congress has led to more explicit laws allowing little leeway for industry, the U. S. Environmental Protection Agency (EPA), or state agencies. A summary of the current regulations and what they mean is given in the following.

10.4.2 REQUIREMENTS

The first step to being taken by a generator of waste is to determine whether that waste is hazardous. Waste may be hazardous by being listed in the regulations, or by meeting any of the four characteristics: ignitability, corrosivity, reactivity, and extraction procedure (EP) toxicity.

Generally: (i) if the material has a flash point less than 140°F it is considered ignitable, (ii) if the waste has a pH less than 2.0 or above 12.5, it is considered corrosive-it may also be considered corrosive if it corrodes stainless steel at a certain rate, (iii) a waste is considered reactive if it is unstable and produces toxic materials, or it is a cyanide or sulfide-bearing waste which generates toxic gases or fumes, (iv) a waste which is analyzed for EP toxicity and fails is also considered a hazardous waste. This procedure subjects a sample of the waste to an acidic environment. After an appropriate time has elapsed, the liquid portion of the sample (or the sample itself if the waste is liquid) is analyzed for certain metals and pesticides. Limits for allowable concentrations are given in the regulations. The specific analytical parameters and procedures for these tests are referred to in 40CRF 261.

The 1984 amendments also brought the owners and operators of underground storage tanks under the umbrella of the Resource Conservation Recovery Act. This can have a significant effect on chemicals production companies (such as refineries) that store products in underground tanks. In addition, chemicals are also regulated by the Resource Conservation Recovery Act, Subtitle I.

Finally, operators of chemicals production companies face stringent regulation of the treatment, storage, and disposal of hazardous wastes. Under recent regulations, a larger number of compounds have been, and are being, studied, and long-time methods of disposal, such as land farming of chemical waste, are being phased out. Thus, many companies are changing their waste management practices.

10.5 GLOBAL ENVIRONMENTAL CHANGE

The environment is in a state of continual change, and not only as a result of modern activities. In fact, ever since life first appeared, the atmosphere has been influenced by the metabolic processes of living organisms. Of which humans have probably caused the greatest changes even though the human tenure on the Earth has been the shortest relative to the terms of other (floral and faunal) species.

When the first primitive life molecules were formed approximately 3.5 billion (3.5 × 109) years ago, the atmosphere was different to the present state of the atmosphere. At that time, the atmosphere is considered to consist primarily of methane, ammonia, water vapor, and hydrogen. These simple molecular species were bombarded by ultraviolet (UV) radiation, which (along with, for example, lightning from the prevalent atmospheric disturbances) provided the energy to cause the chemical reactions that result in the production of relatively complicated molecules, including even amino acids and sugars.

$$CH_4 + NH_3 + H_2O + H_2 + \text{energy} \rightarrow CH_3CH(NH_2)COOH + CH_2OH(CHOH)_nCH_2OH$$

From this rich chemical mixture (a veritable primordial soup or mess of pottage), it is postulated that life molecules evolved. Although, there are alternate theories that need to be given consideration (Britannica, 1969; Christian Bible, 2020).

Initially, these very primitive life forms derived their energy from fermentation of organic matter formed by chemical and photochemical processes, but eventually they gained the capability to produce organic matter by photosynthesis and the stage was set for the massive biochemical transformation that resulted in the production of the majority of the oxygen in the atmosphere.

It is worthy of note that, because of the anaerobic conditions of the primitive atmosphere, the oxygen initially produced by photosynthesis was probably quite toxic to the prevailing life forms. But, eventually, enzyme systems developed that enabled organisms to survive in an oxygen-rich atmosphere and to mediate the reaction of waste-product oxygen with oxidizable organic matter. Later, this mode of waste-product disposal evolved as a means of producing energy for respiration, which is now the mechanism by which non-photosynthetic organisms obtain energy.

As oxygen accumulated in the atmosphere there was the (now widely accepted) additional benefit of the formation of ozone which became a shield against harmful solar UV radiation. With this shield in place, life evolved on the Earth in the form that we now know it insofar as the current environment is much more hospitable to the current flora and fauna (including humans).

There is always the wondering question-related nature of the life forms that would have (or may have) evolved without the protection of the ozone layer. Would the equivalent human form have had skin like a dinosaur? Or perhaps human skin may have, at least, the appearance of the skin of a dried prune, as some continual sunbathers exhibit as they reach the grand old age of 40. Nevertheless, life forms as we now know them were able to venture from the protective surroundings of the sea to the more exposed environment of the land. And these life forms were able to survive.

However, these changes to the primitive atmosphere are not usually considered to be environmentally harmful, even though extinct organisms as yet unknown and yet to be discovered may have suffered as a result of the change. The goal of humankind is to protect the environment as we know it. And yet, at an ever-accelerating pace during the last 200 years, humankind has engaged in a number of activities that are altering the state of the atmosphere.

There are the industrial activities, which emit a variety of atmospheric contaminants, including sulfur dioxide, PM, photochemically reactive hydrocarbon derivatives, chlorofluorocarbon derivatives, and chemicals (such as those that contain toxic heavy metals). These activities include the use (combustion) of fossil fuels which can introduce carbon dioxide, carbon monoxide, sulfur dioxide, nitrogen oxides, hydrocarbon derivatives (including methane), particulate soot, polynuclear aromatic hydrocarbon derivatives, and fly ash into the atmosphere.

Man, a bipedal animal, has learned that self-transportation is slow, ponderous, and energy-consuming. Therefore, in the interests of energy conservation (his or her own, that is), man has developed a variety of mechanical transportation practices hasten his or her movement from one location to another location. However, these practices result in the release of carbon dioxide, carbon monoxide, nitrogen oxides, photochemically reactive (smog-forming) hydrocarbon derivatives (including methane), and polynuclear aromatic hydrocarbon derivatives.

Humans are also adept at altering the surrounding environment by alteration of land surfaces, including deforestation. As part of this alteration of land surfaces, the burning of vegetation, including tropical and subtropical forests and savanna grasses, produces atmospheric carbon dioxide, carbon monoxide, nitrogen oxides, particulate soot and polynuclear aromatic hydrocarbon derivatives. In addition, agricultural practices produce methane (from the digestive tracts of domestic animals and from the cultivation of rice in waterlogged anaerobic soils) and dinitrogen oxide from bacterial denitrification of nitrate-fertilized soils.

These, and related kinds of human activities that are too numerous to note here, have significantly altered the atmosphere, particularly in regard to its composition of minor constituents and trace gases. The major effects have been: (i) an increased acidity in the atmosphere which can result in the deposition of acid rain; (ii) there is also the production of pollutant oxidants in localized areas of the lower troposphere; and (iii) the elevated levels of infrared-absorbing gases (greenhouse gases) that can cause threats to the UV-filtering ozone layer in the stratosphere (Chapters 3 and 8).

Of all environmental hazards, there is little doubt that major disruptions in the atmosphere and climate have the greatest potential for environmental damage. If levels of greenhouse gases and reactive trace gases continue to increase at the projected rates (assuming that the projections are accurate!), potential environmental harm can become real. But the bulk of these emissions is being curtailed by the environmentally-conscious industrialized nations and there are serious efforts to substantially reduce harmful emissions.

It is important to keep in mind that the atmosphere has a strong ability to cleanse itself of pollutant species. Water-soluble gases, including greenhouse gases, acid gases, and fine PM are removed with precipitation. For most gaseous contaminants, oxidation precedes or accompanies removal processes.

The measures to be taken in dealing with the emissions issue can be partly overcome by switching to alternate energy sources, increasing energy conservation, and reversing deforestation. It is especially sensible to use measures that have major effects in addition to reduction of greenhouse warming. However, such a shift is not the complete answer. Indeed, shifting to nuclear-based energy sources just to prevent possible greenhouse warming may give rise to a host of other environmental disadvantages (Pickering and Owen, 1994).

Definite economic and political benefits would also accrue from lessened dependence on uncertain, volatile crude oil supplies. Increased energy efficiency would diminish both greenhouse gas and acid rain production while lowering costs of production and reducing the need for expensive and environmentally disruptive new power plants.

Perhaps the most pertinent response at this time is the introduction of the relevant legislation. However, too much legislation will only be confusing and there must be a balance between the extent of the legislation and the evolution of industrial practices. Old and outdated laws must be taken to task.

As a result, the concept of what seemed to be a good idea at the time no longer holds when the law influences the environment.

10.6 EPILOG

Environmental issues are numerous and often overlap. The macro-issues such as acid rain and global warming involve more than one country while the micro-issues occur on a more localized scale but still need attention. Whether or not the environment was ever pristine is another issue. And, it must be recognized, by those advocating a pristine environment, that some disturbance of the environment will occur if meaningful progress is to be made. On the other hand, the environment must not be raped! Throughout all of this, the major issue is control of the emotionalism that is apt to be the driving force behind many environmental movements. Logic should be the driver.

Perception can also give rise to a powerful form of emotionalism because it can completely alter the real situation. And so, in the environmental arena, we need the true story. The issue is whether or not floral and faunal species really becoming extinct by the dozens, or are they becoming more able to survive in other locales to avoid or to adapt to the presence and effects of humankind.

Darwinism is essentially the survival of the fittest whereby adaptation to change is genetic in origin. The creation of variations in a species that are beneficial to the survival of that species are preserved. Lamarckism is the adaptation that occurs during the lifetime of an organism and which is beneficial to the organism. The Lamarckian change is not usually genetic in character. The ability of the human species to adapt to the environment and the ability of the environment to adapt to the human species is perhaps a combination of both Darwinism and Lamarckism. It is to be hoped that both will be in play during the next decades for the protection of both the human species and the environment.

The formation of various types of emissions (gaseous, liquid, and solids) must be thoroughly understood in order that they may be controlled and even mitigated. However, tolerance is necessary on both sides of the clean environment equation. Industry has taken action to ensure protection of the environment. Although a notable start has been made, further steps are still required. The environmentalists must realize that no one (including themselves) wishes to freeze in the green darkness. A balanced equation is necessary and both sides can make valuable contributions to the future.

There is the realization that human development must move forward with some consideration for its effects on the flora and fauna of the region. Without such consideration, the environment may be doomed. To be followed by human lifestyles as we know them or as we would prefer to have them. Government, industry, and domestic consumers in various countries realize the need for protecting the environment. Like the ceiling of the Sistine that was painted by Michelangelo, the task may never be complete. And it is necessary to be ever-watchful and the overall picture is maintained but not spoiled. Just as it would be an error to focus on one small part of the picture, it would also be an error to focus on one small part of the environment. Also, as many countries criticize the United States for not being a signatory to the Kyoto Accords, some of the countries that have been a signatory to these accords must also (or, at least, take the initiative) clean up their own environment before they venture to criticize any non-signatory countries. It would be a tragedy for the environment to be polluted to such an extent that human existence became impossible. And it would be an equal tragedy if human life ceased when people froze to death in the green darkness. But policies to protect the environment should be based on hard scientific fact and not on rampant emotionalism.

A balance is needed between what should be and what can be. A balance must be struck between human development and the necessary environmental considerations. To swing the pendulum too far to any one side could be a serious error.

KEYWORDS

- **environmental impact assessment**
- **Environmental Protection Agency**
- **flue gas desulfurization**
- **hazardous air pollutants**
- **maximum achievable control technology**
- **National Environmental Policy Act**

REFERENCES

Bahadori, A., (2014). *Natural Gas Processing: Technology and Engineering Design*. Gulf Professional Publishing, Elsevier, Amsterdam, Netherlands.

Britannica, (1969). *Creation, Myths of Encyclopedia Britannica*. William Benton, Chicago.

Christian Bible, (2020). *Genesis*. Chapters 1 and 2.

Dennison, M. S., (1993). *RCRA Regulatory Compliance Guide*. Noyes Data Corp., Park Ridge, New Jersey.

Ingle, G. W., (1983). *TSCA's Impact on Society and the Chemical Industry*. ACS Symposium Series, American Chemical Society, Washington, DC.

Kidnay, A. J., Parrish, W. R., & McCartney, D. G., (2011). *Fundamentals of Natural Gas Processing* (2nd edn.). CRC Press, Taylor & Francis Group, Boca Raton, Florida.

Lazarus, R., (2004). *The Making of Environmental Law*. Cambridge University Press, Cambridge, United Kingdom.

Lunn, G., & Sanstone, E. B., (1994). *Destruction of Hazardous Chemicals in the Laboratory* (2nd edn.). McGraw-Hill.

Majumdar, S. B., (1993). *Regulatory Requirements for Hazardous Materials*. McGraw-Hill, New York.

Mokhatab, S., Poe, W. A., & Speight, J. G., (2006). *Handbook of Natural Gas Transmission and Processing*. Elsevier, Amsterdam, Netherlands.

Nordin, J. S., Sheesley, D. C., King, S. B., & Routh, T. K., (1995). *Environmental Solutions, 8*(4), 49.

Noyes, R., (1993). *Pollution Prevention Technology Handbook*. Noyes Data Corp, Park Ridge, New Jersey.

Pickering, K. T., & Owen, L. A., (1994). *Global Environmental Issues*. Routledge Publishers Inc., Taylor & Francis Group, New York.

Sittig, M., (1991). *Handbook of Toxic and Hazardous Chemicals and Carcinogens* (3rd edn.). Noyes Data Corp., Park Ridge, New Jersey.

Speight, (2013). *The Chemistry and Technology of Coal* (3rd edn.). CRC Press, Taylor & Francis Group, Boca Raton, Florida.

Speight, J. G., (1996). *Environmental Technology Handbook*. Taylor & Francis, Washington, DC.

Speight, J. G., (2014). *The Chemistry and Technology of Petroleum* (5th edn.). CRC Press, Taylor & Francis Group, Boca Raton, Florida.

Speight, J. G., (2017). *Handbook of Petroleum Refining*. CRC Press, Taylor & Francis Group, Boca Raton, Florida.

Speight, J. G., (2019). *Natural Gas: A Basic Handbook* (2nd edn.). Gulf Publishing Company, Elsevier, Cambridge, Massachusetts.

Wang, C. C. K., (1994). *OSHA Compliance and Management Handbook*. Noyes Data Corp., Park Ridge, New Jersey.

Glossary

Abiotic – Nonliving factors or physical factors.

Abiotic Factors – Influences such as light radiation (from the Sun), ionizing radiation (cosmic rays from outer space), temperature (local and regional variations), water (seasonal and regional distributions), atmospheric gases, wind, soil (texture and composition), and catastrophic disturbances.

Acetic Acid – An acid with the structure of CH_3COOH; acetyl groups are bound through an ester linkage to hemicellulose chains, especially xylan derivatives, in wood and other plants; the natural moisture present in plants hydrolyzes the acetyl groups to acetic acid, particularly at elevated temperatures.

Acid – Any of a class of substances whose aqueous solutions are characterized by a sour taste, the ability to turn blue litmus red, and the ability to react with bases and certain metals to form salts; a substance that yields hydrogen ions when dissolved in water and which can act as a proton (H^+) donor.

Acid Deposition – acid rain; also a form of pollution depletion in which pollutants are transferred from the atmosphere to soil or water which is often referred to as atmospheric self-cleaning.

Acid Detergent Fiber (ADF) – Organic matter that is not solubilized after 1 hour of refluxing in an acid detergent of cetyltrimethylammonium bromide in 1N sulfuric acid; includes cellulose and lignin; this analytical method is commonly used in the feed and fiber industries.

Acid Hydrolysis – A chemical process in which acid is used to convert cellulose or starch to sugar.

Acid Insoluble Lignin – Lignin is mostly insoluble in mineral acids, and therefore can be analyzed gravimetrically after hydrolyzing the cellulose and hemicellulose fractions of the biomass with sulfuric acid; standard test method ASTM E1721 describes the standard method for determining acid-insoluble lignin in biomass; see ASTM International.

Acid Mine Water – water which occurs in used/disused mines and an appreciable concentration of free mineral acid.

Acid Rain – the precipitation phenomena that incorporates anthropogenic acids and other acidic chemicals from the atmosphere to the land and water (see "Acid deposition").

Acid Soluble Lignin – A small fraction of the lignin in a biomass sample is solubilized during the hydrolysis process of the acid-insoluble lignin method. This lignin fraction is referred to as acid-soluble lignin and may be quantified by ultraviolet spectroscopy; see Lignin and Acid Insoluble Lignin.

Acidity – The capacity of an acid to neutralize a base such as a hydroxyl ion (OH-).

Actinomycetes – microorganisms that are morphologically similar to both bacteria and fungi.

Activated Sludge – the biologically active sediment produced by the repeated aeration and settling of sewage and/or organic wastes.

Activated Sludge Process – production of a biologically active sludge which is usually brown in color and which destroys the polluting organic matter in sewage and waste.

Add-on Control Methods – the use of devices that remove process emissions after they are generated but before they are discharged to the atmosphere.

Adsorption – The transfer of a substance from a solution to the surface of a solid. Aerobic: in the presence of air.

Aerosol – a colloidal system in which a gas, frequently air, is the continuous medium and particles of solids or liquids are dispersed in it. Usually the particles are less than 100 microns in size; see Micron.

Agitator – A device such as a stirrer that provides complete mixing and uniform dispersion of all components in a mixture; are generally used continuously during the thermal processes and intermittently during fermentation.

Agricultural Residue – Agricultural crop residues are the plant parts, primarily stalks and leaves, not removed from the fields with the primary food or fiber product; examples include corn stover (stalks, leaves, husks, and cobs); wheat straw; and rice straw.

Agricultural Waste – Waste produced at agricultural premises as a result of an agricultural activity.

Air Quality Maintenance Area – Specific populated area where air quality is a problem for one or more pollutants.

Alcohol – The family name of a group of organic chemical compounds composed of carbon, hydrogen, and oxygen. The molecules in the series vary in chain length and are composed of a hydrocarbon plus a hydroxyl group. Alcohol includes methanol and ethanol.

Aldoses – Occur when the carbonyl group of a monosaccharide is an aldehyde.

Algae – Algae are primitive plants, usually aquatic, capable of synthesizing their own food by photosynthesis; currently being investigated as a possible feedstock for producing biodiesel.

Aliphatic – Any non-aromatic organic compound having an open-chain structure.

Alkali – A soluble mineral salt.

Alkali Lignin – Lignin obtained by acidification of an alkaline extract of wood.

Alkalinity – the capacity of a base to neutralize the hydrogen ion ($H+$).

Alkylation – a process for manufacturing high octane blending components used in unleaded petrol or gasoline.

Alluvium – Soil that has been transported by the movement of water; deposits of stream-borne sediments.

Alternative Fuel – As defined in the United States Energy Policy Act of 1992 (EPACT): methanol, denatured ethanol and other alcohols, separately or in blends of at least 10% by volume with gasoline or other fuels; compressed natural gas; liquefied natural gas (LNG), liquefied propane gas, hydrogen, coal-derived liquid fuels, fuels other than alcohols derived from biological materials, electricity, biodiesel, and any other fuel deemed to be substantially not petroleum and yielding potential energy security benefits and substantial environmental benefits.

Ambient Air Quality – The condition of the air in the surrounding environment.

American Society for Testing and Materials (ASTM) – Now known as ASTM International; an international voluntary standards organization that develops and produces technical standards for materials, products, systems, and services.

Amine Washing – a method of gas cleaning whereby acidic impurities such as hydrogen sulfide and carbon dioxide are removed from the gas stream by washing with an amine (usually an alkanolamine).

Anabolic Pathways – See Catabolic pathways.

Anaerobic – Biological processes that occur in the absence of oxygen – (In the absence of air).

Anaerobic Digestion – Decomposition of biological wastes by micro-organisms, usually under wet conditions, in the absence of air (oxygen), to produce a gas comprising mostly methane and carbon dioxide.

Anhydrous – without water; transesterification of biodiesel must be an anhydrous process; water in the vegetable oil causes either no reaction or cloudy biodiesel, and water in lye or methanol renders it less useful or even useless, depending on how much water is present.

Annual Removals – The net volume of growing stock trees removed from the inventory during a specified year by harvesting, cultural operations such as timber stand improvement, or land clearing.

Anthropogenic Acids – those acids which are the result of human activities. Anthropogenic: made by human activities.

Anthropogenic Stress – the effect of human activity on other organisms. Aquasphere: water systems.

API Gravity – a measure of the lightness or heaviness of petroleum that is related to density and specific gravity: °API = (141.5/sp gr @ 60°F) – 131.5.

Aquasphere – Water in various forms; also known as the hydrosphere.

Aquatic Plants – The wide variety of aquatic biomass resources, such as algae, giant kelp, other seaweed, and water hyacinth; certain microalgae can produce hydrogen and oxygen while others manufacture hydrocarbons and a host of other products; microalgae examples include Chlorella, Dunaliella, and Euglena.

Aquiclude – A rock formation that is too impermeable or unfractured to yield groundwater.

Aquifer – A subsurface zone that yields economically important amounts of water to wells.

Arabinan – The polymer of arabinose; can be hydrolyzed to arabinose.

Arabinose – A five-carbon sugar; a product of hydrolysis of arabinan found in the hemicellulose fraction of biomass.

Areic – Regions which lack surface steams because of low rainfall or lithologic conditions.

Aromatics – a range of hydrocarbons which have a distinctive sweet smell and include benzene and toluene that occur naturally in petroleum and are also extracted as a petrochemical feedstock, as well as for use as solvents.

Asbestos – a group of fibrous silicate minerals, typically those of the serpentine group.

Asbestosis – a medical condition caused by inhalation of asbestos fibers into the respiratory system.

Asphalt – A product of petroleum refining used for road construction.

Asphaltene (Asphaltenes, Asphaltene Fraction) – the brown to black powdery material produced by treatment of petroleum, heavy oil, bitumen, or residuum with a low-boiling liquid hydrocarbon.

Asthenosphere (Asthenosphere) – The upper layer of the mantle of the Earth, below the lithosphere, in which there is relatively low resistance to plastic flow and convection is thought to occur.

ASTM International – Formerly American Society for Testing and Materials (ASTM); an international voluntary standards organization that develops and produces technical standards for materials, products, systems, and services.

Atmospheric Aging – The results of the various physical and chemical interactions and transformations of airborne particles.

Atmospheric Pressure – Pressure of the air and atmosphere surrounding us which changes from day to day; equal to 14.7 psia.

Attainment Area – A geographic region where the concentration of a specific air pollutant does not exceed federal standards.

Auger – A rotating, screw-type device that moves material through a cylinder.

Autotrophic Organisms – utilize solar or chemical energy to fix elements from simple, nonliving inorganic materials into complex life molecules that compose living organisms.

Available Production Capacity – The biodiesel production capacity of refining facilities that are not specifically designed to produce biodiesel.

Average Megawatt (MWa or aMW) – One megawatt of capacity produced continuously over a period of one year. Thus: 1 aMW = 1 MW × 8,760 hours/year = 8,760 MWh = 8,760,000 kWh.

B100 – Another name for pure biodiesel.

B Horizon – layer of soil below the A horizon. It receives material such as organic matter, salts, and clay particles leached from the topsoil. Also known as subsoil.

Background Level – The average amount of a substance present in the environment. Originally referring to naturally occurring phenomena; used in toxic substance monitoring.

Backup Electricity – Power or services needed occasionally; for example, when on-site generation equipment fails.

Backup Rate – A utility charge for providing occasional electricity service to replace on-site generation.

Baffle Chamber – In incinerator design, a chamber designed to settle fly ash and coarse particulate matter by changing the direction and reducing the velocity of the combustion gases.

Bagasse – Sugar cane waste.

Baghouse – a filter system for the removal of particulate matter from gas streams; so-called because of the similarity of the filers to coal bags.

Bark – The outer protective layer of a tree outside the cambium comprising the inner bark and the outer bark; the inner bark is a layer of living bark that separates the outer bark from the cambium and in a living tree is generally soft and moist; the outer bark is a layer of dead bark that forms the exterior surface of the tree stem; the outer bark is frequently dry and corky.

Barrel (bbl) – the unit of measure used by the petroleum industry; equivalent to approximately forty-two US gallons or approximately thirty-four (33.6) Imperial gallons or 159 liters; 7.2 barrels are equivalent to 1 ton of oil (metric).

Barrel of Oil Equivalent (boe) – The amount of energy contained in a barrel of crude oil, i.e., approximately 6.1 GJ (5.8 million Btu), equivalent to 1,700 kWh.

Base – A classification of substances which when combined with an acid will form a salt plus water, usually producing hydroxide ions when dissolved.

Baseload Capacity – The power output that generating equipment can continuously produce.

Baseload Demand – The minimum demand experienced by an electric utility, usually 30 – 40% of the utility's peak demand.

Batch Distillation – A process in which the liquid feed is placed in a single container and the entire volume is heated, in contrast to continuous distillation in which the liquid is fed continuously through the still.

Batch Fermentation – Fermentation conducted from start to finish in a single vessel; see Fermentation.

Batch Process – Unit operation where one cycle of feedstock preparation, cooking, fermentation, and distillation is completed before the next cycle is started.

Beer – A general term for all fermented malt beverages flavored with hops; a low level (6 to 12%) alcohol solution derived from the fermentation of mash by microorganisms.

Beer Still – The stripping section of a distillation column for concentrating ethanol.

Benzene – A toxic, six-carbon aromatic component of gasoline; a known carcinogen.

Billion – 1×10^9

Bioaccumulation – The accumulation of a chemical in tissues of an organism to levels greater than in the environment in which the organism lives; a process by which persistent environmental pollution leads to the uptake and accumulation of one or more contaminants, by organisms in an ecosystem.

Biobutanol – Alcohol containing four carbon atoms per molecule, produced from the same feedstocks as ethanol, but with a modified fermentation and distillation process; less water-soluble than ethanol, biobutanol has a higher energy density and can be transported by pipeline more easily.

Biochemical Conversion – The use of fermentation or anaerobic digestion to produce fuels and chemicals from organic sources.

Biochemical Conversion Process – The use of living organisms or their products to convert organic material to fuels, chemicals or other products.

Biochemical Oxygen Demand (BOD) – A standard means of estimating the degree of water pollution, especially of water bodies that receive contamination from sewage and industrial waste; the amount of oxygen needed by bacteria and other microorganisms to decompose organic matter in water-the greater the BOD, the greater the degree of pollution; biochemical oxygen demand is a process that occurs over a period of time and is commonly measured for a five-day period, referred to as BOD5.

Bioconcentration – The amount of a chemical that has accumulated in an organism (including the human organism) through the process of bioaccumulation.

Bioconversion Platform – Typically uses a combination of physical or chemical pretreatment and enzymatic hydrolysis to convert lignocellulose into its component monomers.

Biodegradable – Capable of decomposing rapidly under natural conditions.

Biodegradable Waste – Waste that is capable of being broken down by plants (including fungi) and animals (including worms and microorganisms).

Biodegradation – the conversion of waste materials by biological processes to simple inorganic molecules and, to a certain extent, to biological materials.

Biodiesel – A fuel derived from biological sources that can be used in diesel engines instead of petroleum-derived diesel; through the process of transesterification, the triglycerides in the biologically derived oils are separated from the glycerin, creating a clean-burning, renewable fuel.

Biodiesel Blend – A blend of biodiesel and diesel fuels-the blend can be with Diesel #1, Diesel #2, or JP8; one standard blend that meets the minimum requirements of the federal EPA CAA criteria is B20. The number after "B" indicates the percentage of biodiesel included in the blend-in B20; there would be 20% biodiesel and 80% diesel in the fuel blend; a biodiesel blend can come in any mixture percentage, such as B2, B5, B50, and B85.

Biodiesel Recipe – The most common recipe uses waste vegetable oil (WVO), methanol (wood alcohol), and sodium hydroxide (caustic soda/lye) to produce biodiesel and glycerin; the steps are: (1) cleaning/heating waste vegetable oil, (2) titration of the waste vegetable oil sample, (3) combining methanol and sodium hydroxide in exact amounts, (4) combining (3) with (1) and mixing at 50°C, (5) settling (6) separating the biodiesel from the wastes, (7) washing and drying the biodiesel, (8) disposing of wastes.

Biodiversity (Biological Diversity) – The sum total of all the variety and variability of life in a defined area. See also Ecosystem diversity.

Bioenergy – Useful, renewable energy produced from organic matter-the conversion of the complex carbohydrates in organic matter to energy; organic matter may either be used directly as a fuel, processed into liquids and gasses, or be a residual of processing and conversion.

Bioethanol – Ethanol produced from biomass feedstocks; includes ethanol produced from the fermentation of crops, such as corn, as well as cellulosic ethanol produced from woody plants or grasses.

Biofuels – a generic name for liquid or gaseous fuels that are not derived from petroleum-based fossils fuels or contain a proportion of non-fossil fuel; fuels produced from plants, crops such as sugar beet, rapeseed oil or re-processed vegetable oils or fuels made from gasified biomass; fuels made from renewable biological sources and include ethanol, methanol, and biodiesel; sources include, but are not limited to: corn, soybeans, flax-seed, rapeseed, sugarcane, palm oil, raw sewage, food scraps, animal parts, and rice.

Biofuels – a generic name for liquid or gaseous fuels that are not derived from petroleum

Biogas – A combustible gas derived from decomposing biological waste under anaerobic conditions. Biogas normally consists of 50 to 60% methane. See also landfill gas.

Bioheat – A name sometimes applied to biodiesel when its application is for heating purposes.

Biological Assessment – A specific process required as part of an environmental assessment; an evaluation of potential effects of a proposed project on proposed, endangered, threatened, and sensitive animal and plant species and their habitats.

Biological Oxidation – Decomposition of organic materials by microorganisms.

Biological Waste Treatment – a generic term applied to processes that use microorganisms to decompose organic wastes either into water, carbon dioxide, and simple inorganic substances.

Biomagnification – see Bioconcentration. Biomass: biological organic matter.

Biomass – Any organic matter that is available on a renewable or recurring basis, including agricultural crops and trees, wood and wood residues, plants (including aquatic plants), grasses, animal manure, municipal residues, and other residue materials. Biomass is generally produced in a sustainable manner from water and carbon dioxide by photosynthesis. There are three main categories of biomass-primary, secondary, and tertiary.

Biomass Fuel – Liquid, solid or gaseous fuel produced by conversion of biomass.

Biomass Processing Residues – Byproducts from processing all forms of biomass that have significant energy potential; the residues are typically collected at the point of processing; they can be convenient and relatively inexpensive sources of biomass for energy.

Biomass to Liquid (BTL) – The process of converting biomass to liquid fuels.

Bionaphtha (Bio-Naphtha) – A term used in some eastern European nations for biodiesel.

Biopower – The use of biomass feedstock to produce electric power or heat through direct combustion of the feedstock, through gasification and then combustion of the resultant gas, or through other thermal conversion processes. Power is generated with engines, turbines, fuel cells, or other equipment.

Bioreactor – A vessel in which a chemical process occurs, which usually involves organisms or biochemically active substances derived from such organisms.

Biorefinery – A facility that processes and converts biomass into value-added products. These products can range from biomaterials to fuels such as ethanol or important feedstocks for the production of chemicals and other materials.

Bioremediation – The use of living organisms (primarily microorganisms) to degrade pollutants previously introduced into the environment or to prevent pollution through treatment of waste streams before they enter the environment.

Biosphere – Living organisms and their environments on the surface of the Earth. Biota: living organisms.

Biotic – actions of other organisms.

Biotic Factors – Natural interactions (e.g., predation and parasitism) and anthropogenic stress (e.g., the effect of human activity on other organisms); can also be described as any living component that affects another organism or shapes the ecosystem. This includes both animals that consume other organisms within their ecosystem, and the organism that is being consumed. Biotic factors also include human influence, pathogens, and disease outbreaks. Each biotic factor needs the proper amount of energy and nutrition to function day to day.

Biotransformation – The conversion of a substance through metabolization thereby causing an alteration to the substance by biochemical processes in an organism.

Bitumen – also, on occasion, referred to as native asphalt, and extra heavy oil; a naturally occurring material that has little or no mobility under reservoir conditions and which cannot be recovered through a well by conventional oil well production methods including currently used enhanced recovery techniques; current methods involve mining for bitumen recovery.

Black Liquor – Solution of lignin-residue and the pulping chemicals used to extract lignin during the manufacture of paper.

Bog – A thick zone of vegetation floating on water which lacks a solid foundation. Bottom ash: ash which occurs at the bottom of a (for example, coal) combustor. Bronchogenic carcinoma: cancer originating with the air passages in the lungs.

Boiler – Any device used to burn biomass fuel to heat water for generating steam.

Boiler Horsepower – A measure of the maximum rate of heat energy output of a steam generator; one boiler horsepower equals 33,480 Btu/hr output in steam.

Bone Dry – Having 0% moisture content. Wood heated in an oven at a constant temperature of 100°C (212°F) or above until its weight stabilizes is considered bone dry or oven dry.

Bottoming Cycle – A cogeneration system in which steam is used first for process heat and then for electric power production.

Brewing – Generically, the entire beer-making process, but technically only the part of the process during which the beer wort is cooked in a brew kettle and during which time the hops are added; after brewing the beer is fermented.

British Thermal Unit (Btu) – A non-metric unit of heat, still widely used by engineers; One Btu is the heat energy needed to raise the temperature of one pound of water from 60°F to 61°F at one atmosphere pressure. 1 Btu = 1,055 joules (1.055 kJ).

Brown Grease – Waste grease that is the least expensive of the various grades of waste grease.

Bubble Cap Trays – Crossflow trays usually installed in rectifying columns handling liquids free of suspended solids; the bubble caps consist of circular cups inverted over small vapor pipes-the vapor from the tray below passes through the vapor pipes into the caps and curves downward to escape below the rim into the liquid' the rim of each cap is slotted or serrated to break up the escaping vapor into small bubbles, thereby increasing the surface area of the vapor as it passes through the liquid.

Bubble Wash – A method of final washing of biodiesel through air agitation. Biodiesel floats above a quantity of water; bubbles from an aquarium air pump and air stone are injected into the water causing the bubbles to rise-at the water/biodiesel interface, the air bubbles carry water up through the biodiesel by surface tension, simple diffusion causes water-soluble impurities in the biodiesel to be extracted into the water, as the bubble reaches the surface and breaks, the water is freed and percolates back down through the biodiesel again.

Bunker – A storage tank.

Burden – the amount of a specific pollutant in a reservoir.

Butanol – Though generally produced from fossil fuels, this four-carbon alcohol can also be produced through bacterial fermentation of alcohol.

Byproduct (By-Product) – A substance, other than the principal product, generated as a consequence of creating a biofuel.

C Horizon – The layer of soil composed of weathered rocks from which soil originated.

Canola – A member of the *Brassica* family, which includes broccoli, cabbage, cauliflower, mustard, radish, and turnip; it is a variant of the crop rapeseed, with less erucic acid and glucosinolate derivatives than rapeseed; grown for its seed, the seed is crushed for the oil contained within and, after the oil is extracted, the by-product is a protein-rich meal used by the intensive livestock industry.

Capacity – The maximum power that a machine or system can produce or carry safely; the maximum instantaneous output of a resource under specified conditions-the capacity of generating equipment is generally expressed in kilowatts or megawatts.

Capacity Factor – The amount of energy that a power plant actually generates compared to its maximum rated output, expressed as a percentage.

Capital Cost – The total investment needed to complete a project and bring it to a commercially operable status; the cost of construction of a new plant; the expenditures for the purchase or acquisition of existing facilities.

Capping – A process to cover the wastes, prevent infiltration of excessive amounts of surface water, and prevent release of waste to overlying soil and the atmosphere.

Carbohydrate – A chemical compound made up of carbon, hydrogen, and oxygen; includes sugars, cellulose, and starches.

Carbon Chain – The atomic structure of hydrocarbons in which a series of carbon atoms, saturated by hydrogen atoms, form a chain; volatile oils have shorter chains while fats have longer chain lengths, and waxes have extremely long carbon chains.

Carbon Dioxide (CO_2) – A product of combustion that acts as a greenhouse gas in the atmosphere of the Earth, thereby trapping heat and contributing to climate change.

Carbon Monoxide (CO) – A lethal gas produced by incomplete combustion of carbon-containing fuels in internal combustion engines. It is colorless, odorless, and tasteless.

Carbon Sequestration – The absorption and storage of carbon dioxide from the atmosphere; naturally occurring in plants.

Carbon Sink – A geographical area whose vegetation and/or soil soaks up significant carbon dioxide from the atmosphere. Such areas, typically in tropical regions, are increasingly being sacrificed for energy crop production.

Carbonate Washing – a mild alkali (e.g., potassium carbonate) process for emission control by the removal of acid gases from gas streams.

Carbonization – a high temperature process by which coal is converted into coke.

Carcinogen – A cancer-causing substance.

Carcinogenesis – development of cancer cells within an organism.

Carcinogenic – Cancer-causing.

Caricide – A chemical that is detrimental to mites.

Catabolic Pathways – The metabolic pathways that break down molecules into smaller units that are either oxidized to release energy or used in other reactions. Catabolism breaks down larger molecules into smaller units and is, in fact, the molecular breaking-down aspect of metabolism, whereas anabolic pathways is the building-up aspect.

Catabolism – breaking down complex molecules.

Catalyst – A substance that accelerates a chemical reaction without itself being affected. In refining, catalysts are used in the cracking process to produce blending components for fuels.

Catalytic Cracking – A refinery process in which high-boiling gas oils are converted into products of higher volatility by contacting the higher molecular weight hydrocarbon with the hot catalyst.

Catalytic Oxidation – A chemical conversion process used predominantly for destruction of volatile organic compounds and carbon monoxide.

Cellulase – A biological catalyst; an enzyme.

Cellulose – Fiber contained in leaves, stems, and stalks of plants and trees; most abundant organic compound on Earth; it is a polymer of glucose with a repeating unit of $C_6H_{10}O_5$ strung together by ß-glycosidic linkages – the ß-linkages in cellulose form linear chains that are highly stable and resistant to chemical attack because of the high degree of hydrogen bonding that can occur between chains of cellulose; hydrogen bonding between cellulose chains makes the polymers more rigid, inhibiting the flexing of the molecules that must occur in the hydrolytic breaking of the glycosidic linkages – hydrolysis can reduce cellulose to a cellobiose repeating unit, $C_{12}H_{22}O_{11}$, and ultimately to the six-carbon sugar glucose, $C_6H_{12}O_6$.

Cetane Number – A measure of the ignition quality of diesel fuel; the higher the number, the more easily the fuel is ignited under compression.

Cetane Rating – Measure of the combustion quality of diesel fuel.

Chelating Agents – complex-forming agents having the ability to solubilize heavy metals.

Chemical Disinfection – inactivates bacteria, viruses, and protozoa in waste streams.

Chemical Waste – any solid, liquid, gaseous material discharged from a process and which may pose substantial hazards to human health and the environment.

Chips – Small fragments of wood chopped or broken by mechanical equipment – total tree chips include wood, bark, and foliage while pulp chips or clean chips are free of bark and foliage.

Chlorofluorocarbon – A family of chemicals composed primarily of carbon, hydrogen, chlorine, and fluorine; used principally as refrigerants and industrial cleansers and have the tendency to destroy the protective ozone layer of the Earth.

Clarifier – A tank used to remove solids by gravity, to remove colloidal solids by coagulation, and to remove floating oil and scum through skimming.

Class I Area – Any area designated for the most stringent protection from air quality degradation.

Class II Area – Any area where air is cleaner than required by federal air quality standards and designated for a moderate degree of protection from air quality degradation; moderate increases in new pollution may be permitted in Class II areas.

Clay (Clay Minerals) – Silicate minerals which also usually contain aluminum and have particle sizes are less than 0.002 micron; see Micron.

Clean Air Act (CAA) – US national law establishing ambient air quality emission standards to be implemented by participating states; originally enacted in 1963, the CAA has been amended several times, most recently in 1990 and includes vehicle emission standards regulating the emission of criteria pollutants (lead, ozone, carbon monoxide, sulfur dioxide, nitrogen oxides and particulate matter); the 1990 amendments added reformulated gasoline (RFG) requirements and oxygenated gasoline provisions.

Clean Fuels – Fuels such as E-10 (unleaded) that burn cleaner and produce fewer harmful emissions compared to ordinary gasoline.

Closed-Loop Biomass – Crops grown, in a sustainable manner, for the purpose of optimizing their value for bioenergy and bioproduct uses. This includes annual crops such as maize and wheat, and perennial crops such as trees, shrubs, and grasses such as switchgrass.

Cloud Point – the temperature at which paraffin wax or other solid substances begin to crystallize or separate from the solution, imparting a cloudy appearance to the oil when the oil is chilled under prescribed conditions.

Co-Incineration – Burning different waste materials to produce fuel for energy recovery in furnaces and boilers.

Coal – an organic rock.

Coarse Materials – Wood residues suitable for chipping, such as slabs, edgings, and trimmings.

Cogeneration – The sequential production of electricity and useful thermal energy from a common fuel source.

Coke – The solid product produced by the carbonization of coal; also produced from petroleum during thermal processes.

Coking – a thermal method used in refineries for the conversion of bitumen and residua to volatile products and coke (see Delayed coking and Fluid coking).

Colloid – A stable system of small particles dispersed in another phase; a multi-phase system in which one dimension of a dispersed phase is of colloidal size; colloids are the liquid and solid forms of aerosols, foams, emulsions, and suspensions within the colloidal size class.

Colloidal Size – 0.001 micron to 1 micron in any dimension; dispersions where the particle size is in this range are referred to as colloidal aerosols, colloidal emulsions, colloidal foams, or colloidal suspensions; see Micron.

Colluvium – which is soil that has been transported by gravity.

Colza – Eurasian plant cultivated for its seed and as a forage crop.

Combined Heat and Power – The use of a power station to simultaneously generate both heat and electricity. The steam or hot water generated in the process is utilized either in industrial processes or in community heating.

Combustible Liquid – a liquid with a flashpoint in excess of 37.8°C (100°F) but below 93.3°C (200°F).

Combustion (Burning) – The transformation of biomass fuel into heat, chemicals, and gases through chemical combination of hydrogen and carbon in the fuel with oxygen in the air.

Combustion Gases – The gases released from a combustion process.

Commercial Waste – Waste arising from premises used wholly or mainly for trade, business, sport, recreation or entertainment, excluding municipal waste and industrial waste.

Composting – A resource recovery process where biodegradable waste (such as garden and kitchen waste) is converted, in the presence of oxygen from the air, into a stable granular material which, applied to land, improves soil structure and enriches the nutrient content.

Compound – A chemical term denoting a combination of two or more distinct elements.

Compressed Natural Gas (CNG) – Natural gas that has been compressed under high pressure (typically 2,000 to 3,600 psi).

Compression-Ignition Engine – An engine in which the fuel is ignited by high temperature caused by extreme pressure in the cylinder, rather than by a spark from a spark plug; diesel engine.

Concentrated Acid Hydrolysis – A method of converting biomass into cellulosic ethanol.

Condensation – occurs when the air becomes saturated with moisture (relative humidity = 100%). As temperature falls, the relative humidity of the air rises. The temperature at which condensation begins is termed the dew point. See Dew Point.

Conditional Use Permit – A permit, with conditions, allowing an approved use on a site outside the appropriate zoning class.

Conservation – Efficiency of energy use, production, transmission, or distribution that results in a decrease of energy consumption while providing the same level of service.

Construction and Demolition Waste – Waste arising from the construction, repair, maintenance, and demolition of buildings and structures, including roads. It consists mostly of brick, concrete, hardcore, subsoil, and topsoil, but it can also contain quantities of timber, metal, plastics, and (occasionally) hazardous waste materials.

Containment – The emplacement of physical, chemical, or hydraulic barrier to isolate contaminated areas.

Contaminant – A substance which causes deviation from the normal composition of an environment.

Continuous Fermentation – A steady-state fermentation system that operates without interruption; each stage of fermentation occurs in a separate section of the fermenter, and flow rates are set to correspond with required residence times.

Continuous Flow Process – A general term for any number of biodiesel production processes that involves the continuous addition of ingredients to produce biodiesel on a continual, round-the-clock basis, as opposed to the batch process.

Controlled Waste – The UK term for wastes controlled under the Waste Framework Directive; includes household waste, commercial waste, industrial waste and agricultural waste.

Conventional Biofuels – Biofuels such as bioethanol and biodiesel, which are typically made from corn, sugarcane, and beet, wheat or oilseed crops such as soy and rapeseed oil.

Conventional Crude Oil (Conventional Petroleum) – Crude oil that is pumped from the ground and recovered using the energy inherent in the reservoir; also recoverable by application of secondary recovery techniques.

Conversion Efficiency – A comparison of the useful energy output to the potential energy contained in the fuel; the efficiency calculation relates to the form of energy produced and allows a direct comparison of the efficiency of different conversion processes can be made only when the processes produce the same form of energy output.

Cooker – A tank or vessel designed to cook a liquid or extract or digest solids in suspension; the cooker usually contains a source of heat; and is fitted with an agitator.

Cord – A stack of wood comprising 128 cubic feet (3.62 m^3); standard dimensions are $4 \times 4 \times 8$ feet, including air space and bark. One cord contains approx. 1.2 U. S. tons (oven-dry) = 2,400 pounds = 1,089 kg.

Corn Stover – Residue materials from harvesting corn consisting of the cob, leaves, and stalk; similar to straw, the residue left after any cereal grain or grass has been harvested at maturity for its seed. It can be directly grazed by cattle or dried for use as fodder. Stover has attracted some attention as a potential fuel source, and as biomass for fermentation or as a feedstock for ethanol production from cellulose.

Corrosive Waste – A waste that is capable of corroding metal.

Corrosivity – characteristic of substances that exhibit extremes of acidity or basicity or a tendency to corrode steel.

Cracking – A secondary refining process that uses heat and/or a catalyst to break down high molecular weight chemical components into lower molecular weight products which can be used as blending components for fuels.

Crop Residue – The "backbone" of sugar and starch crops-the stalks and leaves-is composed mainly of cellulose. The individual six-carbon sugar units in cellulose are linked together in extremely long chains by a stronger chemical bond than exists in starch. As with starch, cellulose must be broken down into sugar units before it can be used by yeast to make ethanol. However, the breaking of the cellulose bonds is much more complex and costly than the breaking of the starch bonds. Breaking the cellulose into individual sugar units is complicated by the presence of lignin, a complex compound surrounding cellulose, which is even more resistant than cellulose to enzymatic or acidic pretreatment. Because of the high cost of converting liquefied cellulose into fermentable sugars, agricultural residues (as well as other crops having a high percentage of cellulose) are not yet a practical feedstock source for small ethanol plants.

Cropland – Total cropland includes five components: cropland harvested, crop failure, cultivated summer fallow, cropland used only for pasture, and idle cropland.

Cropland Pasture – Land used for long-term crop rotation. However, some cropland pasture is marginal for crop uses and may remain in pasture indefinitely. This category also includes land that was used for pasture before crops reached maturity and some land used for pasture that could have been cropped without additional improvement.

Cull Tree – A live tree, 5.0 inches in diameter at breast height (d.b.h.) or larger that is non-merchantable for saw logs now or prospectively because of rot, roughness, or species.

Cultivated Summer Fallow – cropland cultivated for one or more seasons to control weeds and accumulate moisture before small grains are planted.

Cultural Eutrophication – changes in species composition, population sizes, and productivity in groups of organisms throughout the aquatic ecosystem.

Curie – unit of measurement of radioactivity.

Darwinism – The theory espoused by Charles Darwin that proposes that an organisms ability to survive is dependent upon hereditary factors.

DDGS (Dried Distillers Grain with Soluble Constituents) – A by-product of dry mill ethanol production that is fed to livestock.

Definition – The means by which scientists and engineers communicate the nature of a material to each other and to the world, either through the spoken word or through the written word.

Dehydrating Agents – substances capable of removing water (drying) or the elements of water from another substance.

Delayed Coking – a coking process in which the thermal reactions are allowed to proceed to completion to produce gaseous, liquid, and solid (coke) products.

Density – the mass (or weight) of a unit volume of any substance at a specified temperature; see also Specific gravity.

Desulfurization – the removal of sulfur or sulfur compounds from a feedstock.

Detoxification – the biological conversion of a toxic substance to a less toxic species, which may still be relatively complex, or biological conversion to an even more complex material.

Devolatilized Fuel – Smokeless fuel.

Dew Point – The temperature at which air becomes saturated with moisture. See Condensation, Humidity (absolute), Humidity (relative).

Dialysis – A process for separating components m a liquid stream by using a membrane.

Diesel #1 and Diesel #2 – Diesel #1 is also called kerosene and is not generally used as a fuel oil in diesel vehicles – it has a lower viscosity (it is thinner) than Diesel #2, which is the typical diesel vehicle fuel. Biodiesel replaces Diesel #2 or a percentage.

Diesel Engine – Named for the German engineer Rudolph Diesel, this internal-combustion, compression-ignition engine works by heating fuels and causing them to ignite; can use either petroleum or bio-derived fuel.

Diesel Fuel – A distillate of fuel oil that has been historically derived from petroleum for use in internal combustion engines; also derived from plant and animal sources.

Diesel, Rudolph – German inventor famed for fashioning the diesel engine, which made its debut at the 1900 World Fair; initially engine to run on vegetable-derived fuels.

Digester – An airtight vessel or enclosure in which bacteria decomposes biomass in water to produce biogas.

Direct-Injection Engine – A diesel engine in which fuel is injected directly into the cylinder.

Dispersion – A stable or unstable system of fine particles, larger than colloidal size, evenly distributed in a medium.

Dispersion Aerosols – Aerosols formed from disintegration of larger particles and are usually above 1 micron in size, e.g., dust; see Micron.

Disposal – the discharge, deposit, injection, or placing of a waste on to, or into, a land facility.

Dissolved Air Flotation – air is dissolved in the suspending medium under pressure and comes out of solution when the pressure is released as minute air bubbles attached to suspended particles, which causes the particles to float to the surface.

Distillate – Any petroleum product produced by boiling crude oil and collecting the vapors produced as a condensate in a separate vessel, for example gasoline (light distillate), gas oil (middle distillate), or fuel oil (heavy distillate).

Distillate Oil – Any distilled product of crude oil; a volatile petroleum product used for home heating and most machinery.

Distillation – The primary distillation process which uses high temperature to separate crude oil into vapor and fluids which can then be fed into a distillation or fractionating tower.

Distillation – The process to separate the components of a liquid mixture by boiling the liquid and then condensing the resulting vapor.

Distillers Grains – Byproduct of ethanol production that can be used to feed livestock; alternatively, distillers dried grains with soluble constituents (DDGS).

Downdraft Gasifier – A gasifier in which the product gases pass through a combustion zone at the bottom of the gasifier.

Dry Deposition – The removal of both particles and gases as they come into contact with the land surface.

Dry Mill – An ethanol production process in which the entire corn kernel is first ground into flour before processing-in addition to ethanol, dry mills also produce dried distillers grains with soluble constituents (DDGS) which is fed to livestock; and carbon dioxide which is used in food processing and bottling; most new ethanol plants are dry mill facilities.

Dry Ton – 2,000 pounds of material dried to a constant weight.

Drying – removal of a solvent or water from a chemical substance; also referred to as the removal of solvent from a liquid or suspension.

Dust Control – particulate matter control.

Dust Explosions – explosions caused by the presence of dust particles m the atmosphere.

Dutch Oven Furnace – One of the earliest types of furnaces, having a large, rectangular box lined with firebrick (refractory) on the sides and top; commonly used for burning wood.

Dystrophic Lakes – shallow, clogged with plant life, and normally contain colored water with a low pH

E Diesel – A blend of ethanol and diesel fuel plus other additives designed to reduce air pollution from heavy equipment, city buses and other vehicles that operate on diesel engines.

E10 – An alcohol fuel mixture containing 10% ethanol and 90% gasoline by volume.

E85 – An alcohol fuel mixture containing 85% ethanol and 15% gasoline by volume, and the current alternative fuel of choice of the government of the United States.

Ecad – A plant species is a population of individuals which although belong to the same genetic stock, but differ markedly in phenotypes such as size, shape, and number of leaves; also known as ecophene.

Ecological Cycles – The cycles involving land systems, water systems, and the atmosphere which are important to life; the processes by which the limited resources of the Earth (water, carbon, nitrogen, and other elements that are essential to sustain life are recycled; an understanding the means by which local cycles fit into global cycles is essential in order to formulate the best possible management decisions to maintain ecosystem health and productivity for now and the future.

Ecology – The branch of science related to the study of the relationship of organisms and their environment; the science of interactions among (floral and faunal) individuals, populations, and communities and it also involves the interrelations between living (biotic) components with their non-living (abiotic) counterparts or environment.

Ecosystem – An ecological community, or living unit, considered together with nonliving factors of its environment as a unit.

Ecosystem Diversity – All of the species, plus all the abiotic factors characteristic of a region. For example, a desert ecosystem has soil, temperature, rainfall patterns, and solar radiation that affect not only what species occur there, but also the morphology, behavior, and the interactions among those species. Ecosystem diversity describes the number of niches, trophic levels and various ecological processes that sustain energy flow, food webs and the recycling of nutrients. See also Biodiversity.

Effective Stack Height – The combination of the physical stack height and plume rise above the stack.

Effluent – The liquid or gas discharged from a process or chemical reactor, usually containing residues from that process.

Electrodialysis – an extension of dialysis which is used to separate the components of an ionic solution by applying an electric current to the solution, thereby causing ions to move in preferred directions through the dialysis membrane.

Electrolysis – a process in which ionic species in solution move to an electrode of opposite electric charge.

Electrostatic Precipitator – A device which operates on the principle of imparting an electric charge to particles in an incoming airstream and which are then collected on an oppositely charged plate across a high voltage field.

Elemental Analysis – The determination of carbon, hydrogen, nitrogen, oxygen, sulfur, chlorine, and ash in a sample as a% w/w of the whole.

Emission Control – The use of gas cleaning processes to reduce emissions.

Emission Offset – A reduction in the air pollution emissions of existing sources to compensate for emissions from new sources.

Emissions – Substances discharged into the air during combustion; waste substances released into the air or water.

Emulsification – To emulsify; to form an emulsion.

Emulsion – A suspension of small drops of 1 liquid in a 2nd with which the 1st will not mix; can be formed either by mechanical agitation or by chemical processes; unstable emulsions will separate with time or temperature but stable emulsions will not separate.

Emulsion Breaking – The settling or aggregation of colloidal-sized emulsions from suspension in a liquid medium.

Encapsulation – A process used to coat waste with an impermeable material so that there is no contact between the waste constituents and the surroundings.

End-of-Pipe Emission Control Methods – the use of specific emission control processes to clean gases after production of the gases.

Endorheic – regions which drain to interior closed basins.

Energy – The output of fuel sources; also the output of living organisms.

Energy Balance – The difference between the energy produced by a fuel and the energy required to obtain it through agricultural processes, drilling, refining, and transportation.

Energy Crops – Crops grown specifically for their fuel value; include food crops such as corn and sugarcane, and nonfood crops such as poplar trees and switchgrass.

Energy from Biomass – The production of energy from biomass.

Energy-Efficiency Ratio – A number representing the energy stored in a fuel as compared to the energy required to produce, process, transport, and distribute that fuel.

Enhanced Recovery – methods that usually involves the application of thermal energy (e.g., steam flooding) to oil recovery from the reservoir.

Environment – The external conditions that affect organisms and influence their development and survival; the sum total of everything that directly influences the animal's chances of survival or reproduction.

Environmental Assessment (EA) – A public document that analyzes a proposed federal action for the possibility of significant environmental impacts-if the environmental impacts will be significant, the federal agency must then prepare an environmental impact statement.

Environmental Impact Assessment (EIA) – assessment of the impact of a proposed change, such as a construction project, on the environment.

Environmental Impact Statement (EIS) – A statement of the environmental effects of a proposed action and of alternative actions. Section 102 of the National Environmental Policy Act requires an EIS for all major federal actions.

Environmental Technology – The application of scientific and engineering principles to the study of the environment with the goal of improving the environment.

Enzymatic Hydrolysis – A process by which enzymes (biological catalysts) are used to break down starch or cellulose into sugar.

Enzyme – A protein or protein-based molecule that speeds up chemical reactions occurring in living things; enzymes act as catalysts for a single reaction, converting a specific set of reactants into specific products.

Eolian Soil – Soil that has been transported by wind.

Eon – A long span of geologic time. In formal usage, an eon is the longest portion of geologic time (an era is the second-longest). Three eons are recognized – the Phanerozoic Eon (dating from the present back to the beginning of the Cambrian Period), the Proterozoic Eon and the Archean Eon. Less formally, the term eon often refers to a time span of 1 billion years (1×10^9 years).

Era – A subdivision of geologic time that sub-divides an eon into smaller units of time.

Esters – Any of a large group of organic compounds formed when an acid and alcohol is mixed; methyl acetate (CH_3COOCH_3) is the simplest ester; biodiesel contains methyl stearate.

Estuary – The place where freshwater (from a land-based water system such as a river) mingles with the saltwater of the ocean.

ETBE – see Ethyl Tertiary Butyl Ether.

Ethanol (Ethyl Alcohol, Alcohol, Grain Alcohol, or Grain-Spirit) – A clear, colorless, flammable oxygenated hydrocarbon; used as a vehicle fuel by itself (E100 is 100% v/v ethanol), blended with gasoline (E85 is 85% v/v ethanol), or as a gasoline octane enhancer and oxygenate (10% v/v).

Ethers – Liquid fuel made from blending an alcohol with isobutylene.

Ethyl Tertiary Butyl Ether (Ethyl t-Butyl Ether) – Ether created from ethanol that can increase octane and reduce the volatility of gasoline, decreasing evaporation and smog formation.

Eutrophic Lakes – Lakes that contain high proportions of more nutrients and in which abnormal growth of organisms occurs (see Eutrophication).

Eutrophication – The deterioration of the esthetic and life-sporting qualities of lakes and estuaries, caused by excessive fertilization from affluent high in phosphorus, nitrogen, and organic growth substances.

Evaporation – The conversion of a liquid to the vapor state by the addition of latent heat of vaporization; the surrounding gas must not be saturated with the evaporating substance.

Evaporites – soluble salts that precipitate from solution *t* the result of evaporation of evaporation.

Exoreic – regions which drain to the sea.

Expanding Clays – Clay minerals which expand or swell on contact with water, e.g., montmorillonite.

Explosives – chemicals which decompose spontaneously, or by initiation/stimulation, with a rapid release of a high amount of energy.

Exposure Route – The way that a released enters chemical enters the flora and fauna as a result of the release.

Extracellular Enzyme – An enzyme which is capable of acting outside an organism.

Extractives – Any number of different compounds in biomass that are not an integral part of the cellular structure-the compounds can be extracted from wood by means of polar and non-polar solvents including hot or cold water, ether, benzene, methanol, or other solvents that do not degrade the biomass structure and the types of extractives found in biomass samples are entirely dependent upon the sample itself.

F-Type Wastes – wastes from nonspecific sources.

FAAE – A term for biodiesel made from any alcohol during its production process.

Fabric Filter – A filter for particulate matter removal from gas streams that is made from fabric materials (see Baghouse).

FAME (Fatty Acid Methyl Ester) – ester that can be created by a catalyzed reaction between fatty acids and methanol; the constituents in biodiesel are primarily FAMEs, usually obtained from vegetable oils by transesterification.

Fast Pyrolysis – Thermal conversion of biomass by rapid heating to between 450 to 600°C (842 to 1,112°F) in the absence of oxygen

Fatty Acid – A carboxylic acid (an acid with a -COOH group) with long hydrocarbon side chains; feedstocks are first converted to fatty acids and then to biodiesel by transesterification.

Fatty Acid Alkyl Ester – see FAAE.

Fatty Acid Methyl Ester – see FAME.

Feedstock – The biomass used in the creation of a particular biofuel (e.g., corn or sugar-cane for ethanol, soybeans or rapeseed for biodiesel).

Fermentation – Conversion of carbon-containing compounds by micro-organisms for production of fuels and chemicals such as alcohols, acids or energy-rich gases.

Fiber Products – Products derived from fibers of herbaceous and woody plant materials; examples include pulp, composition board products, and wood chips for export.

Filtration – the use of an impassible barrier to collect solids but which allows liquids to pass.

Fine Materials – Wood residues not suitable for chipping, such as planer shavings and sawdust.

Fischer-Tropsch Process – process for producing liquid fuels, usually diesel fuel, from natural gas or synthetic gas from gasified coal or biomass.

Fission – Energy from nuclear sources which uses uranium ore or refined uranium-235 as the basic energy source and which splits into lower atomic weight nuclei.

Fixation – a process that binds a waste in a less mobile and less toxic form and is generally included in the definition of stabilization. Also the process by which aerial (molecular) nitrogen is converted to nitrates in the soil (see Nitrogen fixation).

Fixed Bed – use of a stationary bed to accomplish a process (see Fluid bed). Flammability range – the range of temperature over which a chemical is flammable.

Fixed Carbon – The carbonaceous residue remaining after heating in a prescribed manner to decompose thermally unstable components and to distil volatiles; part of the proximate analysis group.

Fixed Hearth Incinerators – incinerators with single or multiple (non-mobile) hearths which are used for the combustion of liquid or solid wastes occurs. ·

Flammable Liquid – A liquid having a flashpoint below 37.8°C (100°F). Flammable: a substance that will burn readily.

Flammable Solid – A solid that can ignite from friction or from heat remaining from its manufacture, or which may cause a serious hazard if ignited.

Flashpoint – The lowest temperature at which a liquid will produce enough vapor to ignite if the vapor is flammable.

Flexible-Fuel Vehicle (Flex-Fuel Vehicle) – A vehicle that can run alternately on two or more sources of fuel; includes cars capable of running on gasoline and gasoline/ ethanol mixtures, as well as cars that can run on both gasoline and natural gas.

Flotation – a process for removing solids from liquids using air bubbles to carry the particles to the surface.

Fluid Bed (Fluid-Bed) – An agitated bed of inert granular material to accomplish a process in which the agitated bed resembles the motion of a fluid.

Fluid Coking – a continuous fluidized solids process that cracks feed thermally overheated coke particles in a reactor vessel to gas, liquid products, and coke.

Fluidized-Bed Boiler – A large, refractory-lined vessel with an air distribution member or plate in the bottom, a hot gas outlet in or near the top, and some provisions for introducing fuel; the fluidized bed is formed by blowing air up through a layer of inert particles (such as sand or limestone) at a rate that causes the particles to go into suspension and continuous motion.

Flushing – a cleanup process in which the soil is left in place and the water is pumped into and out of the soil in order to clean it.

Flux – the rate of transfer of a pollutant from one sphere or domain to another.

Fly Ash – Particulate matter produced from mineral matter in coal that is converted during combustion to finely divided inorganic material and which emerges from the combustor in the gases.

Foam – A dispersion of a gas in a liquid or solid.

Fodder Beets – A promising sugar crop which presently is being developed in New Zealand is the fodder beet. The fodder beet is a high yielding forage crop obtained by crossing two other beet species, sugar beets and mangolds. It is similar in most agronomic respects to sugar beets. The attraction of this crop lies in its higher yield of fermentable sugars per acre relative to sugar beets and its comparatively high resistance to loss of fermentable sugars during storage. Culture of fodder beets is also less demanding than sugar beets.

Forage Crops (e.g., Forage Sorghum, Sudan Grass) – Crops used for ethanol production because, in their early stage of growth, there is only (at best) a small amount of lignin and the conversion of the cellulose to sugars is more efficient. In addition, the proportion of carbohydrates in the form of cellulose is less than in the mature plant. Since forage crops achieve maximum growth in a relatively short period, they can be harvested as many as four times in one growing season. For this reason, forage crops cut as green chop may have the highest yield of dry material of any storage crop. In addition to cellulose, forage crops contain significant quantities of starch and fermentable sugars which can also be converted to ethanol. The residues from fermentation containing nonfermentable sugars, protein, and other components may be used for livestock feed.

Forest Health – A condition of ecosystem sustainability and attainment of management objectives for a given forest area; usually considered to include green trees, snags, resilient stands growing at a moderate rate, and endemic levels of insects and disease.

Forest Land – Land at least 10% stocked by forest trees of any size, including land that formerly had such tree cover and that will be naturally or artificially regenerated; includes transition zones, such as areas between heavily forested and non-forested lands that are at least 10% stocked with forest trees and forest areas adjacent to urban and built-up lands; also included are pinyon-juniper and chaparral areas; minimum area for classification of forest land is 1 acre.

Forest Residues – Material not harvested or removed from logging sites in commercial hardwood and softwood stands as well as material resulting from forest management operations such as precommercial thinning and removal of dead and dying trees.

Fossil Fuel Resources – Natural gas, crude oil (petroleum), heavy crude oil, extra heavy oil, tar sand bitumen, coal, and oil shale.

Fossil Fuel – Solid, liquid, or gaseous fuels formed in the ground after millions of years by chemical and physical changes in plant and animal residues under high temperature and pressure.

Fruit Crops (e.g., Grapes, Apricots, Peaches, and Pears) – Another type of feedstock in the sugar crop category. Typically, fruit crops such as grapes are used as the feedstock in wine production. These crops are not likely to be used as feedstocks for the production of fuel-grade ethanol because of their high market value for direct human consumption. However, the coproducts of processing fruit crops are likely to be used as feedstocks because fermentation is an economical method for reducing the potential environmental impact of untreated wastes containing fermentable sugars.

Fuel Cell – A device that converts the energy of a fuel directly to electricity and heat, without combustion.

Fuel Cycle – The series of steps required to produce electricity. The fuel cycle includes mining or otherwise acquiring the raw fuel source, processing, and cleaning the fuel, transport, electricity generation, waste management and plant decommissioning.

Fuel Oil – A heavy residue, black in color, used to generate power or heat by burning in furnaces.

Fuel Treatment Evaluator (FTE) – A strategic assessment tool capable of aiding the identification, evaluation, and prioritization of fuel treatment opportunities.

Fuelwood – Wood used for conversion to some form of energy, primarily for residential use.

Fugitive Emissions – Emissions that enter the atmosphere from an unconfined area.

Fulvic Acids – A base-soluble fraction of humus.

Fumigants – Volatile substances used as soil pesticides and to control insects in stored products.

Fungi – non-photosynthetic organisms which frequently possess a filamentous structure.

Fungicide – a chemical used to control fungal action.

Furnace – An enclosed chamber or container used to burn biomass in a controlled manner to produce heat for space or process heating.

Fusion – energy produced using the fusion of hydrogen nuclei.

Galactan – The polymer of galactose with a repeating unit of $C_6H_{10}O_5$; found in hemicellulose it can be hydrolyzed to galactose.

Galactose – A six-carbon sugar with the formula $C_6H_{12}O_6$; a product of hydrolysis of Galactan found in the hemicellulose fraction of biomass.

Gas Engine – A piston engine that uses natural gas rather than gasoline-fuel and air are mixed before they enter cylinders; ignition occurs with a spark.

Gas Shift Process – A process in which carbon monoxide and hydrogen react in the presence of a catalyst to form methane and water.

Gas to Liquids (GTL) – The process of refining natural gas and other hydrocarbons into longer-chain hydrocarbons, which can be used to convert gaseous waste products into fuels.

Gas Turbine (Combustion Turbine) – A turbine that converts the energy of hot compressed gases (produced by burning fuel in compressed air) into mechanical power-often fired by natural gas or fuel oil.

Gaseous Emissions – Substances discharged into the air during combustion, typically including carbon dioxide, carbon monoxide, water vapor, and hydrocarbons.

Gasification – A chemical or heat process used to convert carbonaceous material (such as coal, petroleum, and biomass) into gaseous components such as carbon monoxide and hydrogen.

Gasifier – A device for converting solid fuel into gaseous fuel; in biomass systems, the process is referred to as pyrolitic distillation.

Gasohol – A mixture of 10% v/v anhydrous ethanol and 90% v/v gasoline; 7.5% v/v anhydrous ethanol and 92.5% v/v gasoline; or 5.5% v/v anhydrous ethanol and 94.5% v/v gasoline.

Gasoline – A volatile, flammable liquid obtained from petroleum that has a boiling range of approximately 30 to 220°C 86 to 428°F) and is used for fuel for spark-ignition internal combustion engines.

Gel Point – The point at which a liquid fuel cools to the consistency of petroleum jelly.

Genetically Modified Organism (GMO) – An organism whose genetic material has been modified through recombinant DNA technology, altering the phenotype of the organism to meet desired specifications.

Geosphere – the complex and variable mixture of minerals, organic matter, water, and air which make up soil; also known as land systems.

Glacial Soil – Soil that has been transported by the movement of glaciers.

Glassification – encapsulation of a waste in a glass-like coat; vitrification.

Glycerin (CH_2OH. CHOH. CH_2OH) – A byproduct of biodiesel production; each of the hydroxyl (OH) functions is one of the three places where an ester is broken off of the triglyceride molecule (e.g., vegetable oil).

Glycerine (Glycerin; Glycerol) – A liquid by-product of biodiesel production; used in the manufacture of dynamite, cosmetics, liquid soaps, inks, and lubricants.

Grain Alcohol – See Ethanol.

Grassland Pasture and Range – All open land used primarily for pasture and grazing, including shrub and brushland types of pasture; grazing land with sagebrush and scattered mesquite; and all tame and native grasses, legumes, and other forage used for pasture or grazing; because of the diversity in vegetative composition, grassland pasture and range are not always clearly distinguishable from other types of pasture and range; at one extreme, permanent grassland may merge with cropland pasture, or grassland may often be found in transitional areas with forested grazing land.

Grease Car – A diesel-powered automobile rigged post-production to run on used vegetable oil.

Greenhouse Effect – The effect of certain gases in the atmosphere of the Earth in trapping heat from the Sun.

Greenhouse Gases – Gases that trap the heat of the Sun in the atmosphere of the Earth, producing the greenhouse effect. The two major greenhouse gases are water vapor and carbon dioxide. Other greenhouse gases include methane, ozone, chlorofluorocarbons, and nitrous oxide.

Grid – The system used by an electric utility company for distributing power.

Gross Heating Value (GHV) – The maximum potential energy in the fuel as received, considering moisture content (MC).

Growing Stock – A classification of timber inventory that includes live trees of commercial species meeting specified standards of quality or vigor; cull trees are excluded.

GTL (Gas to Liquid) – A refinery process that converts natural gas into longer-chain hydrocarbons; gas can be converted to liquid fuels via a direct conversion or using a process such as the Fischer-Tropsch process.

Habitat – The area where a plant or animal lives and grows under natural conditions. Habitat includes living and non-living attributes and provides all requirements for food and shelter.

Hardwoods – Usually broad-leaved and deciduous trees; one of the botanical groups of dicotyledonous trees that have broad leaves in contrast to the conifers or softwoods-the term has no reference to the actual hardness of the wood-the botanical name for hardwoods is angiosperms; short-rotation, fast-growing hardwood trees are being developed as future energy crops which are uniquely developed for harvest from 5 to 8 years after planting and example include – hybrid poplars (*Populus sp.*), hybrid willows (*Salix sp.*), silver maple (*Acer saccharinum*), and black locust (*Robinia pseudoacacia*).

Hazardous Waste – A broad term for a wide range of waste materials that present different levels of risk. Some present a serious and immediate threat to the population and the environment, for example, those that are toxic, could cause cancer or infectious disease. Others, such as fluorescent tubes or cathode ray tubes in televisions, pose little immediate threat but may cause long-term damage over a period of time.

Heating Value – The maximum amount of energy that is available from burning a substance.

Heavy Oil (Heavy Crude Oil) – Oil that is more viscous thanthan conventional crude oil, has a lower mobility in the reservoir but can be recovered through a well from the reservoir by the application of a secondary or enhanced recovery methods.

Hectare – Common metric unit of area, equal to 2.47 acres. 100 hectares = 1 square kilometer.

Hemicellulose – consists of short, highly branched chains of sugars in contrast to cellulose, which is a polymer of only glucose, hemicellulose is a polymer of five different sugars; contains five-carbon sugars (usually D-xylose and L-arabinose) and six-carbon sugars (D-galactose, D-glucose, and D-mannose) and uronic acid which are highly substituted with acetic acid; the branched nature of hemicellulose renders it amorphous and relatively easy to hydrolyze to its constituent sugars compared to cellulose; when hydrolyzed, the hemicellulose from hardwoods releases products high in xylose (a five-carbon sugar); hemicellulose contained in softwoods, by contrast, yields more six-carbon sugars.

Herbaceous – Non-woody type of vegetation, usually lacking permanent strong stems, such as grasses, cereals, and canola (rape).

Herbaceous Energy Crops – Perennial non-woody crops that are harvested annually, though they may take 2 to 3 years to reach full productivity; examples include: switch-grass (*Panicum virgatum*), reed canary grass (*Phalaris arundinacea*), miscanthus (*Miscanthus x giganteus*), and giant reed (*Arundo donax*).

Herbaceous Plants – Non-woody species of vegetation, usually of low lignin content such as grasses.

Herbicide – A chemical used to control plant growth.

Heteroatom Compounds – Hydrocarbon compounds that contain nitrogen and/or oxygen and/or sulfur and/or metals bound within their molecular structure(s).

Heterosphere – that part of the atmosphere above 60 miles (100 km) in altitude.

Heterotrophic Organisms – organisms which utilize the organic substances produced by autotrophic organisms as energy sources and as the raw materials for the synthesis of their own biomass.

Hexose – Any of various simple sugars that have six carbon atoms per molecule (e.g., glucose, mannose, and galactose).

High-Level Waste – A type of radioactive waste which depends upon the amount of radio-activity (see also Low-level waste).

Higher Heating Value (HHV) – The potential combustion energy when water vapor from combustion is condensed to recover the latent heat of vaporization. Lower heating value (LHV) is the potential combustion energy when water vapor from combustion is not condensed.

History – The means by which the subject is studied, hopefully so that the errors of the past will not be repeated.

Homopause – The boundary between the homosphere and the heterosphere; also known as the turbopause.

Homosphere – The atmosphere below about 60 miles (100 km) in altitude.

Horizon – The top layer of soil, also known as topsoil.

Horizons – The layers of soil which are a more or less parallel to the surface and differing from those above and below in one or more properties such as color, texture, structure, consistency, porosity, and reaction.

Household Waste – Also known as domestic waste; waste from household collection rounds, waste from services such as street sweepings, bulky waste collection, litter collection, hazardous household waste collection and garden waste collection, waste from civic amenity sites and wastes separately collected for recycling or composting through bring recycling schemes and curbside recycling schemes. Household waste is a sub-group of municipal solid waste.

Humic Acid – The base-soluble fraction of humus.

Humidity (Absolute) – A measure of the amount of water in air; varies with temperature–arm air contains more moisture than cold air; saturation represents the point where increasing vapor density results in condensation. See Dew Point.

Humidity (Relative) – A measures the amount of water in air relative to the maximum amount of water the air could hold; increases with increasing water vapor or decreasing temperature.

Humin – An insoluble fraction of humin which is the residue from the biodegradation of plant material.

Humus – A generic term for the water-insoluble material that makes up the bulk of soil organic matter.

Hydrocarbon – A chemical compound that contains a carbon backbone with hydrogen atoms attached to that backbone.

Hydrocarbon Compounds – chemical compounds containing only carbon and hydrogen.

Hydrocarbonaceous Material – A material such as bitumen that is composed of carbon and hydrogen with other elements (heteroelements) such as nitrogen, oxygen, sulfur, and metals chemically combined within the structures of the constituents; even though carbon and hydrogen may be the predominant elements, there may be very few true hydrocarbons (q.v.).

Hydrocracking – A catalytic high-pressure high-temperature process for the conversion of petroleum feedstocks in the presence of fresh and recycled hydrogen.

Hydrodesulfurization – the removal of sulfur by hydrotreating (q.v.).

Hydrologic Cycle – The water cycle.

Hydrology – the study of water.

Hydrolysis – A method used to dispose of chemicals that are reactive with water by allowing them to react with water under controlled conditions.

Hydroprocesses – refinery processes designed to add hydrogen to various products of refining.

Hydrosphere – Water in various forms; also known as the aqua sphere.

Hydrotreating – the removal of heteroatomic (nitrogen, oxygen, and sulfur) species by treatment of a feedstock or product at relatively low temperatures in the presence of hydrogen.

Idle Cropland – Land in which no crops were planted; acreage diverted from crops to soil-conserving uses (if not eligible for and used as cropland pasture) under federal farm programs is included in this component.

Igneous Rock – rock produced from the solidification of molten rock, e.g., granite, basalt, quartz, feldspar, and magnetite.

Ignitability – characteristic of substances that are liquids whose vapors are likely to ignite in the presence of ignition source; also characteristic of non-liquids that may catch fire from friction or contact with water and which bum vigorously.

Impact Assessment – the interpretation of the significance of anticipated changes related to the proposed project.

***In-Situ* Immobilization** – A process used to convert waste constituents to insoluble or immobile forms that will not leach from a disposal site.

***In-Situ* Treatment** – Waste treatment processes that can be applied to wastes in a disposal site by direct application of treatment processes.

Incinerator – Any device used to burn solid or liquid residues or wastes as a method of disposal.

Inclined Grate – A type of furnace in which fuel enters at the top part of a grate in a continuous ribbon, passes over the upper drying section where moisture is removed, and descends into the lower burning section. Ash is removed at the lower part of the grate.

Indirect Liquefaction – Conversion of biomass to a liquid fuel through a synthesis gas intermediate step.

Indirect-Injection Engine – An older model of diesel engine in which fuel is injected into a pre-chamber, partly combusted, and then sent to the fuel injection chamber.

Industrial Wood – All commercial round wood products except fuelwood.

Insecticide – A chemical which interferes in the life cycle of certain insects to control insect populations; the chemical may also kill the insect.

Iodine Value – a measure of the number of unsaturated carbon-carbon double bonds in a vegetable oil molecule-double bonds can allow polymerization, leading to the formation of lacquers and possibly blockage and damage to engine or fuel train components; in liquid biofuel applications the iodine value gives a lower cold filter plugging point (CFPP) or cloud point.

Ion Exchange – A means of removing cations or anions from solution on to a solid resin.

Ionizing Radiation – Cosmic rays from outer space.

Ionosphere – The layer of the atmosphere at altitudes of about 30 miles (50 km) and up.

Jatropha – A non-edible evergreen shrub found in Asia, Africa, and the West Indies; the seeds contain a high proportion of oil which can be used for making biodiesel.

Jerusalem Artichoke – A member of the sunflower family, this crop is native to North America and well-adapted to northern climates. Like the sugar beet, the Jerusalem artichoke produces sugar in the top growth and stores it in the roots and tuber. It can grow in a variety of soils, and it is not demanding of soil fertility. The Jerusalem artichoke is a perennial; small tubers left in the field will produce the next seasonal crop, so no plowing or seeding is necessary.

Joule – Metric unit of energy, equivalent to the work done by a force of one Newton applied over a distance of one meter (= 1 kg m^2/s^2). One joule (J) = 0.239 calories (1 calorie = 4.187 J).

K-Type Waste – Waste from a specific source.

Kármán Line – An attempt to define a boundary between the atmosphere of the Earth and outer space.

Karst System – Formed from the dissolution of soluble rocks such as limestone ($CaCO_3$), dolomite ($CaCO_3$. $MgCO_3$), and gypsum ($CaSO_4$) and is characterized by underground drainage systems with sinkholes and caves. Subterranean drainage may limit surface water, with few to no rivers or lakes. However, in regions where the dissolved bedrock is covered (perhaps by debris) or confined by one or more superimposed non-soluble rock strata, distinctive karst features may occur only at subsurface levels and may not be noticed above ground.

Kerosene – A light middle distillate that in various forms is used as aviation turbine fuel or for burning in heating boilers or as a solvent, such as white spirit.

Kilowatt (kW) – A measure of electrical power equal to 1,000 watts. 1 kW = 3,412 Btu/hr = 1.341 horsepower.

Kilowatt Hour (kWh) – A measure of energy equivalent to the expenditure of one kilowatt for one hour. For example, 1 kWh will light a 100-watt light bulb for 10 hours. 1 kWh = 3,412 Btu.

Klason Lignin – Lignin obtained from wood after the non-lignin components of the wood have been removed with a prescribed sulfuric acid treatment; a specific type of acid-insoluble lignin analysis.

Knock – Engine sound that results from ignition of the compressed fuel-air mixture prior to the optimal moment.

KOH – see Potassium hydroxide.

Land Farming – Solid-phase bioremediation.

Land Systems – Components that form the Earth.

Landfill – a site where waste is placed in or on land (often in separate trenches depending upon the waste and to prevent contact of reactive, waste materials) and which may/should be lined to prevent leakage and to prevent run-off of the contaminated surface water.

Landfill Gas – A type of biogas that is generated by decomposition of organic material at landfill disposal sites. Landfill gas is approximately 50% methane. See also biogas.

Landfill Licensed Facilities – Facilities where waste is permanently deposited for disposal into land. According to the waste hierarchy, the final disposal of waste through landfill is the least preferred way of managing waste.

Leaching – washing chemicals out of the soil.

Life Cycle Assessment (LCA) – The systematic identification and evaluation of all the environmental benefits and disbenefits that result, both directly and indirectly, from a product or function throughout its entire life from extraction of raw materials to its eventual disposal and assimilation into the environment. LCA helps to place the assessment of the environmental costs and benefits of these various options, and the development of appropriate and practical waste management policies, on a sound and objective basis.

Light Radiation – Radiation from the Sun.

Lignin – Structural constituent of wood and (to a lesser extent) other plant tissues, which encrusts the walls and cements the cells together; energy-rich material contained in biomass that can be used for boiler fuel.

Lignocellulose – Plant material made up primarily of lignin, cellulose, and hemicellulose.

Limnology – the branch of science dealing with the characteristics of freshwater, including biological properties as well as chemical and physical properties.

Lipid – Any of a group of organic compounds, including the fats, oils, waxes, sterols, and triglycerides, that are insoluble in water but soluble in nonpolar organic solvents, are oily to the touch, and together with carbohydrates and proteins constitute the principal structural material of living cells.

Liquid Particulate Matter – mist which includes raindrops, fog, and other chemical mists.

Lithosphere – The rigid, outermost shell of the Earth that is composed of the crust and the portion of the upper mantle that behaves elastically on time scales of thousands of years or greater; defined on the mineralogy.

Live Cull – A classification that includes live cull trees; when associated with volume, it is the net volume in live cull trees that are 5.0 inches in diameter and larger.

Logging Residues – The unused portions of growing-stock and non-growing-stock trees cut or killed logging and left in the woods.

Low-Level Waste – a type of radioactive waste depending upon the amount of radioactivity (see also High-level waste).

Lower Heating Value (LLV, Net Heat of Combustion) – The heat produced by combustion of one unit of a substance, at atmospheric pressure under conditions such that all water in the products remains in the form of vapor; the net heat of combustion is calculated from the gross heat of combustion at 20°C (68°F) by subtracting 572 cal/g (1,030 Btu/lb) of water derived from one unit mass of sample, including both the water originally present as moisture and that formed by combustion-this subtracted amount is not equal to the latent heat of vaporization of water because the calculation also reduces the data from the gross value at constant volume to the net value at constant pressure and the appropriate factor for this reduction is 572 cal/g.

LVOCs – Low-volatile organic compounds.

Lye – See Sodium hydroxide.

M85 – An alcohol fuel mixture containing 85% methanol and 15% gasoline by volume. Methanol is typically made from natural gas, but can also be derived from the fermentation of biomass.

Macronutrients – The elements that occur in standard levels in plant materials or in fluids in the plants.

Mafic – An adjective describing a silicate mineral or igneous rock that is rich in magnesium and iron.

Magma – Molten rock, as occurs in volcanoes.

Megawatt (MW) – A measure of electrical power equal to 1 million watts (1,000 kW).

Mercury (Hg) – A chemical element commonly known as quicksilver, a liquid under standard conditions of temperature and pressure.

Mesosphere – The part of the atmosphere immediately above the stratosphere at an altitude of 53 miles (85 km).

Mesothelioma – a tumor of the mesothelial tissue lining the chest cavity adjacent to the lungs.

Methanol – A fuel typically derived from natural gas, but which can be produced from the fermentation of sugars in biomass; also derived from wood and known as wood alcohol.

Methoxide (Sodium Methoxide, Sodium Methylate, $CH_3O^- Na^+$) – An organic salt, in pure form a white powder; in biodiesel production, *methoxide* is a product of mixing methanol and sodium hydroxide, yielding a solution of sodium methoxide in methanol, and a significant amount of heat; making sodium methoxide is the most dangerous step when making biodiesel.

Methyl Alcohol – see Methanol.

Methyl Esters – see Biodiesel.

Mica – A complex aluminum silicate mineral, which is transparent, tough, flexible, and elastic.

Micron – A unit pf size; 1 micron (1 μm, 1 meter × 10^{-6}) also called 1 micrometer is equivalent to 1 meter × 10^{-6} in diameter, i.e., 1 millionth of a meter or one thousandth of a millimeter, 0.001 mm, or approximately 0.000039 of an inch.

Micronutrients – The elements that are essential only at low levels and which are generally required for the functioning of essential enzymes.

Microorganisms – Organisms such as algae, bacteria, and fungi.

Mill Residue – Wood and bark residues produced in processing logs into lumber, plywood, and paper.

Million – 1×10^6

Minerals – Naturally-occurring inorganic solids with well-defined crystalline structures.

Mitigation – The identification, evaluation, and cessation of potential impacts on the environment.

Mobile Emissions Sources – anthropogenic air emissions produced by the various forms of transportation.

Modified/Unmodified Diesel Engine – Traditional diesel engines must be modified to heat the oil before it reaches the fuel injectors in order to handle straight vegetable oil. Modified, any diesel engine can run on veggie oil; without modification, the oil must first be converted to biodiesel.

Mohs Scale – measurement of hardness of a mineral ranging from 1 to 10 with 10 being the hardest.

Moisture Content – The weight of the water contained in wood, usually expressed as a percentage of weight, either oven-dry or as received.

Moisture Content, Dry Basis – Moisture content expressed as a percentage of the weight of oven-wood, i.e.: [(weight of wet sample-weight of dry sample)/weight of dry sample] × 100.

Moisture Content, Wet Basis – Moisture content expressed as a percentage of the weight of wood as-received, i.e.: [(weight of wet sample-weight of dry sample)/weight of wet sample] × 100.

Moisture Free Basis – Biomass composition and chemical analysis data is typically reported on a moisture-free or dry weight basis-moisture (and some volatile matter) is removed prior to analytical testing by heating the sample at 105°C (221°F) to constant weight; by definition, samples dried in this manner are considered moisture-free.

Monosaccharide – A simple sugar such as a five-carbon sugar (xylose, arabinose) or six-carbon sugar (glucose, fructose); sucrose, on the other hand is a disaccharide, composed of a combination of two simple sugar units, glucose, and fructose.

MTBE – Methyl tertiary butyl ether is highly refined high octane light distillate used in the blending of petrol.

Municipal Solid Waste (MSW) – Household waste and other wastes collected by a waste collection authority or its contractors, such as municipal parks and gardens waste, beach cleansing waste and any commercial waste and industrial waste for which the collection authority takes responsibility.

Municipal Waste – Residential, commercial, and institutional post-consumer waste that contain a significant proportion of plant-derived organic material that constitutes a renewable energy resource; waste paper, cardboard, construction, and demolition wood waste, and yard wastes are examples of biomass resources in municipal wastes.

Mutagenic – Tending to cause mutations.

Naft – pre-Christian era (Greek) term for naphtha.

Naphtha – the volatile fraction of petroleum which is used as a solvent or as a precursor to gasoline.

Natural Gas – The gaseous components that often occur in reservoirs with petroleum.

Neutralization – a process for reducing the acidity or alkalinity of a waste stream by mixing acids and bases to produce a neutral solution. Also known as pH adjustment.

New Chemical Substance – A chemical that is not listed in the Environmental Protection Agency's Inventory of Chemical Substances or is an unlisted reaction product of two or more chemicals.

Nitrogen Fixation – The transformation of atmospheric nitrogen into nitrogen compounds that can be used by growing plants.

Nitrogen Oxides (NOx) – Products of combustion that contribute to the formation of smog and ozone.

Non-Attainment Area – Any area that does not meet the national primary or secondary ambient air quality standard established (by the Environmental Protection Agency) for designated pollutants, such as carbon monoxide and ozone.

Non-Forest Land – Land that has never supported forests and lands formerly forested where use of timber management is precluded by development for other uses; if intermingled in forest areas, unimproved roads and non-forest strips must be more than 120 feet wide, and clearings, etc., must be more than 1 acre in area to qualify as non-forest land.

Non-Industrial Private – An ownership class of private lands where the owner does not operate wood processing plants.

NVOCs – Non-volatile organic compounds.

Oceanography – The science of the ocean and its physical and chemical characteristics.

Octane Number – Measure of the resistance of a fuel to self-ignition; the octane number of a fuel is indicated on the pump-the higher the number, the slower the fuel burns; bioethanol typically adds two to three octane numbers when blended with ordinary petroleum-making it a cost-effective octane-enhancer; see Knock.

Oil from Tar Sand – synthetic crude oil (q.v.).

Oil Mining – application of a mining method to the recovery of bitumen.

Oligotrophic Lakes – deep, generally clear, deficient in nutrients, and without much biological activity.

OOIP (Oil Originally in Place or Original Oil in Place): The quantity of petroleum existing in a reservoir before oil recovery operations begin.

Open Windrow – a type of composting.

Open-Loop Biomass – Biomass that can be used to produce energy and bioproducts even though it was not grown specifically for this purpose; include agricultural livestock waste, residues from forest harvesting operations and crop harvesting.

Organic Sedimentary Rocks – Rocks containing organic material such as residues of plant and animal remains/decay.

Overturn – The disappearance of thermal stratification causing an entire body of water to behave as a hydrological unit.

Oxidation – A process which can be used for the treatment and removal of a variety of inorganic and organic wastes.

Oxygenate – A substance which, when added to gasoline, increases the amount of oxygen in that gasoline blend; includes fuel ethanol, methanol, and methyl tertiary butyl ether (MTBE).

Oxygenated Fuels – Ethanol is an oxygenate, meaning that it adds oxygen to the fuel mixture-more oxygen helps the fuel burn more completely, thereby reducing the amount of harmful emissions from the tailpipe; a fuel such as ethanol-blended gasoline that contains a high oxygen content is called *oxygenated*.

Ozone (O_3) – A pale blue gas with a distinctively pungent smell; formed from oxygen by the action of ultraviolet light and (UV) light and electrical discharges (lightning) within the atmosphere of the Earth; present in low concentrations throughout the atmosphere; the highest concentration is in the ozone layer of the stratosphere.

P-Type Waste – A hazardous waste; usually a specific chemical species

Palm Oil – A form of vegetable oil obtained from the fruit of the oil palm tree; widely used feedstock for traditional biodiesel production; the palm oil and palm kernel oil are composed of fatty acids, esterified with glycerol just like any ordinary fat.

Particulate – A small, discrete mass of solid or liquid matter that remains individually dispersed in a gas or liquid emissions.

Particulate Emissions – particles of a solid or liquid suspended in a gas, or the fine particles of carbonaceous soot and other organic molecules discharged into the air during combustion.

Particulate Matter – particles in the atmosphere or on a gas stream which may be organic or inorganic and originate from a wide variety of sources and processes; the sources can be natural or anthropogenic.

Pay Zone Thickness – the depth of a tar sand deposit from which bitumen (or a product) can be recovered.

Peat Soils – Soils containing as much as 95% w/w organic material.

Pedosphere – The soil on the earth.

Perennial – Plant that does not have to be planted every year like traditional row crops.

Permeability – The ease of flow of the water through the rock.

Permeability (Soil) – The rate at which water and air move from upper to lower soil layers.

Persistent Organic Chemicals (POPs) – Organic compounds that are resistant to environmental degradation by chemical, biological, and photolytic processes.

Persistent Pollutants – Chemicals that are not easily degraded in the environment due to their stability and low decomposition rates and, thus, have a long life in various ecosystems; these types of pollutants typically require other forms of removal such as physical or chemical methods of cleanup as well as the addition of non-indigenous microbes for cleanup.

Petrodiesel – Petroleum-based diesel fuel, usually referred to simply as diesel.

Petroleum – A hydrocarbon-based substance comprising of a complex blend of hydrocarbons derived from crude oil through the process of separation, conversion, upgrading, and finishing, including motor fuel, jet oil, lubricants, petroleum solvents, and used oil.

Petroleum refining – a complex sequence of events that result in the production of a variety of products.

pH – A measure of acidity and alkalinity of a solution on a scale with 7 representing neutrality; lower numbers indicate increasing acidity, and higher numbers increasing alkalinity; each unit of change represents a tenfold change in acidity or alkalinity.

pH Adjustment – Neutralization.

Photochemical – Light-induced chemical phenomena.

Photosynthesis – Process by which chlorophyll-containing cells in green plants concert incident light to chemical energy, capturing carbon dioxide in the form of carbohydrates.

Pipestill – The refinery distillation unit.

Pipestill Gas – The most volatile fraction which contains most of the gases which are generally dissolved in the crude. Also known as pipe still light ends.

Pitch – The non-volatile product of the thermal decomposition of coal.

Plasma Incinerators – Incineration by use of an extremely hot plasma of ionized air injected through an electrical arc.

Podzol-Soil – Soil formed in temperate-to-cold climates under relatively high rainfall conditions under coniferous or mixed forest or heath vegetation.

Point Sources – sources which emit air emission through a confined vent or stack.

Polishing – The treatment of a waste product for safe discharge.

Pollutant – A substance or energy introduced into the environment that has undesired effects, or adversely affects the usefulness of a resource. A pollutant may cause long- or short-term damage by changing the growth rate of plant or animal species, or by interfering with human amenities, comfort, health, or property values. Some pollutants are biodegradable and therefore will not persist in the environment in the long term but can still cause damage to the environment during the lifetime of the pollutant. A pollutant can also be a naturally-occurring material that is reintroduced into the environment in an amount that exceeds the naturally occurring amount in the environment.

Pollutant (Degradable or Non-Persistent) – A pollutant that can be rapidly broken down by the natural processes; also known as biodegradable as the biological agents are involved in the process of the degradation.

Pollutant (Non-Degradable) – A pollutant that cannot be degraded by natural processes; once are released into the environment, the pollutant is difficult to eradicate and will continue to accumulate; examples are toxic elements such as lead or mercury.

Pollutant (Primary) – A chemical which is emitted directly from an identifiable source; examples include the oxides of sulfur (SOx), the oxides of carbon (COx), and the oxides of nitrogen (NOx).

Pollutant (Secondary) – A pollutant produced by the combination of primary pollutants in the environment. For example, peroxy-acetyl nitrate (PAN) and ozone (O_3) which are formed due to photochemical reactions between nitrogen oxides, oxygen, and hydrocarbon derivatives present in the atmosphere.

Pollutant (Slowly Degradable, Persistent Pollutant) – A pollutants that remain in the environment for many years in an unchanged condition and take decades or longer to degrade; examples include certain pesticides such as DDT and most plastics.

Pollution – The introduction into the land water and air systems of a chemical or chemicals that are not indigenous to these systems or the introduction into the land water and air systems of indigenous chemicals in greater-than-natural amounts; the introduction of indigenous (beyond the natural abundance) and non-indigenous (artificial) gaseous, liquid, and solid contaminants into an ecosystem.

Polycyclic Aromatic Hydrocarbon Derivatives (PAHs) – Chemical compounds containing only carbon and hydrogen in the form of multiple fused aromatic rings; the simplest such chemicals are naphthalene and phenanthrene, having two fused aromatic rings, and the three fused aromatic rings, respectively; sometimes referred to as polynuclear aromatic hydrocarbon derivatives (PNAs).

Pond – A small body of freshwater with shallow and still water, marsh, and aquatic plants that could dry up on a seasonal basis.

Porosity – The percentage of rock volume available to contain water or other fluid.

Porosity (Soil) – A measure of the volume of pores or spaces per volume of soil and the average distances between these pores.

Possible Reserves – The reserves where there is an even greater degree of uncertainty but about which there is some information.

Potassium Hydroxide (KOH) – used as a catalyst in the transesterification reaction to produce biodiesel.

Potential Reserves – based upon geological information about the types of sediments where such resources are likely to occur, and they are considered to represent an educated guess.

Pour Point – The lowest temperature at which oil will pour or flow when it is chilled without disturbance under definite conditions.

Premanufacture Notification (PMN) – requirement of any manufacturer to notify the Environmental Protection Agency at least 90 days prior to the production of a new chemical substance.

Primary Emissions – Pollutants which are directly emitted to the atmosphere.

Primary Oil Recovery – oil recovery utilizing only naturally occurring forces; recovery of crude oil from the reservoir using the inherent reservoir energy.

Primary Pollutants – Pollutants which are emitted directly from the sources.

Primary Waste Treatment – preparation for further treatment, although it can result in the removal of byproducts and reduction of the quantity and hazard of the waste.

Primary Wood-Using Mill – A mill that converts round wood products into other wood products; common examples are sawmills that convert saw logs into lumber and pulp mills that convert pulpwood round wood into wood pulp.

Probable Reserves – The reserves of the mineral that are nearly certain but about which a slight doubt exists.

Process Heat – Heat used in an industrial process rather than for space heating or other housekeeping purposes.

Producer Gas – Fuel gas high in carbon monoxide (CO) and hydrogen (H_2), produced by burning a solid fuel with insufficient air or by passing a mixture of air and steam through a burning bed of solid fuel.

Protein – A protein molecule is a chain of up to several hundred amino acids and is folded into a more or less compact structure; in the biologically active state, proteins function as catalysts in metabolism and to some extent as structural elements of cells and tissues; protein content in biomass (in mass percentage) can be estimated by multiplying the mass percentage nitrogen of the sample by 6.25.

Proved Reserves – The reserves of the mineral that have been positively identified as recoverable with current technology.

Proximate Analysis – The determination, by prescribed methods, of moisture, volatile matter, fixed carbon (by difference), and ash; the term proximate analysis does not include determinations of chemical elements or determinations other than those named and the group of analyzes is defined in ASTM D3172.

Pulpwood – Round wood, whole-tree chips, or wood residues that are used for the production of wood pulp.

Pump and Treat – The extraction of contaminated groundwater from the aquifer with subsequent treatment at the surface and disposal or reinjection.

Pyrethroids – Nerve poisons, acting through interference of ion transport along the axonal membrane.

Pyrolysis – exposure of waste to high temperatures in an oxygen-poor environment. Pyrolysis: The thermal decomposition of biomass at high temperatures (greater than 400°F, or 200°C) in the absence of air; the end product of pyrolysis is a mixture of solids (char), liquids (oxygenated oils), and gases (methane, carbon monoxide, and carbon dioxide) with proportions determined by operating temperature, pressure, oxygen content, and other conditions. In contrast to incineration, pyrolysis is the thermal degradation of a substance in the absence of oxygen. This process requires an external heat source to maintain the temperature required. The products produced from pyrolyzing materials are a solid residue and a synthetic gas (syngas). The solid residue (sometimes described as a char) is a combination of noncombustible materials and carbon. The syngas is a mixture of gases (combustible constituents include carbon monoxide, hydrogen, methane, and a broad range of other volatile organic compounds). A proportion of these can be condensed to produce oils, waxes, and tars. If required, the condensable fraction can be collected by cooling the syngas, potentially for use as a liquid fuel.

Pyrolysis Oil – A bio-oil produced by fast pyrolysis of biomass; typically a dark brown, mobile liquid containing much of the energy content of the original biomass, with a heating value about half that of conventional fuel oil; conversion of raw biomass to pyrolysis oil represents a considerable increase in energy density and it can thus represent a more efficient form in which to transport it.

Pyrophoric – substances that catch fire spontaneously in air without an ignition source

Quad – One quadrillion Btu (10^15 Btu) = 1.055 exajoules (EJ), or approximately 172 million barrels of oil equivalent.

Radiation – The mechanism by which heat is transported away from the earth, usually after conduction and convection effects have transported the heat to atmosphere.

Radioactive Waste – See High-level waste and Low-level waste.

Radioactive Waste Management – the treatment and containment of radioactive waste materials that arise from such industry.

Radionuclide – An atom that has excess nuclear energy, making it unstable. The excess energy can be used in one of three ways: emitted from the nucleus as gamma radiation transferred to one of the electrons to release it as a conversion electron or used to create and emit a new particle (referred to as an alpha particle) from the nucleus. During those processes, the radionuclide is said to undergo radioactive decay; sometimes spelled radio nucleide and referred to as radioactive nuclide, radioisotope, or radioactive isotope.

Radon – A gas product of radium decay.

Rapeseed (*Brassica napus*; Rape, Oilseed Rape or Canola: A bright yellow flowering member of the family *Brassicaceae* (mustard or cabbage family); a traditional feedstock used for biodiesel production; canola is a name taken from *Canada oil* due to the fact that much of the development of the oil was performed in Canada; see Colza.

Rapeseed Oil – Food grade oil produced from rapeseed is called Canola oil; see Colza.

Raw Materials – Minerals as extracted from the earth prior to any refining or treating.

Reactive Substances – Chemicals which undergo rapid or violent reaction under certain conditions.

Reactive Waste – Waste unstable under ambient conditions.

Receptor – An object or location that is affected by a pollutant.

Recovery Boiler – A pulp mill boiler in which lignin and spent cooking liquor (black liquor) is burned to generate steam.

Recycling – The or reuse of chemical waste as an effective substitute for a commercial product or as an ingredient or feedstock in an industrial process.

Reformulated Gasoline (RFG) – gasoline designed to mitigate smog production and to improve air quality by limiting the emission levels of certain chemical compounds such as benzene and other aromatic derivatives.

Refractory Lining – A lining, usually of ceramic, capable of resisting and maintaining high temperatures.

Refuse-Derived Fuel (RDF) – Fuel prepared from municipal solid waste; non-combustible materials such as rocks, glass, and metals are removed, and the remaining combustible portion of the solid waste is chopped or shredded.

Regulated Pollutants – Pollutants which have been singled out for regulatory control. Renewable energy sources – solar, wind, and other non-fossil fuel energy sources.

Renewable Fuels Standard (RFS) – Legislation enacted by United States Congress as part of the Energy Policy Act of 2005, requiring an increasing level of biofuels be used every year.

Reserves – Well-identified resources that can be profitably extracted and utilized with existing technology.

Reservoir – A domain where a pollutant may reside for an indeterminate time.

Residual Waste – Waste remaining to be disposed of after re-use, recycling, composting, and recovery of materials and energy.

Residues – Bark and woody materials that are generated in primary wood-using mills when round wood products are converted to other products.

Residuum (pl. Residua, also Known as Resid or Resids) – the non-volatile portion of petroleum that remains as residue after refinery distillation; hence, atmospheric residuum, vacuum residuum.

Resource – The total amount of the mineral that has been estimated to be ultimately available.

Reverse Osmosis – separation of components in a liquid stream by applying external pressure to one side of the membrane so that solvent will flow in the opposite direction.

Rotary Kiln – a versatile large refractory-lined cylinder capable of burning virtually any liquid or solid organic waste. The unit is rotated to improve turbulence in the combustion zone.

Rotation – Period of years between establishment of a stand of timber and the time when it is considered ready for final harvest and regeneration.

Rotenone – An electron-transport inhibitor.

Round Wood Products – Logs and other round timber generated from harvesting trees for industrial or consumer use.

RTFO (Renewable Transport Fuels Obligation) – A United kingdom K policy that places an obligation on fuel suppliers to ensure that a certain percentage of their aggregate sales is made up of biofuels.

Run-of-the-River Reservoir – A reservoir with a large rate of flow-through compared to the volume.

Sand – Soil with particle sizes between 0.06 – 2 micron; see Micron.

Sandstone – A sedimentary rock formed by compaction and cementation of sand grains; can be classified according to the mineral composition of the sand and cement.

Saponification – The reaction of an ester with a metallic base and water (i.e., the making of soap); occurs when too much lye is used in biodiesel production.

Saturated Steam – Steam at boiling temperature for a given pressure.

Screening – a process for removing particles from waste streams used to protect downstream pretreatment processes.

Second Generation Biofuels – Biofuels produced from biomass or non-edible feedstocks.

Secondary Emissions – Pollutants which are formed in the atmosphere as the result of the transformation of primary emissions.

Secondary Oil Recovery – application of energy (e.g., water flooding) to recovery of crude oil from a reservoir after the yield of crude oil from primary recovery diminishes.

Secondary Pollutants – Pollutants produced by interaction of primary pollutants and with another chemical or by dissociation of a primary pollutant or other effects within a particular ecosystem.

Secondary Wood Processing Mills – A mill that uses primary wood products in the manufacture of finished wood products, such as cabinets, moldings, and furniture.

Sedimentary Strata – typically consist of mixtures of clay, silt, sand, organic matter, and various minerals.

Silt – A granular material of a size between sand and clay that originated from quartz; may occur as a soil (often mixed with sand or clay) or as a sediment mixed in suspension with water (also known as a suspended load) and soil in a body of water such as a river; may also exist as soil deposited at the bottom of a water body, like such as mudflows and landslides.

Slurry Phase Reactor – A tank into which the wastes, nutrients, and microorganisms are placed Sodium hydroxide (lye, caustic soda, NaOH).

Smog – a description of the combination of smoke and fog.

Softwood – Generally, one of the botanical groups of trees that in most cases have needle-like or scale-like leaves; the conifers; also the wood produced by such trees; the term has no reference to the actual softness of the wood; the botanical name for softwoods is gymnosperms.

Soil – finely divided rock-derived material containing organic matter and capable of supporting vegetation.

Soil Heaping – piling wastes in heaps of several feet high on an asphalt or concrete pad.

Solid Phase Bioremediation – treatment of wastes using conventional soil management practices to enhance the microbial degradation of the wastes. Also known as land farming.

Solid Waste – garbage refuse, sludge from a waste treatment plant, water supply treatment plant, or air pollution control facility and other discarded material, including solid, liquid, semisolid, or contained gaseous material resulting from industrial, commercial, mining, and agricultural operation, and from community activities.

Solubility – The maximum quantity of solute that can dissolve in a specified quantity of solvent or quantity of solution at a specified temperature or pressure (in the case of gaseous solutes).

Solvent Extraction – A process for separating liquids by mixing the stream with a solvent which is immiscible with part of the waste but which will extract certain components of the waste stream.

Sorghum Bicolor – This crop has been cultivated on a small scale in the past for the production of table syrup, but other varieties can be grown for the production of sugar. The most common types of sorghum species are those used for the production of grain.

Source Reduction – the reduction or elimination of chemical waste at the source, usually within a process.

Soy (Soy Oil) – A vegetable oil pressed from soybeans.

Soy Diesel – A general term for biodiesel which accentuates the renewable nature of biodiesel; popular in soy producing regions.

Soybean – A bushy, leguminous plant, Glycine max, native of South-East Asia that is grown for the beans, which are used widely in the food industry, for protein in cattle feed and for oil production.

Specific Gravity – the mass (or weight) of a unit volume of any substance at a specified temperature compared to the mass of an equal volume of pure water at a standard temperature.

Spontaneous Ignition – ignition of a fuel, such as coal, under normal atmospheric conditions; usually induced by climatic conditions.

Stabilization – the addition of chemicals or materials to a waste to ensure that the waste is maintained in their least soluble or least-toxic form.

Stabilized Waste – Waste that has been treated so that it is chemically stable.

Stand (of Trees) – A tree community that possesses sufficient uniformity in composition, constitution, age, spatial arrangement, or condition to be distinguishable from adjacent communities.

Starch – A molecule composed of long chains of a-glucose molecules linked together (repeating unit $C_{12}H_{16}O_5$); these linkages occur in chains of a-1,4 linkages with branches formed as a result of a-1,6 linkages; widely distributed in the vegetable kingdom and is stored in all grains and tubers (swollen underground plant stems) – this polymer is highly amorphous, making it more readily attacked by human and animal enzyme systems and broken down into glucose; gross heat of combustion: Qv(gross) = 7,560 Btu/lb.

Starch Crops – Crops in which most of the six-carbon sugar units are linked together in long, branched chains (starch). Yeast cannot use these chains to produce ethanol. The starch chains must be broken down into individual six-carbon units or groups of two units. The starch conversion process, described in the previous chapter, is relatively simple because the bonds in the starch chain can be broken in an inexpensive manner by the use of heat and enzymes, or by a mild acid solution.

Steam Turbine – A device for converting energy of high-pressure steam (produced in a boiler) into mechanical power which can then be used to generate electricity.

Stover – See Corn stover.

Straight Vegetable Oil (SVO) – Any vegetable oil that has not been optimized through the process of transesterification; using this type of vegetable oil in a diesel engine requires an engine modification that heats the oil before it reaches the fuel injectors.

Strata – Layers including the solid iron-rich inner core, molten outer core, mantle, and crust of the earth.

Stratosphere – The atmospheric layer directly above the troposphere.

Stripping – a means of separating volatile components from less volatile ones in a liquid mixture by the partitioning of the more volatile materials to a gas phase of air or steam.

Sugar Beet – A plant whose root contains a high concentration of sucrose and which is grown commercially for sugar production; grown in many areas of the United States; must be rotated with non-root crops (1 beet crop per 4 year period is the general rule). While beet by-products cannot provide fuel for the distillery, the beet pulp and tops are excellent feed in wet or dry form. Or the tops may be left on the field for fertilizer and erosion control.

Sugar Cane (Sugarcane, or Simply Cane) – Several species of tall perennial true grasses of the genus Saccharum, tribe *Andropogoneae*, used for sugar production. The plant is six to twenty feet tall with stout, jointed, fibrous stalks that are rich in sucrose; belongs to the grass family *Poaceae*, an economically important seed plant family that includes maize, wheat, rice, and sorghum, and many forage crops. It is native to the warm temperate to tropical regions of South Asia, Southeast Asia, New Guinea, and South America. Only a minority of states cultivate sugar cane, but there are hybrids (such as *Saccharum spontaneum*) which can be grown further north. High yields per acre of both sugar and crop residue are strong points of sugar cane production. The crop residue, called *bagasse*, is used in Brazil to provide heat for the distilleries.

Superheated Steam – Steam which is hotter than boiling temperature for a given pressure.

Surface Impoundment – use of natural depressions, engineered depressions, or diked areas for treatment, storage, or disposal of chemical waste.

Surface water – water in lakes, streams, and reservoirs.

Suspension – A dispersion of a solid in a gas, liquid, or solid.

Sustainable – An ecosystem condition in which biodiversity, renewability, and resource productivity are maintained over time.

SVO – Straight vegetable oil; burns well in many diesel engines but does not start the engine and will coke in the injectors as a hot engine cools; a separate tank of petro-diesel or biodiesel is often used during starting and stopping engine, and an electric valve allows transfer to the straight vegetable oil tank.

Swamp – a wetland where trees and shrubs are an important part of the vegetative association.

Sweet Sorghum – The name given to varieties of a species of sorghum.

Switch Grass – Prairie grass native to the United States and known for its hardiness and rapid growth, often cited as a potentially abundant feedstock for ethanol.

Syngas (Synthesis Gas) – A mixture of carbon monoxide and hydrogen; see Synthesis gas.

Synthesis Gas – A mixture of carbon monoxide (CO) and hydrogen (H_2) which is the product of high temperature gasification of organic material such as biomass; after clean-up to remove any impurities such as tars, syngas) can be used to synthesize organic molecules such as synthetic natural gas (SNG, methane (CH_4)) or liquid biofuels such as gasoline and diesel fuel via the Fischer-Tropsch process.

Synthetic Crude Oil (Syncrude) – a hydrocarbon product produced by the conversion of coal, oil shale, or tar sand bitumen that resembles conventional crude oil; can be refined in a petroleum refinery.

Synthetic Ethanol – Ethanol produced from ethylene, a petroleum by-product.

Tallow – Produced by meat rendering which evaporates the moisture and enables the fat, known (the tallow) to be separated from the high-protein solids (the greaves), which are pressed, centrifuged or subjected to a process of solvent extraction to remove more tallow, before being ground into meat and bone meal (MBM); pure tallow is a creamy-white substance.

Tar – a volatile liquid or semi-solid product from the thermal decomposition of coal (see also Pitch).

Tar Sand (Bituminous Sand) – A formation in which the bituminous material (bitumen) is found as a filling in veins and fissures in fractured rocks or impregnating relatively shallow sand, sandstone, and limestone strata; a sandstone reservoir that is impregnated with a heavy, extremely viscous, black hydrocarbonaceous, petroleum-like material that cannot be retrieved through a well by conventional or enhanced oil recovery techniques; (FE 76 – 4): The several rock types that contain an extremely viscous hydrocarbon which is not recoverable in its natural state by conventional oil well production methods including currently used enhanced recovery techniques.

Teratogenic – tending to cause developmental malformations.

Terminology – The means by which various subjects are named so that reference can be made in conversations and in writings and so that the meaning is passed on.

Texture (Soil) – The relative amounts of the different sizes and types of mineral particles.

Thermal Conversion – A process that uses heat and pressure to break apart the molecular structure of organic solids.

Thermal Treatment – The treatment of waste using elevated temperatures as the primary means to change the chemical, physical, or biological character or composition of the waste. Examples of thermal treatment processes are gasification, incineration, and pyrolysis.

Thermochemical Conversion – Use of heat to chemically change substances from one state to another, e.g., to make useful energy products.

Thermochemical Platform – Typically uses a combination of pyrolysis, gasification, and catalysis to transform wood into syngas-the gaseous constituents of wood- and then into fuels or chemicals.

Timberland – Forest land that is producing or is capable of producing crops of industrial wood, and that is not withdrawn from timber utilization by statute or administrative regulation.

Tipping Fee – A fee for disposal of waste.

Titration – Applied to biodiesel, titration is the act of determining the acidity of a sample of waste vegetable oil by the drop-wise addition of a known base to the sample while testing with pH paper for the desired neutral reading (pH =7); the amount of base needed to neutralize an amount of waste vegetable oil determines how much base to add to the entire batch.

Ton (Short Ton) – 2,000 pounds.

Tonne (Imperial Ton, Long Ton, Shipping Ton) – 2,240 pounds; equivalent to 1,000 kilograms or in crude oil terms about 7.5 barrels of oil.

Topping and Back Pressure Turbines – Turbines which operate at exhaust pressure considerably higher than atmospheric (non-condensing turbines); often multistage with relatively high efficiency.

Topping Cycle – A cogeneration system in which electric power is produced first. The reject heat from power production is then used to produce useful process heat.

Toxic Wastes – Wastes such as chemicals that are harmful or fatal when ingested or absorbed.

Toxicity – Defined in terms of a standard extraction procedure followed by chemical analysis for specific substances.

Toxicity Characteristic Leaching Procedure (TCLP) – a test designed to determine the mobility of both organic and inorganic contaminants present in liquid, solid, and multiphasic wastes.

Trace element – An element that occurs at a low levels in a given system.

Transpiration – The process by which water enters the plant's leaves from the atmosphere Transesterification – The chemical process in which an alcohol reacts with the triglycerides in vegetable oil or animal fats, separating the glycerin and producing biodiesel.

Traveling Grate – A type of furnace in which assembled links of grates are joined together in a perpetual belt arrangement. Fuel is fed in at one end and ash is discharged at the other.

Treatment – any method, technique, or process that changes the physical, chemical, or biological character of any chemical waste so as to neutralize the effects of the constituents of the waste.

Trillion – 1×10^{12}

Tropopause – layer at the top of the troposphere.

Troposphere – The lowest layer of the atmosphere extending from sea level to an altitude of 635 miles (1,016 km).

Turbine – A machine for converting the heat energy in steam or high temperature gas into mechanical energy. In a turbine, a high velocity flow of steam or gas passes through successive rows of radial blades fastened to a central shaft.

Turbopause – See Homopause.

Turbulence – air disturbances which result from such factors as the friction of the land surface, physical obstacles to wind flow, and the vertical temperature profile of the lower atmosphere.

Turn Down Ratio – The lowest load at which a boiler will operate efficiently as compared to the maximum design load of the boiler.

U-Type Waste – A hazardous waste which is a specific compound.

Ultimate Analysis – The determination of the elemental composition of the organic portion of carbonaceous materials, as well as the total ash and moisture; a description of the elemental composition of a fuel as a percentage of the dry fuel weight as determined by prescribed methods; see American Society for Testing and Materials.

Ultra-Low Sulfur Diesel (ULSD) – Ultra-low sulfur diesel describes a new EPA standard for the sulfur content in diesel fuel sold in the United States beginning in 2006-the allowable sulfur content (15 ppm) is much lower than the previous US standard (500 ppm), which not only reduces emissions of sulfur compounds (blamed for acid rain), but also allows advanced emission control systems to be fitted that would otherwise be poisoned by these compounds.

Unsaturated Zone – the region in which water is held in the soil.

Unstable Atmosphere – marked by a high degree of turbulence.

Uronic Acid – A simple sugar whose terminal -CH_2OH group has been oxidized to an acid, COOH group; uronic acids occur as branching groups bonded to hemicelluloses such as xylan.

Vacuum Distillation – A secondary distillation process which uses a partial vacuum to lower the boiling point of residues from primary distillation and extract further blending components.

Vadose Water – Water that is present in the unsaturated zone.

Van der Waals' Forces – The residual attractive or repulsive forces between molecules or atomic (functional) groups that do not arise from a covalent bond, or electrostatic interaction of ions or of ionic groups with one another or with neutral molecules; the resulting van der Waals' forces can be attractive or repulsive.

Viscosity – a measure of the ability of a liquid to flow or a measure of its resistance to flow; the force required to move a plane surface of area 1 square meter over another parallel plane surface 1 meter away at a rate of 1 meter per second when both surfaces are immersed in the fluid; the higher the viscosity, the slower the liquid flows.

Vitrification – a phase conversion process in which a chemical waste (or constituents of the waste) are melted at high temperature to form an impermeable capsule around the remainder of the waste; also known as glassification.

VOCs – see Volatile Organic Compounds.

Volatile Organic Compounds (VOCs) – Organic compounds that have a high vapor pressure at room temperature; the high vapor pressure results from a low boiling point, which causes large numbers of molecules to evaporate (passes from liquid to gas) or sublimes (passes from solid to gas) and enter the surrounding air-often expressed as volatility.

Volatility – Propensity of a fuel to evaporate.

Washout – The uptake of particles and gases by water droplets and snow and their removal from the atmosphere when rain and snowfall on the ground.

Waste Management – An organized system for waste handling in which chemical pass through appropriate pathways leading to elimination or disposal in ways that protect the environment.

Waste Piles – Piles that are used to contain an accumulation of solid waste and which also may be used for final disposal or for temporary storage.

Waste Streams – Unused solid or liquid by-products of a process.

Waste Vegetable Oil (WVO) – Grease from the nearest fryer which is filtered and used in modified diesel engines, or converted to biodiesel through the process of transesterification and used in any old' diesel car.

Water Pollution – Any change in natural waters which may impair their further use, caused by the introduction of organic or inorganic substances or a change in temperature of the water.

Water-Cooled Vibrating Grate – A boiler grate made up of a tuyere grate surface mounted on a grid of water tubes interconnected with the boiler circulation system for positive cooling; the structure is supported by flexing plates allowing the grid and grate to move in a vibrating action; ash is automatically discharged.

Watershed – The drainage basin contributing water, organic matter, dissolved nutrients, and sediments to a stream or lake.

Watt – The common base unit of power in the metric system; one watt equals one joule per second, or the power developed in a circuit by a current of one ampere flowing through a potential difference of one volt. One Watt = 3.412 Btu/hr.

Wet Deposition – Particle deposition with precipitation of airborne water particles.

Wet Flatlands – areas where mesophytic vegetation is more important than open water and which are commonly developed in filled lakes, glacial pits, and potholes, or in poorly drained coastal plains or flood plains.

Wet Mill – An ethanol production facility in which the corn is first soaked in water before processing; in addition to ethanol, wet mills have the ability to produce co-products such as industrial starch, food starch, high fructose corn syrup, gluten feed and corn oils.

Wet Scrubbers – devices in which a counter-current spray liquid is used to remove impurities and particulate matter from a gas stream.

Wetlands – flooded areas in which the water is shallow enough to enable growth of bottom-rooted plants.

Wheeling – The process of transferring electrical energy between buyer and seller by way of an intermediate utility or utilities.

Whole Tree Chips – Wood chips produced by chipping whole trees, usually in the forest and which contain both bark and wood; frequently produced from the low-quality trees or from tops, limbs, and other logging residues.

Whole-Tree Harvesting – A harvesting method in which the whole tree (above the stump) is removed.

Wind – Horizontally moving air.

Wood – A solid lignocellulosic material naturally produced in trees and some shrubs, made of up to 40 to 50% cellulose, 20 to 30% hemicellulose, and 20 to 30% lignin.

Wood Alcohol – see Methanol.

Wort – An oatmeal-like substance consisting of water and mash barley in which soluble starch has been turned into fermentable sugar during the mashing process-the liquid remaining from a brewing mash preparation following the filtration of fermentable beer.

Xylan – A polymer of xylose with a repeating unit of $C_5H_8O_4$, found in the hemicellulose fraction of biomass-can be hydrolyzed to xylose.

Xylose – A five-carbon sugar $C_5H_{10}O_5$; a product of hydrolysis of xylan found in the hemicellulose fraction of biomass.

Yarding – The initial movement of logs from the point of felling to a central loading area or landing.

Yeast – Any of various single-cell fungi capable of fermenting carbohydrates; bioethanol is produced by fermenting sugars with yeast.

Yellow Grease – A term from the rendering industry; usuall y means used frying oils from deep fryers and restaurants' grease traps; can also refer to lower-quality grades of tallow from rendering plants.

Zone of Incorporation – the top foot of soil. Zone of aeration: unsaturated zone.

Zone of Saturation – at lower depths, in the presence of adequate amounts of water, all voids are filled.

Index

B

C

D

E

I

N

O

P

Q

R

S

X

Y

Z

Printed in the United States
by Baker & Taylor Publisher Services